DARE TO INVENT THE FUTURE

Global South Cosmologies and Epistemologies

Clapperton Chakanetsa Mavhunga

Dare to Invent the Future: Knowledge in the Service of and through Problem-Solving, Clapperton Chakanetsa Mavhunga

DARE TO INVENT THE FUTURE

KNOWLEDGE IN THE SERVICE OF AND THROUGH PROBLEM-SOLVING

CLAPPERTON CHAKANETSA MAVHUNGA

The MIT Press
Cambridge, Massachusetts
London, England

The MIT Press would like to thank the anonymous peer reviewers who provided comments on drafts of this book. The generous work of academic experts is essential for establishing the authority and quality of our publications. We acknowledge with gratitude the contributions of these otherwise uncredited readers.

This book was set in Stone Serif and Stone Sans by Westchester Publishing Services. Printed and bound in the United States of America.

Library of Congress Cataloging-in-Publication Data

Names: Mavhunga, Clapperton Chakanetsa, 1972– author.
Title: Dare to invent the future : knowledge in the service of and through problem-solving / Clapperton Chakanetsa Mavhunga.
Description: Cambridge, Massachusetts : The MIT Press, [2023] | Series: Global South cosmologies and epistemologies | Includes bibliographical references and index.
Identifiers: LCCN 2022059079 (print) | LCCN 2022059080 (ebook) | ISBN 9780262546867 (paperback) | ISBN 9780262376723 (epub) | ISBN 9780262376716 (pdf)
Subjects: LCSH: Knowledge economy. | Information technology—Economic aspects. | Industrial productivity. | Philosophy, Black.
Classification: LCC HC79.I55 M3728 2023 (print) | LCC HC79.I55 (ebook) | DDC 338.9—dc23/eng/20221213
LC record available at https://lccn.loc.gov/2022059079
LC ebook record available at https://lccn.loc.gov/2022059080

10 9 8 7 6 5 4 3 2 1

For my children, Nyasha, Chitsidzo, and Dadiso, to whom the future belongs.
You must dare to invent the future!

You cannot carry out fundamental change without a certain amount of madness. It took the madmen of yesterday for us to be able to act with extreme clarity today. I want to be one of those madmen. . . . We must dare to invent the future. . . . Everything man is capable of imagining, he can create.

—Thomas Sankara, 1985

While revolutionaries as individuals can be murdered, you cannot kill ideas.

—Thomas Sankara, October 1987, one week before his assassination

I am just an ordinary African who wants to repay his debt to his people.

—Amílcar Cabral, 1973

(Amílcar Cabral)

Never be content with waiting for the bus
  Get on the bus!
Never be content with being a passenger or conductor on the bus
  Drive the bus!
Never be content with driving the bus
  Own the bus!
Never be content with owning the bus
  Own the road!
Never be content with not traveling because there is no path
  Make the path!

Never be content with being a messenger
  Make your own journey!
Never be content with elaborating the visions of others
  Be the visionary!
Never have visions that end as visions
  Make them realities!
Never be just a critic
  Be the builder!

Never limit your horizon
  Have no horizons

CONTENTS

INTRODUCTION

> Always bear in mind that the people are not fighting for ideas, for the things in anyone's head. They are fighting to win material benefits, to live better and in peace, to see their lives go forward, to guarantee the future of their children.
>
> —Amílcar Cabral, "Tell No Lies, Claim No Easy Victories," 1965

The problem of Africa's poverty is that, besides the minerals it exports raw, it is rich in critical thinkers who are not doers, and doers who are not critical thinkers, which is the legacy of an unreformed colonial education, wherein students come to class, the professors pour all the information they have into their ears, the students memorize, write exams, and pass, and the school head and teachers, vice-chancellor and professors, congratulate themselves on the year's pass rate when results are published.

Little has changed from my primary and secondary school days from 1978 to 1992. Then as now, there is no integration of the school and the home as complementary educational spaces. The school remains the quintessential place of education, whereas the home, located in the villages or townships, is simply where schoolchildren come to school from before the bell rings for assembly or resumption of classes and return to when the school bell rings to mark dismissal time. Despite that students live off the family agriculture, livestock, and forestry and learn from their parents, brothers, kin, and peers when heading cattle, working in the fields, or doing chores in the home, none of this practiced, lived, everyday knowledge is integrated into the school curriculum or considered educative. Similarly, little to none of the knowledge learned in class is applied—even applicable—to everyday activities and problems. Two different worlds. The schoolteacher becomes the only teacher.

When students who pass the exams go to college or university, usually in the city or isolated campus, *there are no deliberate mechanisms to instill an ethic of responsibility to community so that they carry their community to campus, and bring the university back to the community*. Everything they have grown up doing every day to live is even more alienated; they become proper artifacts of the college or university. The life sciences do not include their everyday life, so whose life is it accounting for? The physics describes things too abstract from the lived reality of society, examples plucked from alien worlds, things some of us who have been fortunate to travel see in Europe and North America on our journeys, most of which are things seldom relevant to the local situation. The environmental sciences talk of climate and sustainability whose values and priorities seldom speak to locally experienced ones. Nothing on the rich countries' planned obsolescence and fad culture of tech-savvy where people change their smartphones with each new Apple or Samsung model, exchanging the "old" one that is barely scratched, without thought to where it goes to die, and the posthumous lives of the deceased technological, including as toxic e-waste in the environments and lives of the marginalized and poor. When the curriculum talks about e-waste, the missing chapter is that of the various everyday innovations the poor and polluted are engaging in to deal with such waste, giving it new forms of posthumous lives, as part of new things contrived by them.

Nothing on the local conditions of dwindling rains, bad engineering decisions that create environmental disasters—like the housing stands allocated in flood plains by

FIGURE 0.1
Collectors, who live on site, comb through unsegregated waste at Bulawayo's Richmond Landfill, 2021. Photo: Personal archives/e-waste.

FIGURE 0.2
My crop in Chihota rural area, thriving from mid-November to late December, begins to wilt as the January dry spell takes its toll, January 2017. *Source*: Personal archives.

corrupt local politicians; bridges that empty out flooded rivers but drain out water and leave dry riverbeds and valleys come winter. More and more fields moving toward the rivers and valleys, squeezing livestock pastures into pockets, as drylands become drier. The rains, locally, becoming erratic, and yet, even when more rains fall in a season, they are concentrated in one sustained period. The three weeks or so of no rains at critical months destroy the crop or halve the potential harvest, because our agriculture continues to be rain dependent.

The syllabus, couched in the name of universal knowledge, or its local dynamics, seldom starts with the everyday and the priorities of ordinary people; it's the naivety of assuming what is being taught is what the whole world should know, unaware that this is merely extending things that matter to the demographic dominant in the United States' and Europe's learning institutions. Lectures on environment focus on conservation and partnerships, participation and management; they do not talk about ownership. They talk about tourism, remote pleasure islands for white elites, the black participation reduced to selling arts and crafts, entertaining the visitors with "local culture"—that would be ancient dances, the scarier the better, the more authentically primitive, truly African. Long after independence we continue to say "The scientific name for musasa is *Brachystegia spiciformis*." The African professor feels so good about himself pronouncing that, never asks why *musasa* is not scientific but this jawbreaker is. To know is to speak the colonizer's language. Nobody ever asks whose science, what is science? Our goal is not to interrogate, master, create, and turn theory into everyday tools and outcomes. Knowledge starts in the book, ends in the book. We wait for the bus, get on it, and take our seats as the sheepish passengers that we are, and the driver drives the bus following the prescribed colonial route of knowledge, to wherever ancestor-colonial said it will take us.

Civil engineering is supposed to be civil. At the basic level, in *chidzimbahwe* civil means *kuva nehunhu* (to be well-mannered, well-raised, cultured, polite, caring, and embodying the values of being among others, what the Zulu call *ubuntu*). Uncivil means the opposite: *kushaya hunhu* (to lack manners); all told, an engineering without manners is not engineering. *Kuti ive nehunhu* (for it to have manners), it must be concerned with ordinary people and their needs, which include water, roads, bridges, housing, communication systems, energy, agricultural technologies, and other basic needs and infrastructures of their time. How many roads have our graduates built, how many boreholes have they drilled, let alone civil engineering techniques invented, companies founded, and public works undertaken? In other countries, mechanical engineering has resulted in students and graduates designing new combustion engines, pumps, irrigation systems, hydraulic applications to cranes, earthmovers, and compactors. All that our graduates are good for is assembling, repairing, and operating automobiles, locomotives, irrigation systems and so on made by others. They are nothing but glorified mechanics; they do not invent or innovate anything. The same for our electrical engineers. In North America, Europe, South Korea, Japan, India, and China, electrical and electronic engineers design computers, cellular phones, photovoltaics, battery systems—things that power life today. Ours just install and repair. When spares run out, they wait for the next shipment.

In the end, it all boils down to the knowledge, knowing, and its ends. *Why do we know?* For me, the answer is: *To solve problems*. That asks serious questions regarding the relationship between my work station, campus, and my community, especially as a black man who is minority in the United States, majority in Africa, but away from home. *How do we come to campus? Who do we come to campus as?* Answer: As people who know nothing and are looking to be educated. We bring nothing; we are sheep. We come as individuals, in search of an individual certification that we know. We know individually, for our individual selves. We must leave everything, every fiber of our being—what constitutes us—at home, to become proper academic product. The pulse of everyday challenges, opportunities, experience, and knowledge we leave at home. It was never part of knowledge since creche; it can't start now, when the rough stone is being polished into gem. This is rubbish education!

And because the leadership and followership constituting the citizenry is a product of a useless education, this translates into the conception, conduct, and perception of development. Our projects don't need to be sustainable. They just need to be seen to be happening. Especially toward election time—the graders and all their cousins come out. Huge trees are rudely awoken from their serene sleep, the soils parted, and roads begin to take shape. People see the development. They vote. The graders start slackening on us. Then one by one they disappear. The road has snaked a few more villages further, then like all snakes do, hibernate in the winter, until another election beckons. We are in permanent election campaign mode. People are nothing but votes; they are not human beings. Just numbers.

Development is what governments say it is. If they say it is sustainable, believe them. Engineering is what the government engineer says it is. Here is our college-educated engineer, pompously insisting to be addressed as Engineer Dutururu, coming to tell us that for the area to develop we need a big dam, to be located upstream of the village, waving away the community's concern that it will flood the pastures. What people want is water, the rivers to flow and hold water perennially again, as they used to when they were still growing up, but not at the expense of waters shaving off half their pasturelands, or boreholes they complain spout water mixed with grease and rust, which upon breaking down or drying up, government abandons, and the community has neither the expertise nor resources to maintain.

The bridges with large openings that the good engineer is replacing the smaller-mouthed bridges of the colonizer with have the hydrological effect of trading impassable flooded bridges for rapid drainage, passable bridges for dry riverbeds, once-flourishing springs for dry patchy valleys, and perennial rivers for drained-out arteries and dwindling water tables. Having run out of answers, our engineer now blames climate change—everybody's alibi now. Politicians in power blame it for their failures in agricultural production. NGOs use it to raise funding to prepare Africa for climate catastrophe or "resilience" (Africa is not a baby!), and then spend the money lavishly on overheads and SUVs. Scientists use it to draw money to their research. Corporations use climate and sustainability to spruce up their image, despite contributing the most to pollution and global warming. The good engineer is in quite some company.

Let's assume Engineer Dutururu is right, and development looks like boreholes, the big dam, and big-mouthed bridges, and communities now have piped water irrigation systems, powered by the national grid (thanks to another "development-must-be-seen-to-be-believed" icon, rural electrification). They still export their agricultural and forest harvests raw to the city, where very basic manufacturing still takes place. Like pressing sunflower, peanuts, and other grain into vegetable oil, milling maize and wheat into flower, cooking, drying, packaging, and labeling; separating ore from earth, and smelting to facilitate export, but seldom turning the minerals into end-use products like electricals, electronics, automotive parts, farm machinery, and so forth. All of this to make other countries rich while we remain perennial primary producers! Corporates call it the "supply chain," or "feedstock," to make it sound less exploitative. African politicians tout the sector-by-sector export contributions in their annual budget statements. When the rural exports its products raw without processing, it is creating jobs for cities that add that value, and to countries that import our minerals. When I was in Dubai in 2019, a Zimbabwean-based relative told me that our unprocessed aluminum comes here, is processed in those big factories into multiple end products or value-added exports, which we then import at exorbitant prices. The other day I imported a barbed wire–making machine from China (which our useless engineers should be making to start with!), only to discover that even the galvanized wire for the strands and barbs was no longer made locally. It has to be imported from China, to which we export our iron and chrome. All our diamonds from Chiyadzwa are sent raw, sometimes all the earth containing it, to China! And the list goes on and on.

Meanwhile, we, Africa's academics who know this through research, are busy writing our books, teaching, and doing consultancy work for NGOs that are applying bandages or even causing more wounds to our eroding self-reliance! Our undecolonized curriculum is swelling the ranks of the unemployed, their critical thinking and theoretical skills useless to turn the surfeit of problems, the raw potential, and opportunities into wealth. To even see the opportunities in the problems! To even see the problem to lie not just in government or politicians, but in themselves! The community, country, and continent count on us, but we take refuge and comfort from distance. We research, publish, get tenure, title, and accolades. We measure our impact according to bibliometrics. And we love to blame. "Our governments are corrupt." "They are dictators." "They are puppets of the West." "They are puppets of China." We expect the politics to be right before we come in to innovate. Entire lifetimes wasted. A life spent waiting. Until we are too old to be useful even to ourselves, let alone to Africa. Or our communities. We only return home to retire. To die. Be buried. We can't even become ancestral spirits anymore. We murdered our own belief systems and embraced those of others. Perhaps we fear jeopardizing our chances of "going to heaven" if we embraced who we are, our ancestry, for it might convert us away from Jesus and embrace our own ancestors if we set aside a Judeo-Christian or Muslim lens and saw our past with deeply African eyes. Not for the beliefs—whether one way only or many ways to this force white people called God with a capital, it matters not—but the knowledges, the sciences, the technology, the engineering, and so on that they produced, which we have foolishly abandoned because we can't embrace the ancestor in spirit.

In our energetic years, all we do is criticize what others are doing while just sitting there! What have our most famed African scholars ever built in Africa? And I am not just talking of the distinguished professoriate in US or European academies; I mean our vice-chancellors, deans, department chairs, and professors in our African universities. All we do is write from the comforts of our high offices and cities.

This book is not the end of the story but just the beginning of a trilogy. Subsequent books will turn to the invention of a knowledge architecture to correct the problem as outlined above, at the level of the knowledge itself (one book) and the physical (including human) infrastructure already emerging to turn the knowledge into products and services beyond the writing, art, or audiovisual.

Nothing represents the uselessness of Africa's education and knowledge production in recent history more than the Covid-19 pandemic. A whole continent frozen in waiting mode. First it was the personal protective equipment, ventilators, and sanitizers; Africa could not even produce its own until months later. Even then, in most countries the fabric was imported from China! Covid arrived well before Africa had sufficient 3D printing capacity, so it has continued importing N95 (medical-grade) masks. Next, the vaccines: while Europe, China, India, and North America were busy researching the virus to develop a vaccine, Africa was again waiting to receive donations, reducing itself to one vast clinical trial field. We have to wait on others to find a vaccine or cure; all we can do is follow the instructions of the WHO. The vaccine is being developed only in the Global North and Asia while we just sit there! For this reason, the partnership between Rwanda, Ghana, and BioNTech to establish a vaccine manufacturing factory in Rwanda is a welcome one, hoping it builds conditions to not just attract innovators and manufacturers from the Global North and China and India, but puts faith in and invests in African creatives home and abroad well qualified to lead the charge. The curriculum also has to align with that "can do" spirit, instead of always waiting for "foreign investors."

At present that thinking has not trickled down to our training institutions on the continent. Our medical schools train applicators of therapies, not developers of medical technologies, vaccines, and drugs. Our labs are for identifying (some) microbes and determining the nature and cause of illnesses, not making the antidotes or equipment. Often our equipment is donated (the right word is disposed) by Western and Eastern hospitals or manufacturers switching to new generations of technology. It is a glorified form of medical electronic waste! As part of a review of one of the top private hospitals in Harare, Zimbabwe, I and four African colleagues based in the United Kingdom, the United States, and South Africa saw that even where such equipment was purchased, it had no warranty, could only be maintained by the technicians of the supplier (at great expense in flights, hotel, and labor fees), and some of the equipment was imported without checking something as basic as voltage standards. The United States and Japan use 110 to 130V/60Hz voltages, EU 220 to 240V/50 to 60Hz, and the United Kingdom and

FIGURE 0.3
Presidents Akufo Addo of Ghana (left) and Paul Kagame of Rwanda (middle) with BioNTech SE CEO Uğur Şahin, at the groundbreaking ceremony for the construction of the company's vaccine manufacturing plant in Rwanda, set for completion by December 2022, with manufacturing commencing end of 2023. *Source*: Office of the President, Rwanda.

South Africa 230V/50Hz, and the local importer used to local voltage just plugs them equipment in without checking, and their transformers burn up. Piles and piles of this "dead" medical equipment can be found at these hospitals, despite having technicians in-house.

We find in Africa a degree in business management, business administration, business studies, accountancy, and commerce. It teaches the student to manage a business, but whose business are they going to manage? To manage one, you have to create one! Whose books are you going to keep? As we see later, the problem is that Africa's universities are training for employment, not employers. The models we are drilling into our children in the name of entrepreneurship are all Western-and-white; in a continent so thoroughly dominated by informal(ized) economic activities, we never stop to consider everyday life as a business school characterized by hustling, social and political lubricating (bribing), queueing for everything, shortages of almost everything any time, these transient workspaces of fleeting customers created and accessed on the move, involving our own and others' movements. We read what we get from the famous business schools at Harvard, Stanford, Columbia, MIT, London School of Economics, and so forth, and jacket-and-tie these methods to our African situation at the total ignorance of *the street, village, and everyday experience as distinguished professor in strategy and creative resilience.* I want to think of business out of the realities of the local everyday and the local inventions too, and mix with everything else that is useful from those other places, which is ours by discovery, is a product of our global hunt, and which we reassemble and purpose into new forms. Our own. Why would we want to lose the advantage of mastery over our indigenous ingredients that those that privileged Western-and-white values do not know? I know what they know; they do not know what I know as an African. That's a lot of power.

Let me define what I mean by Western-and-white values as impediments to knowledge in the service of and through problem-solving. "Western-and-white" does not refer to people, or individuals, but to the racialized system of knowledge that currently governs the world, or *knowledge racism.* Namely, *a system or order of knowledge that is explicitly Western-and-white in what it privileges, in the composition of the people, institutions, concepts, or spaces it takes place in, in its premises, methods, processes, and ends, even where it is imposed upon or freely sought and deployed by non-Western-and-white. Always, it has the effect of including and excluding based on race, even through inclusive language, a Western-and-whitewash given the semblance of disgust for racism in words when an incident happens, meanwhile doing nothing about, actively participating in, writing, teaching, building, and even strengthening such exclusionary knowledge. By doing nothing they are doing something.*

Western-and-white also stipulates the values developed in and by people who historically come from a very specific region of the world (western Europe) and are of a very specific color (white), who are not tethered to one region but carry their ways of being, seeing, thinking, knowing, and doing with them. Equally important, Western-and-white enables us to separate geographies of solidarity with racially discriminated peoples from those that use racism to demarcate and colonize. For example, some black people, usually for personal material benefits, may behave and even disfigure themselves in order to be Western-and-white (through skin lighteners or plastic surgery), while white people in the western and northern hemisphere can be anti-Western-and-white. History has examples of white people who risked their lives in the cause of black freedom, not least Quaker abolitionists like Thomas Garrett of Wilmington, Delaware, who worked with Harriet Taubman in the Underground Railroad to smuggle enslaved Africans from the south to the north.

In the struggle for African independence, Swedish prime minister Olof Palme was assassinated by the apartheid regime for his unwavering support for the liberation movement. Then there are countries and individuals that are white-but-not-Western, which came to Africa's assistance to fight for independence against Western-and-white imperialism in the twentieth-century.—Like the Union of Soviet Socialist Republics (Russia and eastern Europe), which fall outside the West but are majority white. These countries and (some of) their citizens may have harbored racist ideas but their work, if only to advance their own interests, ended up advancing black struggles for freedom where even black governments and people did not. I have not seen any black struggles that did not have that good white person that transcended color and lived life as a human being, not a white person. I too have benefitted from a few such.

Both the anti-slavery and anti-colonial struggles have important lessons for us, not least that solidarities built through not just talking and writing (academia) but actually solving problems that threaten the lives of others, are more likely to have sincerity and depth compared to mere academic or civic discourse. Where such struggles have overcome, the challenge still outstanding is to render freedom fruitful—those are the Underground Railroads and liberation struggles of our own time.

Whereas science, engineering, and business are charged with making and moving things, we African academics have allowed our humanities and social sciences to simply study and write. Our history is the study of the past. Sterile, frozen in time, too empirical, full of dates, quotation marks, references, and all those things that make us proper scholars in the eyes of our Western-and-white peers. The study of history is a very colonial practice: it is the greatest obstacle to telling our own stories. We are asked to write for an academic audience so that they can quote us, reference us, cite us, and as such show that we are "making an impact." "What is your genre?" I am always asked. All I care about is "why" I am writing, "to whom" I am writing, and "why" it matters. "What is your archive?" To which I answer: "What is and is not an archive? According to who?" Now that I am here, no longer object of study, what does that presence, my story, how I see things, how I think, how I know, how I make, how I think about consequences before or after, must matter. I can't just start with definitions, formats, reasons, instruments, and expected outcomes inherited from exclusionary spaces, practices, and practitioners.

Knowing toward what end? Well, just knowing and being knowledgeable is not enough on its own, never mind the urgencies and emergencies our historically marginalized and overresearched communities face. That was my early problem with philosophy as a graduate student at the University of Zimbabwe in 1993–1994 (which I found later reflected in African philosophy written by African scholars), which like other humanities and social sciences was never intended to walk out of its home in the book pages, elite cafes, and seminar rooms to enter and impact people's material everyday life. The African philosophers were trying very hard, too hard, to imitate the Greeks, to stay true to what the latter meant and intended by this or that term, and to dismiss what did not conform, especially that which seemed troubled by privileging of form and content to fit into the Greek way, whereas what they sought was to explain African situations—"the purpose of knowing." It was the same with religious studies or theology, chopped off from the everyday languages, now called "African languages," and from philosophy, which was now put on a scale and measured for difference or similarity with Socrates, Aristotle, Plato, Kant, Marx, and so forth. And before we know it, we are always squirreled back to the Western-and-white referent!

The question for some of us in the twenty-first century is this: *Can we trust and rely for our self-rehumanization as a people those orders of knowledge that reduced black people to human machines since late August, 1619, when the first enslaved Africans landed at Point Comfort (Fort Monroe in Hampton, Virginia), as objects of study under colonial rule, and continue to constitute themselves based on methods that reduce Africans and other peoples of color worldwide to informants or simply sources of information? Can such orders of knowing remain legitimate, moral, ethical, and even legal where such peoples are extracting our knowledge and publishing it in their own name for their own benefit, leaving nothing tangible for us?* The challenge is that the reduction of indigenous peoples into informants is transactional; it reduces knowledge into information and, worse, just data, and places the honor, skill, and benefits of producing knowledge in the hands of the anthropologist, while stripping the knower-designated-informant of the capacity and attribute of producing knowledge. Now rendered data or information, knowledge becomes commodifiable as a book, as data, individually owned and transactable. The exact role and usefulness of anthropology to everyday life has eluded reimagination.

So too history as a subject. Some universities in Africa, like the University of Zimbabwe, are phasing anthropology out altogether, at least on paper. Increasingly "history" is being supplanted by "knowledge systems and heritage" as the country moves to take ownership of and localize its narrative, which by and large has been written by white outsiders. In 2007 Nigeria banned "history" as a subject in schools, and it would only be reinstated in 2019—a sign of unsureness about its usefulness. And that is precisely the problem: relevance and purpose to Africans themselves.

GRADUATES OF A USELESS EDUCATION

Then comes Twitter and other social media. The products of our disciplines are always writing and talking about problems. They are always tweeting; and to tweet you must have your more powerful, usually writing, hand on the keyboard; the hand that is supposed to be working to solve challenges that won't be tweeted or critiqued away is spending too much time "protexting" (protesting through texting) and demanding that somebody else—government, the world—come and solve the problems, instead of running toward them, to be first to turn them into opportunities and solutions. Ours is an education that is all mind, no hand. No heart for—no responsibility to—the community. Just me, my book, and self. It is as if being educated absolves us from physical work; as if physical work, the work of sweating, is for uneducated people, a lesser art, a mechanic art, as distinct from the finer arts of education that, once acquired, absolve one from physical work.

Knowledge must reward us, our communities, with happiness, livelihood, wealth, prosperity, or it dies. Western-and-white society invested its system of formalized knowledge production with individual material rewards and credentialing. With knowledge-ability (hence knowledgeability). They called it a school, a college, a university, and designated its contents disciplines, and when they came to our shores, consigned our own ways of knowing into "Bantu studies" and later "area studies" and "African studies," simply raw empirical materials to be studied and disciplined by the "real" disciplines. That process of designing and designating the school/college/university was an exclusionary process, an exclosure of other knowledges; knowing could not be anything else besides what was in this enclosure. That was not all: a whole economy and bureaucracy was created to employ graduates of this funnel-tunnel, and that monoskilled funnel-tunnel exclosed any other grounds of competence but its disciplines, and proof of disciplined knowing was codified in the certificate, the diploma. Without that piece of paper there could be no other claim to knowing.

In our stampedes to get certified, we asked no questions why that should be, when our ancestors had built pyramidal structures in old Egypt, tamed the Nile and diverted its waters, dry-bonded stone buildings at many *madzimbahwe* (houses of stone) with such mathematical precision, turned dirt into metals and ceramics, tree bark and wild cotton into cloth, and coexisted with deadly insects and microorganisms without poisoning their way to disease-free environments. *Then again, why would we have been bothered?* All we wanted was to get a certificate with which to look for a job. With a job, we could fulfill those narrow, individual materialistic chores to family and self. We were placed on a narrow tunnel of vision; the purpose of education was to turn us inward, to have no ambition beyond the household, the individual, and the material. Beyond knowing how molecules and electric current functioned, not to deploy that knowledge as an ingredient to invent something of our own. It was never the business of Western-and-white education to prepare black people for self-government, to be inventors, to generate and govern wealth, to create viable self-sustaining communities, to encourage independent thinking that defies received Western-and-white versions of knowledge, and to come to knowledge from wherever we want, including asking what the relevance is to our specific pasts, presents, and futures. To speak to our own challenges and opportunities. The famous universities of today earned their name addressing the needs of the nineteenth and twentieth centuries (i.e., Western-and-white definitions and forms of scientific and industrial revolution, colonial resource exploitation, and satisfying the thirsts of burgeoning middle and upper classes in North America, Europe, and the colonies). Those in Africa were merely local branches of these European universities. Today, a century later, US higher education institutions are following suit, setting up branches in Africa and Asia. At least these universities are being open about their educational imperialism. In Ghana, Mauritius, Rwanda, and Kenya, in particular, private colleges and innovation start-ups are supposedly "owned" or "founded" by Africans, but the real power is in Silicon Valley. I often look beyond that to a more strategic question: what these (usually young) entrepreneurs intend to do, their reasons, not the intentions of Silicon Valley and white monopoly capital.

A PARTNERSHIP BETWEEN RIDER AND HORSE

There are those Western and traditionally white universities that claim to be "partnering" local institutions, with their professoriate as riders enjoying the ride (parading their global profiles, getting tenure and positions of responsibility off it, and much funding that they control) while locals are the horses carrying the burden of research, risky clinical and technology trials, and receiving only bread crumbs from the budget.

In any "joint" project, the Western-and-white partners own and control the budget, disburse money, and do the accounting. It is in very rare cases that the controller of the purse is black, given that only 4 percent of full professors are black and only 7.4 percent of any professors at all in the United States are black. Of these, just 4.9 percent of African Studies professors are black, compared to 70.8 percent white; we are even less than Asian professors (12.2 percent) and Latinos (7.2 percent) (Zippia 2023a, 2023b). Out of these, even fewer do research on Africa, illustrating the specific connection between system racism in the United States and its direct effect on global partnerships, not just with Africa, but the non-white world in general. Black professors constitute 6.4 percent of engineering faculty compared to 64.2 percent white in the United States, and 7.1 percent compared to 66.4 percent white science professors (Zippia 2023c, 2023b). "Black" includes African Americans who may not focus on continental Africa at all, hence the 4.9 percent figure for African Studies professors.

As I discussed in *The Mobile Workshop* (Mavhunga 2018), African Studies emerged out of Bantu Studies, the specific intent being to study 'the native.' An analysis of the founding charter and mechanism for funding African Studies in the United States is tied closely to its utility to the country's national interests, which may circumscribe to what purposes the money is put. Then again, it is not the responsibility of the United States, Europe, Asia, or any outside country to establish research on Africa that is invested in Africa's priorities. If Africa's intellectuals think it is a legitimate enterprise to simply "study" Africa regardless of whether such pursuit defines or meets self-determined African values and priorities, or whether the research leads to anything made, built, or solved by such intellection, they are complicit in the coloniality of knowledge production about the continent and themselves. At least the "Africanist" is doing their job—their brief is to study Africa, and provided they stay within the federal funding charter that supports area studies, they cannot be counted upon as keepers of keys to the gates protecting or advancing what is in Africa's interests. *They do not have to build bridges; they are there to only walk them. We, by contrast, are counted upon by a continent to show how to build bridges, to build them ourselves; instead we just talk.*

We do not even own the venues; we are just invited guests to conferences about our own continent. We the professoriate are not immune to Rwandan President Paul Kagame's pointed and welcome rebuke to his peers in the African Union:

> African leaders, we don't need to be invited anywhere to go and address our problems without first inviting ourselves to come together to tell each other the actual truth that we must tell each other about our serious problems. For me, when I am watching on television, and I find our leaders, who should have been working together all along to address these problems that commonly affect their countries wait until they are invited to go to Europe to sit there and just . . . you know, it's like they are made to sit down and address their problems. Why does anybody wait for that? What image does it even give about us? About Africa? In fact, the image it gives is that we are not there to even address the problems we are there for a photo opportunity. We are happy to sit there in Paris with the president of France and address. It doesn't make sense that our leaders cannot get themselves together to address problems affecting our people. It doesn't make sense. (Kagame 2014)

We are just happy to be invited, to take our place, and parrot our substantiation of the theories of others, *kupupa furo* (froathing at the mouth) to show we are better applicators of Western-and-white theory more than the Europeans themselves.

Kubinduka kusakurira minda yevamwe	Toilsomely weeding the fields of others
yedu yakati tunguzuzu sora	While ours are submerged in weeds
Kusadharara kukohwa mhunga dzevamwe,	Cumbersomely harvesting the millet of others
edu mapfunde achikuyiwa nezvimokoto	While our own sorghum is being ground by quelea birds
Kudida here kufambira dzavamwe,	Traveling far and wide on the errands of others
dzedu dzakati tumbi	While ours are on hold
Kutobuda tsinga dzekumusana, nekudetemba nhetembo dzaamwe	Veins bulging on our backs, from (the effort of) reciting the poetry of others with such melody
dzedu dzichiziizirwa nemazizi	While we abandon ours to the hoarse hooting of owls
Many'ana endongamahwe dzedzimbahwe	The nestlings of the stone-builders of houses of stone
Kubvuma kuitwa guyo nehuyo isu tiri vene	Accepting to be made grinding stone and grinding pebble when we are the owners
Kukwikwidza here nezviguyakuya	Competing with geckos
Kugutsurira savanamudengumuneyi	Nodding like what's-in-the-calabashes?
Vakaenda zvavakati wani "Kudya nyemba dzeshamwari, dzako dziri pamoto"?	Did not the departed say "To eat the boiled beans of friends, yours are cooking on the fire"?
Saka dzedu dziripi?	So where are our own (beans)?

And so here we are, keeping the fire burning for others to cook their own food. The headmaster in the North, the class monitor or research assistant in the South—are the proverbial partnership of rider and horse. I have seen how even programs that are emerging, that purport to be democratic about ensuring each research partner has conceptual and budgetary say—and pretty much runs their own budget—soon frustrate the more egalitarian younger (usually female) and white faculty into leaving when the matriarchs and patriarchs revert to their colonialistic mode. I have seen it in Switzerland. Germany. France. Belgium. United Kingdom. The United States. Friends of mine whose genuine, humane intentions were squashed once the money applied for was in the account.

The partition of Africa continues in conceptual, financial, and intellectual property form what the colonial occupation did through physical settler occupation: for some odd reason European and North American institutions doubt that African faculty can be trusted with money. They will blame their governments and the funders. Well, tell them it is colonialistic! The African institution—its faculty and students—is reduced to do all the empirical work; the Western partner does all the theoretical framework and secondary literature stuff. The knowledge labor is weighted heavily against the former, but the most resources are reserved for the Western-and-white partner's per diem and other "fieldwork" over the Western-and-white summer. I have attended conferences organized by these Western-and-white partners to discuss "their" work. In attendance will be the Western-and-white partners and their African faculty counterparts and their students. The relationship resembles that of the German professor and his postdocs and PhD students: the Western-and-white speak first, and the African research assistants (the faculty and students both, what's the difference?) then come in to simply reinforce or polish off. They cannot disagree; it is the master and servant. You do not want to bite the hand or kill the goose. The rider and the horse.

I do not feel sorry for the horses, because they give the rider so much power. I have seen broncos and stallions throw off their burden. Mules and donkeys too. In Mozambique and Zimbabwe, for example, hundreds of hectares of university-owned land, lakes, and (in the former) ocean waters lie fallow while these institutions beg for state allocations and demean themselves by sending delegation after delegation with begging bowls to traditionally Western-and-white universities, asking for the fish, not tools to explore, mine, farm, or process aquatic and terrestrial resources that take longer to make locally. It would appear the horses have chosen to be slaves of traditionally Western-and-white institutions. How many ever visit historically black universities and colleges to share black-to-black community experiences? Why do we not see that these riders are attracted like flies to the smell of problems in our countries and communities that our unreformed education has blinded us from seeing and turning into opportunities? Then when the outsider comes in, sees what we cannot, takes risks we cannot, and succeeds where we cannot, we turn around and blame them for taking our opportunities and jobs, like South Africans are doing to non-South Africans now, especially Zimbabweans! *The chickens of an unreformed colonial apartheid education have now hatched and are coming home to roost.*

To be clear: no journalist, researcher, or investor comes to a place without crises; reality or stereotype, an Africa with wars, famine, pandemics, corruption, and dictatorships attracts bad press. In Africa's history, the so-called "investor" is typically an "exploiter." Bad press sells papers, hits, tweets, viewing, airtime, and data bundles, attracts more ads, and makes more money for the new global click-bait economy; problems mean demand, a market, investment opportunity for corporate outsiders. Our university vice-chancellors form beelines to treasury for increased operational funding from government; why not recraft the research agenda to focus on looking for problems, understanding them, designing solutions to them? Why not recraft university financial policy so as to engage government (most African universities are state universities), industry, and commerce for mutually beneficial partnerships and loans to turn this research into products and services? What contributions are our students making to society in terms of quantifiable production and exchange value? And meanwhile we just continue to invest in training more?

DARING TO INVENT A NEW KNOWLEDGE FUTURE

Whether we center knowing or knowledge is not a spot of bother for me; the concern is the same—method, process, meanings, infrastructures, and ends. *Dare to Invent the Future* is inspired by African ancestors who grappled with the question of knowledge in the service of and through problem-solving, which entails an education that deliberately trains students to organize for a happy, equal, and productive society. Presidents breaking ground to establish vaccine factories is a trickle-down model that, while positively uplifting, needs to be firmly anchored in an innovative education system that does not simply consume or copy and paste things from outside. Such an education must encourage students, staff, and faculty to bring their community onto campus, and to take campus back to the community, with them acting as the bridge connecting the knowledges and needs of community and university, harmonizing them into a critical thinker-doer mode of knowing-how for the betterment of the world. *The community, when brought to campus, becomes the star that gives the student and faculty member bearings when choosing what classes to take or teach, so that subject choice becomes a search for ingredients to craft tools for diagnosing and solving the challenges our communities face. Further: the student and faculty member can begin to imagine a role for the university in their own community and draw it there as a partner, in a way that reinvents the institution into a university for, among, and of the people.*

If we all remembered to go back, this responsibility to community has the potential to make the students and faculty potentially the largest global force for community self-development, where the current international development terrain is run by non-governmental organizations, the World Bank and International Monetary Fund (IMF), the United Nations Development Program, and national governments. Without centering the relevance and purpose of education and the knowledge, the present system of international development has reduced educational institutions and the students who come from communities that most need development as self-redevelopment (after colonial underdevelopment) to bystanders, volunteers, and aid workers who are in it for employment. The "quest for relevance," while acknowledged (wa Thiong'o 1985), has eluded the technological and scientific imagination precisely because we as Africans have borrowed without questioning what these two terms mean, or shouldn't, according to our own worldview, experience, and knowledge as a people.

The relevance of our education systems was the subject of two of Tanzania's founding president Julius Kambarage Nyerere's most seminal lectures and speeches in the 1960s, summarized in his question: "What kind of society are we trying to build?" (Nyerere 1966c). His calls do not appear to have landed on fertile soils. As we explore in detail in chapter 8, the continuing African obsession with STEM (science, technology, engineering, and mathematics) without stamping our own meanings not only marginalizes the humanities, arts, and social sciences (and African priorities–driven research) in terms of funding, but also strips our STEM graduates of the critical thinking and social skills prerequisite to their success. We pressure our students, our kids, to be good in chemistry, biology, physics, accounts, and math—especially in high school—so that they can be engineers, doctors, accountants, lawyers, and pharmacists, instead of being makers or builders of things and problem-solvers with multiple skill sets. They end up being just glorified mechanics only good at applying spanners, syringes, and calculators, who cannot design or build public health systems, develop vaccines, roads, and equipment tailored to local context or to lift their disadvantaged communities. In the post-2000 era, our governments, through easy consumption of Western-and-white paradigms as well as conditions of funding from the IMF and the World Bank, forced their education ministries to prioritize STEM at the expense of the humanities, arts, and social sciences. Fellowships were introduced to incentivize a bias toward a specific discipline or disciplines that say true knowledge only comes from spatulas and test tubes, buildings, and benches, rather than the deep exercising of heart, hand, and heart (see chapter 7), perspective, and cosmology, which could, if steered toward problem-solving, instill in students a sense of responsibility (heart) toward communities, which Africa urgently needs.

In effect, without knowing, we were disqualifying any other ways of knowing, except those that were brought in and monopolized by the colonizer, who had specialized far further than us in that particular way of knowing and ordering knowledge and was ignorant or contemptuous of any other, including our own. In his book *Remembering the Dismembered Continent* (2010), Ayi Kwei Armah defined this crisis of our education in Africa while pointing forward to alternative knowledge futures where we think as Africans, home and abroad, and not in the colonial brackets masquerading as nationalities:

> Africans who learn to access the available information directly will know that centuries and millennia ago, people like ourselves, our ancestors, thought seriously about the central issues of life here. Some of the ideas and procedures they originated are sharply relevant to this day. . . . Such knowledge frees us from the crippling, cretinising superstition that we are Ghanaians, Nigerians, Senegalese, Malawians, Chadians and Sierra Leoneans, and liberates us to think as Africans.

> Thinking as Africans, we are free, mentally, to take in the vast range of possibilities open to the entire continent, as we look for ways of solving problems confronting us all living here. . . . The academy as brought to Africa in the wake of Arab and European invaders is very different from the ancient African learning institution that was the house of life. . . . *In many ways, our universities and schools in Africa today are intended not to help our society live, but to fix it in a quasi-permanent state of half-life, half-death, as its vital resources get steadily drained away. Our universities, set to help us vegetate, are national universities; to help us live, they would have to become, or to be replaced by, African universities. . . . We need to retrieve our murdered memory, to revive our starved recollection of our potential.* (Armah 2010, 35–36; see also Armah 1968; my own emphasis)

The theme of murdered memory articulates well with what Ganesh Devy (1995) calls "after amnesia"—an inability to remember or resurrect ways not just of thinking, but seeing and doing that existed before or in lieu of the European colonizer. It is what necessitates "decolonizing the mind" (wa Thiong'o 1986; Ibekwe 1987), but the effect and symptom of colonial trauma tends to be what D. A. Masolo (2017) calls "self-mortification"—the hapless mindset that we can only come to science and technology as its consumers, that we have no capacity to be scientific and technological, that the colonizer already thought for us, and we are dead to knowing and inventing. Masolo was drawing attention to "the place of science and technology in our lives."

The challenge has been explained in terms of the relevance of the racialization of disciplines, whose structures and practices of knowledge production remain undecolonized since their inception by the West under slavery, colonialism, the Cold War, and apartheid. The disciplines study the Western-and-white experience as a universal, human, experience; area studies focus on the experience of nonwhite people as an ethnic one. For Ngugi wa Thiong'o (1986, 95), what needs to happen is to put Africa in the center, to make education "a means of knowledge about ourselves," and then, "after we have examined ourselves, we radiate outwards and discover peoples and worlds around us [and thus be] better able to embrace and assimilate other thoughts without losing [our own] roots." To assimilate with being assimilated. To put Africa at the center is to make the everyday the focus, echoing a much earlier suggestion by the Indian intellectual Rabindranath Tagore in 1902. Following Tagore, in 2003 Ranajit Guha rejects Hegel's concept of world history as reducing human history to states, empires, great men, and clashing civilizations and calls for a return to "the quotidian experience of ordinary people." Individual researchers, institutions, and communities are implored instead to develop methods that capture and are based on the idioms of everyday life, and to approach issues, needs, and capacities that all human beings share from our own cosmologies and cultures, from who we are, and who we could be and who we certainly do not wish to be. This combative return to everyday life was the central theme of Mamadou Diouf's seminal keynote address at

ECAS calling for a revisioning of Africa as a subject of research, in which he invoked Tagore and Guha when rejecting that "history is the science of the nation" systematized by Hegel's "world history" (Diouf 2019).

The second archive, language, is what African intellectuals like Ngugi wa Thiong'o (1986) and Henry Odera-Oruka (1983) deferred to in order to position Africa in the center, to rethink knowledge as an archive for understanding the dynamics of everyday life. What was problematic to D. A. Masolo (1991, 1005) was how these scholars deployed African languages as simply raw materials that the academic mines for knowledge. Masolo therefore urges us to position African knowledge as a joint outcome of conversations between scholars, cultural practitioners or experts, and "the social actors of everyday life."

One of the problems with such brilliant ideas is that they too often end up like the clever rabbit that hides behind the book pages. In *chidzimbahwe* we call it *tsuro yemubhuku* (the book rabbit). It is flat. It never leaves the pages. It behaves like its creator, the academic, who never leaves the academy, who creates knowledge for the sake of publishing for readers, to solve academic, not everyday, problems. "It's a division of labor, Chakanetsa." A good friend checked my frustration. Some will create knowledge; others will turn it into projects and programs.

That is exactly how the Western-and-white humanities and social sciences project is designed. That is how it has been since its foundations; it studies society, each branch its own method. Often, to develop a critical mind is enough; always to advance the values implicated within, defined by, specific societies. Because, after all, the critical mind is always particular, not universal. It is a creature that lives within its habitat; the fish that exercises adventure by exiting the pond dies. In the current order of Western-and-white humanities and social sciences, to think critically with that mind is enough. To research evidence and analyze it in ways that enable one to write a narrative—that is what we teach. We do not have to build anything; we write books, articles, give papers at conferences. We don't go around building stuff—physical buildings, infrastructures, equipment—(and I don't mean metaphorical ones); that's for engineers.

The problem is, for Africa and marginalized communities more generally, those that are charged with designing, making, and building are as theoretical (do not translate their academic skills) as the humanities and social sciences, which are not even constructively critical or doing social research relevant to Africa's needs or, better yet, conducting research for the express purpose of solving a problem. The culture of knowledge for the sake of knowing, never translating. In most cases the only knowledge that translates is the consultancy report these local academics do for nongovernmental organizations, for their overseas partners who literally fund them, and who often end up being their masters because they are funding the research. They conceive the project.—Select who to 'partner.'—Approach them with an already set agenda.—Assign Africans roles.—Timelines.—Set deadlines.—Budgets.—Hypotheses derived from western-and-white scholars.—Address only those questions relevant to western institutions.—The National Science Foundation.—The National Endowment for the Humanities.—The Social Sciences Research Council.—Department for International Development.—Deutsche Akademischer Austauschidienst.—Swiss National Science Foundation.—European Research Council. Tellingly, most of these are taxpayer-funded, they sponsor their citizens (individual and institutions) to do research relevant to their countries, and do not allow their tax dollars, pounds, and Euros to be used to build or benefit the "fieldwork sites" (Africa).

Science is what *they* call science.
  Not what *we* call science.
Technology is what *they* mean by technology.
  Not what *we* think, experience, and purpose as such.
Art is what *they* call art
  Not what *we* mean.
And innovation is what *they* say it is.
  Not what *we* say it is.

All the money is controlled by *them.*—
  The budgeting.—The how much, how, when.—*Them.*—

The outcomes, applications.—*All them.*
  Them Them Them. T.H.E.M.

As bad as that seems, we black people must carry our own cross without expectation that systemic racism and racist attitudes will wake up gone one day soon. We must know what we can control. Some of these things happen in Africa and in our communities because we allow them to. We allow them to because we are not self-reliant. We are not self-reliant not because we don't have resources to be so. No. We have so much mineral wealth, oil, gas, fertile land, water, a very creative, hardworking, and resilient people. But very bad education systems that are based on just studying and passing exams. That confuse qualifications for certificates where a better measure is application of the certified knowledge. Very little to no real-life extension of academic-learned skills to everyday problems.

A useless education and knowledge.

OUR HISTORY AS A TOOL FOR SELF-INSPIRATION

And yet, throughout black history home and abroad, there have been moments where Africa's children have shown what knowledge in the service of and through problem-solving can be. It is those Africans that have inspired me in my own efforts as an intellectual creating new spaces for transhemispheric conversations between knowledges murdered by Western-and-white enslavement, colonization, and continuing systemic racism as ingredients—along with deimperialized Western-certified knowledges that we find relevant to local situations—in addressing the problems of Africa and its children. I selected these knowledges for their practical value to my own project, whose objective is *to invent an education or knowledge at the intersection of the university and everyday (or community), that equips young people to organize for a happy, equitable, and productive society.*

Dare to Invent the Future draws its title from one of the most famous speeches by one of the most inspirational African figures of the twentieth century: Thomas Isodore Noël Sankara, president of Burkina Faso, assassinated in a hail of bullets aged just thirty-eight, on October 15, 1987. In this particular speech in 1985 and indeed his short-lived presidency (1983–1987), in word and deed, Sankara bravely confronted the realities of his country and Africa, and the world context within which it had to exist. Realizing the impossibility of freedom as a gift, Sankara would declare later that freedom can only be conquered (struggled for). Such a struggle, he said, was possible only through an act of daring and the taking of calculated risks.

The purpose of this book is to look for moments in Africa's story where precedents of critical thought and knowledge in service of and through problem-solving are evident, in order to inspire ourselves to create and deploy knowledge in service of problem-solving in our own time. In other words, not always or necessarily privileging academic enquiry, writing, or seeking to know as the objective or primary goal or purpose, but with that as outcome, formula, or byproduct of solving problems through making and building.

I am only interested in reading the critical thinker-doer in these texts, as instantiations of knowledge or theory in the service of and through problem-solving. I resist being enlisted into arguments and judgments that others have made about these personalities. There is no one Blyden, Césaire, Booker T. Washington, Fanon, Cabral, Nyerere, or Sankara; we can all find a little piece of each that interests us.

We write for different reasons. Some are just looking for a good story; I don't write for that audience. Some are fascinated by these African figures; I am serious about what they mean, their words, their whole life's work. Some write for an audience; I write in aspirational, anticipative mode with an eye on the long future, especially beyond my living years. I do not write for fashion or flavor. I write necessary knowledge, for who it is necessary. Any other description of my work are your own labels.

We read for different reasons, look out for different things. Some will be excited about these "men of ideas"; but people do not eat ideas—only products of such, acted upon. Some will search for the individual; I just prefer to see how these black men, privileged to have gone to school as high as they did, in a time when only black men were allowed to go to school or that far with formalized education, extended and strategically deployed that knowledge to fight for and institute inclusion of everybody. Consequently, for their time, it was inevitable that these men led the struggles of their time: their education was not simply for men, but everybody under the knee of oppression. Born of a mother and father, having many sisters, a wife, and a daughter, as an African I relate to women not according to their biology; always relationally, socially, fates intertwined and interdependent. I chose these specific leaders not because they make "fascinating" reading, but as examples of a selfless duty to the oppressed race in total and the role of knowledge or education as a self-acquired tool for organizing everybody to conquer freedom. It is a tribute to ourselves as black people that more women have since attended university and dedicated the knowledge they have acquired toward the service of their own communities than the colonizer ever allowed. The rationale for picking the critical thinker-doers discussed lies in their being far ahead of their times in shuttering Western-and-white-imposed barriers to education not only for women, but black people in general. Lest we get carried away, I don't simply mean access to education; I am talking about knowledge in the service of and through problem-solving. Do novelists, poets, distinguished academics qualify? To which I answer: yes if engaged beyond writing, the academy, and critique; building or making, not simply writing.

I mean, as in my case, writing knowledge that aids the strategy and implementation of work I am doing on the ground, with the people, learning what they know, sharing what I have learned, for I have traveled and seen more than they have been privileged to, and they have produced more knowledge of/in the local while I was away. Hence there is no learning without sharing, no sharing without learning. Sharing may mean sweaty clothes and blistered palms, to build a better future with the people, while generating new knowledge through working. The figures in this book inspired in me the excitement to glue my academic knowledge and physical labor with sweat, thereby leading through the power of example. *I write as—I am—a catalytic space.*

Some will see in some of these figures great statespersons and stalk the ire of those who see in them ruthless rulers who imprisoned their opponents. Others will see Uncle Toms who compromised where confrontation was necessary; and they may be

FIGURE 0.4
Electrifying our own village, Ward 17, Marondera West, Zimbabwe, 2017–2018. *Source*: Personal archives.

justified if the subject of judgment is the specific milieu in which these figures lived and acted. I only read them from now, a century later, where suddenly industrial education, responsibility to community, and entrepreneurship are the necessary but missing elements in black struggles for self-emancipation, especially in Africa, and Booker T. Washington's Tuskegee model, when read with an open mind, provides precedent and clarity on potential futures. At the same time, figures like W. E. B. Du Bois, whose works may have been critical, inspirational, and instrumental in another time, are dear to me and continue to inspire, but I may find little that is useful of them as ingredients for what I want to build with the people on the ground. I am not looking from a militant, cultural or political revolution as Césaire, Fanon, Cabral, Nyerere, or Sankara did; I am talking about rendering African independence and civil rights gains in the United States fruitful.

Knowledge or theory in the service of/through problem-solving entails a pragmatic choice of archive and a strategic deployment of the material, which becomes raw material for the design and making of entirely new tools. *It will not always conform nicely to the conventions of the academy because the academy itself comes with deep historical baggage and a violence by method and discipline that may be, in fact is, the problem.* And the problem is that, for an African or an African American, while we fight and wait for the long-dragging demands for reparations against the injury to our being (performed through race) to be answered, we have a responsibility to our communities because of the success we have become. And what Geri Augusto calls "race duty"—responsibility to community, to remember through critical thinking-doing—is not accounted for in the systems that govern the production of knowledge and the current stations of our work. The irony is that we are potentially the best resource for our institutions to impact and solve the racism and poverty problem highest in our communities. *We are a living example of creative resilience—the spirit that, where death is certain, and people's backs are against the wall, they do not just wait for fate or raise their hands in surrender, but will fight in order to live and, if it has to come to it, die fighting.*

Let's now be clear. The Booker T. Washington of this book is not the "compromised" nemesis of Du Bois or the figure of the Uncle Tom but a black person who deployed a self-acquired education to dare to invent the future against impossible odds, in his case an industrial education based on *hand, head, and heart*. The commitment to community development, to restoring the dignity of work destroyed by slavery, to knowledge in the service of/through problem-solving, and business acuity—the idea that knowledge must bear fruits—is simply without equal. The Césaire and Fanon of this book are not the theorists of colonialism, nor the national heroes of Martinique and Algeria claimed for their intellectual contributions by France but (r)ejected for their rebellion against Paris. Rather, I walk tall as a black man because of their Négritude and its daring. How could these two black men and their colleagues, women and men, even start to imagine something worth rescuing out of the ultimate symbol of their very own malign: their skin color? And if I focus on Amílcar Cabral, it is not the son of Guiné and Cape Verde, the domestic figure and hero, but a living example of what he called "the weapon of theory," from whence came my own philosophy of knowledge and education. The Nyerere of this book is not the Tanzanian head of state, the African statesman, architect of the liberation of Mozambique, Angola, Zimbabwe, Namibia, and South Africa, or the teacher, *mwalimu*. He is, rather, the architect of self-reliance and the role of education in the service of and through self-reliance, whose fundamental questions and possible answers not just Africa, but the twenty-first century need to know about and think-do with. Finally, my Thomas Sankara is not the founder of Burkina Faso but the eternal spirit of his youth, which lives through us in daring to invent the future. In our time.

OUTLINE OF THE BOOK CHAPTERS

The black people whose inspirational deeds are explored in this book embody that daring and creative resilience, a spirit for the world within our own hands to make, shape, build, and enjoy. This book serves as a mind-reengineering tool for a generation that will make and live in that future, and my audience depends on who needs it. All will share one thing in common: at best a commitment to, at the very least curiosity in, the title. And when that future, fair world finally happens, I want to make sure that these black people are on the time stamp.

In essence, *Dare to Invent the Future* is a careful rereading of critical Africa-centered texts showing how the disciplines have been, and could be, deployed in the service of and through problem-solving, building on what people are doing and know. That happens not because the curriculum is set up to achieve that, but because of creative resilience and strategic deployments by those who find themselves without jobs or facing threats to their existence. At core, knowledge in the service of and through problem-solving derives from reading the past for and asking it new questions, performing serious due diligence on the present, and contriving an anticipative approach to futures, in search of the best ingredients and instruments to craft and execute recipes for building a better, more verdant and sustainable planetary existence.

I write this book in search of exemplary moments in knowledge in the service of and through problem-solving. Moments in which the enterprise of knowledge among black folk promised something fresh and different. Call it a rebellion from the Western-and-white enlightenment tradition of disciplining (and among black folk that always comes with the colonizer-slaver's whip bludgeoning our own orders of knowledge and ways of knowing into submission). A rejection of knowledge-in-rooms and an embrace of open spaces of knowing, found anywhere were people and animals exist. A humility before knowledge, wherein my ancestors opened themselves to being educated even by animals. A revisitation of the history of African knowledge production as knowledge in the service of and through problem-solving, to ask it of serious questions, to mix it with what we know from other cultures of knowing, on the way to creating something useful to us.

The book does this in five sections. Part I, "Mindset Re-engineering," takes us back to black experiences of Western-and-white enslavement and its legacy of systemic racism today in and outside Africa that necessitates daring to invent a future without systemic racism. It is composed of three chapters. Chapter 1, "Systemic Racism: Starting Out from Our Dehumanization," explores the origins of racism against black people as an idea, practice, and experience. It first traces the roots of dehumanization, reading closely the writing of the Jamaican laureate, Sylvia Wynter, concerning the question of "the human," before turning to the experiences of race and the psychological effects of racism, as experienced by Aimé Césaire and his student, Frantz Fanon. The next two chapters turn to the task, begun in *The Mobile Workshop*, of self-rehumanization and self-reintellectualization after, and it the face of, the dehumanizing experience of slavery, colonialism, and racism. Chapter 2 focuses on the pioneering work of Edward Wilmot Blyden, a black man from whom can be traced the origins of a tradition of black people strategically deploying historical methods to restore the personality of the African animalized and thingified by the racist portrait. Chapter 3 turns to the deployment of poetry in restoring the African personality, this time turning to the black experiences under French rule, through the life, writings, and interviews of Aimé Césaire. All three parts of the chapter enable us to see clearly that the world we live in is ordered according to a colonial, racist, and unsustainable formulation of the white man as the normative human, along with his values and biases. Being black isn't something one sees or observes; it can only be felt, experienced, and lived. The failure to be in the body and mind of another, to be black, is the source of indifference.

Emphasis is on "self-rehumanization" because rehumanization is not a gift; it is a burden we the dehumanized must carry, must lead in carrying, the question being: What kind of knowledge is needed to conquer freedom? And what knowledge does the process of self-rehumanization produce? Well before Césaire, there was Edward Wilmot Blyden, one of the most published black authors of all time, statesman, and historian of African and the Africans. It was he who introduced the terms "African for the Africans," "Pan-Africanism," and the "African personality." The objective of the chapter is to examine how Blyden assigned to historical research and writing the purpose of engineering and restoring personality to the African deintellectualized, dehumanized, and depersonalized by the slaver, colonizer, and racist. Césaire, Senghor, and Damas would continue this restoration in the French context through Négritude, an early twentieth-century cultural-intellectual movement that marshalled and strategically deployed poetry and history as instruments of self-rehumanization. I am inspired by how this movement, led by Césaire, Leopold Sedar Senghor, and Leon Damas, inverted the blackness (*noir, négre*) that the white man had deployed as a weapon of dehumanization into the core of a black attitude, hence Négritude, that was assertive, confident, proud, and solidly aware of its great pasts that the white man had falsified as primitive and worthless. Everywhere in Africa and abroad, as black people awoke to their blackness as self-worth, they developed a Négritude, a pride in themselves, chin up, eyes front, and started walking tall, to demand their independence (Africa) and civil rights (the United States).

Self-rehumanization and self-reintellectualization are neither a gift nor always bloodless. In most African countries and even in the United States, neither the liberation struggles nor the civil rights struggle were entirely peaceful. From Algeria to Guinea Bissau and Cape Verde to Angola and Mozambique to Zimbabwe and Namibia to South Africa, it took armed struggle to attain independence, with self-pride (be it African personality or Négritude), solidarity (Pan-Africanism), theory (Marx, Lenin, Mao), and guns (the AK-47, bazooka, and landmine) as the preferred tools of struggle. Typically such struggles are confined to the realm of the political and military history, often by political elites to shoeshine their credentials and right to stay in power forever.

The purpose of Part II, "Freedom Struggles as Innovation," is to position not only liberation struggles or wars of independence but also sites of protest and protest movements as sites of innovation, with struggling itself as innovation. The next three chapters are devoted to the sites and actions of struggling—specifically national liberation struggles—as sites of knowledge in the service of and through problem-solving and creative resilience. Chapter 4 draws on Frantz Fanon's *A Dying Colonialism*, specifically three chapters on the veil, radio, and medicine, as examples of strategic deployments illustrative of creative resilience. Chapter 5 turns to the production and extension of theory in the service of and through problem-solving, what the Guinéan and Cape Verdean freedom fighter and agricultural engineer Amílcar Cabral famously called "The Weapon of Theory." Theory is deployed as a method or tool for investigating and understanding the reality, problematizing it for solving, with the intention of designing a solution or strategy that we do not hand over to others to execute, but which we, as critical thinker-doers, are duty bound to execute. We are not just theorists that leave it to others to build, nor are we makers without a deliberate plan; above all we are the catalysts right in the thick of critical thinking in service of and through problem-solving. The chapter therefore revisits Cabral's speeches, interviews, and lectures to explore how theory has been positively weaponized in seminal moments of African innovation. In that "Weapon of Theory" lecture, Cabral implored the educated elite to "commit suicide as a class in order to be reborn as revolutionary workers, completely identified with the deepest aspirations of the people to which they belong." Nobody who has not yet done that can produce knowledge in the service of and through problem-solving. Elements of this phenomenon spill over into chapter 6, where I discuss the liberation struggle in Zimbabwe as an innovative space where guerrillas and villagers drew upon both allopathic and indigenous medical practices, techniques, tools, and materials to build healthcare systems in the bush. The chapter takes the conversation away from the political leadership to the ground level, among the masses that provide the water (information, camouflage, food, and other logistics) for the fish (guerrillas with guns) to swim (engage in mobile warfare against the Rhodesia regime's forces). All of these efforts were innovations in the service of freedom, which could only be realized through independence. The next section thus turns to the moment after emancipation (the United States) and independence (Africa).

The question animating Part III, "To Render Freedom Fruitful," is this: What kind of knowledge is needed to render successful struggles against slavery and dehumanization fruitful, to render such gains permanent, and to build a nonracist—equal—society? It focuses on the two moments after the edict of emancipation from slavery (the United States, 1865) and the declarations of independence from colonial rule (Africa, 1957–1994), in which black people now had to invent tools to render their hard-won struggles

FIGURE 0.5
A poster of showing a female and a male freedom fighter of the Zimbabwe African People's Liberation Army (ZANLA), 1976. *Source*: African Activist Archive, Michigan State University.

fruitful and to defend them against erosion from within and without. What kind of knowledge or education was needed to build a new society, given the legacies of slavery and colonial education that did not equip black people with hand, head, and heart skills? The first two chapters focus on the founding and emergence of Tuskegee Institute under Booker T. Washington in 1881 and the echoes of the Tuskegee model in the emergence of Black Wall Streets in the United States, as well as Washington's impact among proteges who extended his model to international business and the establishment of educational institutions. The next chapter on the role of education and the university in independent Tanzania under President Julius Kambarage Nyerere shows Washington's ideas as one of the many ingredients that newly independent Africa's leaders acquired to design educational systems to render freedom fruitful. The other critical ingredient, discussed in chapter 9, is African communality or community, which they deployed to define African socialism, the philosophy of governance within which Nyerere's notion of education for self-reliance can best be understood. Nyerere defined the purpose of education, knowledge and, more specifically, the university through a question: What kind of a society are we trying to build? The role of the university, Nyerere argued, must be applied, not pure or basic, scientific research to produce *necessary knowledge* that solves our particular and urgent problems, not luxury knowledge that panders to theory without application.

The last two chapters focus on the aftermath of independence, wherein the founding fathers' promises fell far short or no attempt was made to deliver. Some were failed—by US-Soviet Cold War meddling, by ethnic conflict, by continuing European colonial powers, and so on. Others were distracted by power and greed. At that point, in the late 1970s–1980s, the younger generation saw it fit to escape the ancient regime by daring to invent the future. The face of that generation of inspirational thinker-doers is Thomas Sankara (today it is Paul Kagame). The spirit of that generation, and that type of leadership, is summarized in Thomas Sankara's famous 1985 (232) statement: "You cannot carry out fundamental change without a certain amount of madness. . . . It took the madmen of yesterday for us to be able to act with extreme clarity today. I want to be one of those madmen. . . . *We must dare to invent the future*" ("Dare to Invent the Future" 1985, 232; my own emphasis). It is fitting that right after this chapter, the conclusion follows up with "Why We Must Dare," devoted to the present moment and the African struggles of our own time and that define our own "generational mandate," looked at from deep historical trends.

In its place, some of us are turning the village and community into the catalytic space for training students; they do not have to leave for the city-based university to acquire an education. (That story deserves its own book on knowing in order to solve and through solving; another book will focus on the knowledge itself. That is why the two had to be deferred, with the current book laying the foundation.) The everyday becomes not just the open class site, but a professor in its own right. As a member of my marginal community, I have long since returned as a catalyst and leader in my own community's self-upliftment. To offer myself up as a visible role models to the youth, especially during childhood. To inspire in them a "responsibility to community," so that when they come to campus one day, they remember to bring their community with them, as a prismatic compass guiding them in their choices of classes to take, each problem back home serving as purpose shaping subject choices. Forging networks to turn problems into new opportunities. I write this book no longer in speculative mode, but someone who is doing what I am saying, and thoroughly enjoying it.

My community speaks through my contributions on campus; my university speaks through the catalytic space I have established. This is an outcome of a realization that, for all my contributions to academia, the many lectures I have delivered, conferences attended, and the world traveled, up until 2016 the "impact" was completely unfelt or known in Mavhunga Village and the parent ward, Ward 17, let alone the country or continent. The children of my village are a metaphor of my having not done anything (barring the close family), a sad indictment of the knowledge I produce and the education I received and offer. I see them every time I am back in the village, starting for school early, around 6 or 6:30 a.m. I see them when driving from my village, from Border Church to Ten Miles—trudging to and from school. I always stop to let them hop onto the truck if empty or when not rushing to Marondera or Harare for business. To and from, every weekday, often carrying *mangai* (dry maize grains boiled thoroughly), *mutakura* (usually of *nyimo* or *nzungu*, our varieties of nuts), the signs of extreme poverty.

They arrive weary and sweaty and hungry.
Just as I did in the late 1970s–1980s.
Parents continue to toil in the vegetable gardens and fields for meager earnings to feed their hungry children.
Just like mine did.
So what is our educating busy doing?

I MINDSET REENGINEERING

1 SYSTEMIC RACISM: STARTING FROM OUR DEHUMANIZATION

> I hear the storm. They talk to me about progress, about "achievements," diseases cured, improved standards of living.
>
> *I* am talking about societies drained of their essence, cultures trampled underfoot, institutions undermined, lands confiscated, religions smashed, magnificent artistic creations destroyed, extraordinary possibilities wiped out.
>
> They throw facts at my head, statistics, mileages of roads, canals, and railroad tracks.
>
> *I* am talking about thousands of men sacrificed to the Congo Ocean. I am talking about those who, as I write this, are digging the harbor of Abidjan by hand. I am talking about millions of men torn from their gods, their land, their habits, their life—from life, from the dance, from wisdom.
>
> *I* am talking about millions of men in whom fear has been cunningly instilled, who have been taught to have an inferiority complex, to tremble, kneel, despair, and behave like flunkeys.
>
> They dazzle me with the tonnage of cotton or cocoa that has been exported, the acreage that has been planted with olive trees or grapevines.
>
> *I* am talking about natural economies that have been disrupted—harmonious and viable economies adapted to the indigenous population—about food crops destroyed, malnutrition permanently introduced, agricultural development oriented solely toward the benefit of the metropolitan countries; about the looting of products, the looting of raw materials.
>
> They pride themselves on abuses eliminated.
>
> *I* too talk about abuses, but what I say is that on the old ones—very real—they have superimposed others—very detestable. They talk to me about local tyrants brought to reason; but I note that in general the old tyrants get on very well with the new ones, and that there has been established between them, to the detriment of the people, a circuit of mutual services and complicity.
>
> They talk to me about civilization, I talk about proletarianization and mystification.
>
> —Césaire (1955/2000, 44)

The "*I*" (italicized) in this poetic passage from Aimé Césaire's *Discourse on Colonialism* is the felt category of the black enslaved. Elsewhere in the context of Zimbabwe, I have said that all words derived from the "colon-" (colonial, colonialism, colonizer, colony, decolonial, noncolonial) must be viewed as imposed categories or responses thereto, and that those who suffered the colon-actions describe their experience as slavery/*hunhapwa* (Mavhunga 2018). When Césaire juxtapositions what they (colon-) are talking about with what the "*I*" feels and then vocally expresses, it doesn't matter how it was meant or intended or how the colon- interprets it. The "*I*" feels and experiences the effects as "*my*" essence drained.—Culture trampled.—Institutions undermined.—Lands robbed.—Religions denigrated and trivialized.—Artistic creations destroyed.—And self-determined futures rendered impossible. The colon- may cite all sorts of facts, statistics, and cement and steel (infra)structures built, the agricultural and industrial production, but "*I*" have experienced the pain of thousands of black lives and energy sacrificed to its construction, torn from "*my*" gods, thrown off "*my*" land to make way for such, how it feels to be made to live in fear, in a complex of inferiority, to tremble, kneel, and despair, to lose confidence in myself. The "*I*" have felt their indigenous knowledge-based economies, harmonious and viable, destroyed; .—Food crops destroyed.—Permanent malnutrition introduced.—Cash crop economies to feed the colon- introduced.—Natural resources extracted to Europe as raw materials, and all the value addition located on the other side of the Atlantic and Mediterranean. The colon- beats his chest on how many abuses he has eliminated—ending slavery, liberating African women, the girlchild, tyranny and bringing "light" to "darkness," hence "enlightenment"—yet Césaire, without dismissing the "very real" ills Africa had before the colon- arrived, says the colon- added new and far worse ones, including the European tyrants forming an alliance with the African ones to oppress black people.

The relationship between the colon- and the black person, for Césaire, is one of thingification, the reduction of black people into a thing, an object, a nonhuman as distinct from the human, who is the white man. This dehumanization is simultaneously a deintellectualization, wherein black people were now studied as objects of science rather than producers of science in their own right. They were now turned into beasts of burden and tools of mass production on the plantations and laboratory samples, their bodies and skulls acquired in places like Namibia and Tanzania through very gruesome methods to feed the medical anthropology that churned out very racist statements sanitized as science (phrenology, eugenics), as revealed by the skulls in Germany and New York, which now constitute embarrassment for their erstwhile collectors (Mavhunga 2018; Gross 2018; Zimmermann 2001).

Every society has a way of defining its humanity; every society has its own meanings of the human. The notion of a global humanity ignores that even at the bipedal level, the biological, let alone the color line, the definition of the human starts at the level of the colon- and that of the felt/experienced/lived. In this chapter we have three writings that deal with the human at the level of the colon-, in the first instance, and how that hegemonic order, when imposed, is felt by those under the knee. The first is the work of Sylvia Wynter on the origins of the racist notion of the human in Western-and-white society, which then gets extended with Western-and-white imperialism from 1500 to the present. The other two are the seminal writings of Aimé Césaire and his student/mentee, Frantz Fanon, who we will meet again in a later chapter.

The point I make is that those who have felt/experienced/lived under the knee do not take the status of being human for granted and thus declare it the most important struggle of the ages: to be human among, with, and like others. The condition of being denied a self-defined humanity through being enslaved or subject to racism can grind the enslaved under in despair, resignation, and acceptance of the right of the colon- to keep the knee on the neck, or it may also inspire those under the knee to fight, to understand that freedom cannot be given, but must, instead, be conquered.

ORIGINS OF WHITE-ON-BLACK RACISM

The Jamaican scholar Sylvia Wynter has traced the deep historical origins of white systemic racism more than anyone I know. After the original Adamic sin, in which Adam and Eve connived to eat the forbidden fruit, a theocratic map of heaven and earth emerged. The earth was vile, base, mortal, and the heavens perfect, divine, eternal. The earth was condemned to fixity and immobility (the unmoving). Earth was the post-Adamic "fallen" mankind. Hence the legitimacy of the church to rule over the world. From the European renaissance to the eighteenth century and thereafter to present, that definition has been based on natural causality, with nature as an autonomous force that is governed by its own laws. The European physical and biological sciences were based on this despiritualized, so-called "natural" conception (Wynter 2003, 264, 267). With the collapse of the Roman Empire and the rise of a Judeo-Christian Europe, this classical Greco-Roman (i.e., Ptolemaic) astronomy was now Christianized in ways that shifted the division between the divine and heavenly and the evil. Now the difference was between the redeemed spirit (represented by the celibate clergy) and the fallen flesh condemned by the original Adamic sin (laymen and laywomen). Medieval Latin-Christian Europe's idea of the world and order was mapped according to spiritual perfection versus imperfection (Wynter 2003, 274), a break between the supralunar and the sublunar, heaven and earth. The spiritual was the clergy's domain; that of earth became a physical reality whose "facts" must be explored and known (discovered).

The break from the Ptolemaic to Judeo-Christian, Wynter says, made possible the emergence of a natural-scientific "objective set of facts" view of physical reality

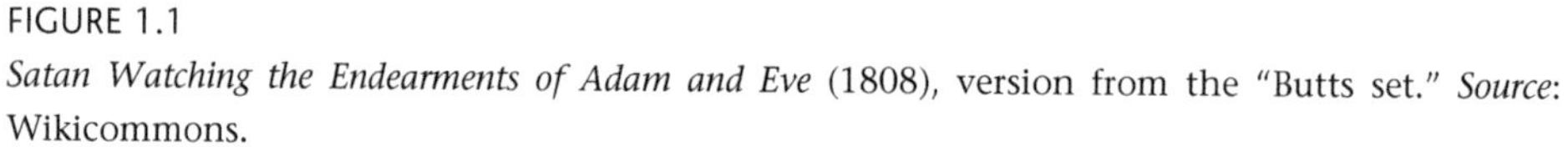

FIGURE 1.1
Satan Watching the Endearments of Adam and Eve (1808), version from the "Butts set." *Source*: Wikicommons.

concerning what was happening "out there." This is what now guided the Portuguese voyages of the fifteenth century to and around Africa to the East and Christopher Columbus's across the Atlantic. Up until then, Europeans had been totally ignorant about and never been to Africa and the Americas; in fact, western Europeans speculated the Atlantic to be unnavigable, Black Africa and the Americas uninhabitable and "outside God's grace," the one "too hot," the other "under water" (2003, 275). Until Columbus, therefore, anything beyond "Cape Bojador" (upper coast of Africa) was assumed to be too hot for habitation, and the Americas under water, both enslaved to original sin. The shock of discovering that Africa south of the Sahara was habitable and the Americas not underwater commenced the process that led to the rise of the physical sciences, and eventually, the gradual freeing of physical reality from the Spirit (Wynter 2003, 280).

Europe invented race as a construct to replace the earlier mortal/immortal, natural/supernatural, human/ancestors, and gods/God distinction. It needed racial difference as a tool for global expansion and domination based on a human/subhuman distinction of colonizer/colonized. The sixteenth-century dispute between Bartomé de Las Casas, the missionary priest, on the one hand, and the humanist royal historian and apologist for the Spanish settlers of then Santo Domingo, Ginés de Sepúlveda, on the other, was about who/what could be human or not. The expansion of the Spanish state followed that of the Christian evangelizing mission and its theocentric conception of the human (Christian) as a subject to the Christian state. The colonizer of the day, the Spanish state, appended race to a "natural" difference between Spaniards and the indigenous peoples of the Americas, a difference almost akin, Sepúlveda insisted in the sixteenth century, to that between monkeys and men or true humans (Wynter 2003, 270, citing Padgen 1982).

The Las Casas/Sepúlveda debate clarified and justified the terms under which the colonizer and colonized would relate. Namely, indigenous peoples as slaves by nature, hence justifying the colonizer's occupation and enslavement of land that belonged to these othered. The Spaniards designated indigenous peoples as pagans needing salvation, as inferior "children are to adults, . . . almost as monkeys are to men," thus justifying salvation and civilization by the adults and men—the white man! Las Casas rejected this notion and proposed that Africans be imported as slaves "in order to liberate the Indians" (Wynter 2003, 283–285).

The indigenous peoples of the Americas/the Caribbean (now simply called [red] Indians/Indios) and the enslaved peoples of Africa transported across the Atlantic (classified as Negroes, *negros/Negras*) were through the discipline of what later became anthropology now constructed as the physical specimen of the irrational or subrational human other, with the blacks being constructed as the ultimate referent of the "racially inferior" human other. Other non-European, dark-skinned peoples were also classed as "natives" (as savages), a register later given ultimate "scientific" fodder by Charles Darwin on the basis of a biocentric model. Race was now naturalized. White became natural man/human, completely evolved/civilized, while black was on the way or not yet started: the case for the black's legitimate subordination, indeed the white colonization and civilization of the world, was complete. Colonial curriculum then systematically entrenched the negative representation of the black and their ancestral pasts in the colonies, to self-hate, self-doubt, even self-mortify, to seek closeness to whiteness and its traps and trappings as being the marker of civilized, successful, modern.

This is the context in which reason replaced the medieval redeemed spirit, with reason versus sensuality and rationality versus irrationality now the designations for colonizer versus colonized. The Las Casas versus Sepúlveda clash was a theocentric Christian versus ratiocentric (rational) colonizer-colonized relation. Enslavement of othered humankind was no longer the consequence of Adamic Original Sin but irrational human nature, with the secularization of the "plan of salvation" now the white man's burden, a civic humanist one at that! One had Higher Reason, the othered was irrational and sensual—what better justification for taking their lands by claiming it as part of some civilizing mission (Wynter 2003, 288–290).

The Portuguese arrival in Senegal in 1444 and their commencement of enslavement of Non-Christians was justified in theological terms. The pope, who was supposed to be a "holy" man, in full conscience gave the Portuguese and Spanish kings papal bulls designating these lands *terra nullius* (the lands of no one) and available to Christian seizure. They could plunder black labor, and the intellect and knowledge black people brought to perform it, for free, extracting the ores from the earth and growing crops in the ground: this is how western Europe began building the wealth that gave it a liftoff and global expansion from the 1400s onward. The truth guiding it was a Christian versus non-Christian order of humanity, pitching the peoples of Africa and the New World as "pagan idolators," and "Enemies-of-Christ," defined by lack rather than worth, measured against self-anointed and publicized true "Man"—the white man. Peoples of these lands saw European incomers as "abnormal," ghostly, monsters, and spirit-human manifestations. The Spanish and Portuguese states used papal bulls as tools for imperial expansion; the pope was a spiritual, temporal, and territorial sovereign using the state power of both to expand the evangelizing mission, and the evangelizing mission to control the world through spiritual obedience.

The theological grounds for the legitimacy of Spain and its settlers over the indigenous peoples' lands and enslaving them became complicated because they were not "Enemies of Christ": after all, Christ's apostles had not crossed the Atlantic to make the gospel available to them. Therefore, they had not "refused" the Word, they were definitely not Christ-Refusers. How then could the colonizer take their lands and enslave them with "just title"? Las Casas proposed a solution: that the enslaved African, "whom he then believed to have been acquired with a just title, should be brought in limited numbers as a labor force to replace the Indians" (Wynter 2003, 293). By the time Las Casas became aware of the brutality of the transatlantic slave trade, it was too late to withdraw his proposal. A way had been devised to sidestep the "Christ Refusers" requirement; a notary accompanying each colonizing or enslaving force simply read out a judicial document called "The Requisition" (*Requirement*) to forcibly assembled indigenous people, who now "literally heard the Word of the Christian Gospel, so that they could then be later classified as having refused it," and therefore in that state became Enemies of Christ, not based on the Adamic Original Sin but irrationality. "The Requisition" said that Christ was king of the world and had anointed the pope as his earthly representative, and the pope had delegated the king of Spain to occupy the lands of the barbarians, and the present company was there to enforce the king's authority. Accept and you shall be unharmed; refuse Christ and you shall die or become slaves and lose the land too (Wynter 2003, 294).

The black was not to be put in the same bracket of subrationality as the Indian. Well before Darwin, the black was designated the last link in the Chain of Being—a hierarchy linking living organisms worldwide during the European Enlightenment, from the tiniest insect, to birds, animals, Man, the angels, and, ultimately, God at the top. Lower beings existed to serve the "higher beings," plants at the animals' service, animals as meat and pets to man, and men on their knees to God. And the black was placed somewhere between rational humans and irrational animals—without a soul or reason, and (by the late 1530s) legitimate for labor when Indians were declared free de jure. The popularized monotheistic religions—Judaism, Christianity, and Islam—had the nonblack as the normative, universal value and their god(s) the only and true gods, with blacks assigned roles as infidels (*kafirs*) and pagans, "not far removed from the apes, as Man made degenerate by sin." The origins lay in the biblical tradition that all three major monotheisms shared: "that the sons of Ham were cursed with blackness, as well as being condemned to slavery." Black became a diabolical (from Spanish *el diablo*, the devil), sinful color. The black became not just ape, but degenerate descendants of true man (Adam). Descendants of Ham replaced Enemies of Christ the moment that enslaving Africans was packaged as a way to "save their souls." It fitted snuggly into the Judeo-Christian "order of existence," where being a Christian meant being saved from sin and degeneracy, and being a Christian or "being saved" was the only universal mode of being human, comparable to being a Muslim in Islam (Wynter 2003, 302–303, citing Fernandez-Armesto 1987).

Aped, irrationalized, adjudged "significant ill" (mad or lunatic in Europe, replacing the medieval leper), black was no longer just "degenerate by sin" and supernaturally determined via Noah's curse to Ham's descendants but closest to ape—the missing link between true human (rational) and ape (irrational). There was only one problem: the absence of a being that would slot in between animal and man. So the chimpanzee was brought into Britain in 1699—the perfect link, of course! Not exactly. By 1677, William Petty had already found the elusive link: "the savage." Then in 1735 the Swedish biologist Charles Linnaeus identified the faculty of reason that he said distinguished primates from humans to be a result of varying climates and cultures. There was *Americanus* (the

Native American), "red" with black hair and sparse beard, stubborn, angry, "free" and bound by "tradition." *Americanus* was inferior and uncivilized. There was *Asiaticus*, "yellowish, melancholy, endowed with black hair and brown eyes . . . severe, *conceited*, and stingy. He puts on loose clothing. He is governed by opinion." *Asiaticus* was a mediocre species. Next, *Africanus*: unmistakable by his skin color, facial structure, and kinky hair; cunning yet passive, inattentive and guided by impulse, the female "shameless" and "lactat[ing] profusely"! Then there was *Europeaus*, towering gracefully above them all like a giant elephant or giraffe, "changeable, clever, and inventive," with "tight clothing," and "governed by laws." Europeans started identifying with the color white, in their eyes and minds the color of perfection (Hossain 2008).

In 1837, Charles Robert Darwin embarked on an imperially funded expedition to the Galápagos Islands of Ecuador on board the *HMS Beagle*, his mission to observe and document the flora and fauna, their history, and their variety. In case one thinks the British have been unique at this, Bennetta Jules-Rosette reminds us that Jean Louis Armand de Quatrefages de Bréau, the first French scholar to teach anthropology, had in 1856 described anthropology as a branch of zoology wherein human anatomy, physiology, and customs of their development was studied in the complete evolutionary sense. "Fossils" and "savages" were appropriate objects of anthropological inquiry; anthropology was a "branch of zoology or mammalogy [in which] man must be studied as an insect or a mammal" so that the "hygienic conditions necessary for each race" would be understood (Jules-Rosette 1991, 945).

Darwin made biocentric man the "native," an object of fascination for evolutionary theory and natural selection, and not just Shem's descendant or savage nature, but racist science. Says Wynter: "It was to be the trope of 'purity of blood,' together with that of its threatening 'stain' (itself a 're-troped' form of the matrix negative construct of the 'taint' of Original Sin) that, once re-troped as 'racial purity,' would come to be attached to peoples of Black African hereditary descent" (2003, 308). The black became the "bottom marker" on a universalized human "scale of being." Both in the Enlightenment's "nature" and later in Darwin's "evolution," the blacks were permanently placed bottom in what mattered to white Man. Early sixteenth-century writing began to classify the Indians as "white" in contrast to "black" Ethiopians, white or "light" (hence the Enlightenment) being the "normal" race, black being the "abnormal" (hence Darkness) with people born of white and black classed "mestizo" and white/black "mulatto"—the nonwhite being the pollutant to pure white stock. The blacker the less pure—mestizo more human than mulatto—and less human. Having started from the Adamic Original Sin, via Noah's curse to Ham, by the time of Darwin, white craziness and fascination with the black body had entered the very center of natural-scientific knowledge at the biological and physical levels of reality (Wynter 2003, 309).

FIGURE 1.2
Da Vinci's *Last Supper*. *Source*: Wikicommons.

FIGURE 1.3
Michelangelo's *The Last Judgment*. *Source*: Wikicommons.

From 1495 to 1498 Leonardo da Vinci would depict Judas Iscariot unmistakably with darker, kinkier hair than other disciples in *The Last Supper*. The black comes in as a figure of betrayal. If that picture leaves room for ambiguity, there is more clarity with the figure of Jesus, who becomes white in the same picture and in Michelangelo's *Last Judgment*, and whose father, therefore, must also be white. And white becomes the color of purity; white values become the marker of excellence. The black has therefore no soul and plays no positive part in the narrative of the divine.

This is the context within which the disciplinary divide between the humanities and the sciences emerged. In *The Souls of Black Folk*, Du Bois (1903) called it the "Color Line," one drawn between the lighter and darker peoples of Earth, enforced by laws and social domination (Wynter 2003, 310). And guns. And death. . . . And the knee on the neck. And racist policing. In the post-1950s era, the new divide shifted to a postcolonial division of the world between the "developed" First World versus "underdeveloped" third and fourth worlds and a "developing" world in between—the evolutionist language of dehumanization before and since Darwin. If the natural sciences managed to develop a descriptive statement of Man in its "second, purely secular, biocentric, and overrepresented modality of being human," Wynter asks, how come the social sciences and humanities have failed to do so?

THINGIFICATION: OR HOW IT FEELS, HOW IT IS EXPERIENCED[1]

In the lines we opened with, Aimé Césaire reminds Europe not of what it built, brought, or planted, but what it destroyed, took away, or killed. He restores erased histories and realities that Europe rendered invisible.—The Madagascans who had been poets, artists, and administrators.—The Africans who had built the Sudanic empires.—Carved Shango sculpture.—Alloyed and manufactured products out of bronze. Inventors that Europe reduced to mere savages.

And in prophetic mode, Césaire consoles himself: "Periods of colonization pass . . . nations sleep only for a time, and peoples remain" (Césaire 1955/2000, 44). Neither a black return to an idyllic past nor a pastless future, for we are not an "either or" people but those who "go beyond," not revivers of dead societies or prolongers of a present colonial one. "It is a new society that we must create . . . a society rich with all the productive power of modern times, warm with all the fraternity of olden times" (Césaire 1955/2000, 52). *Not founded on hating those who have deeply hurt us through enslavement, colonization, or systemic racism, but remembering in order to re-member ourselves after their past and continuing dismemberment of us.*

Rejecting the argument that justified colonization as bringing development (civilization) to Africa, Césaire's reflections enable us to ask: *Who is to say where an uncolonized*

Africa pursuing a self-determined path would have been by now? African societies had already been participating in deliberate technological contacts with Europe well before actual colonization. Case in point Japan, where Europeanization was "even slowed down [and] distorted by the European takeover." After World War II, peoples of Africa and Asia demanded the development of ports and roads, but colonialist Europe was not just dragging its feet but slowing down indigenous peoples keen to develop their countries. Europe's contact with Africa was a contact of inequality.—Savagery.—Genocide. And it would boomerang in the form Adolf Hitler and Nazism. Europeans were shocked. Césaire wasn't. They had supported Nazism.—Absolved it.—Been indifferent to it.—Until Hitler "applied to Europe colonialist procedures which until then had been reserved exclusively for the Arabs of Algeria, the 'coolies' of India, and the 'niggers' of Africa." Hitler was important as an analytical tool for Césaire to see capitalist society in its larger context, its incapacity to establish a concept of rights for *all* humans—to even consider nonwhites human. Whatever entry point into the narrative of Europe, there is, will always be, Adolf Hitler (Césaire 1955/2000, 36–46).

And yet the whole concept of ruling humankind and making the world obey, did not come with Hitler. The relegation of nonwhite peoples to the white man's menial tool, created solely to serve "a race of masters and soldiers, the European race"; the European colonial enterprise based on "an alleged right to possess the land one occupies"; the strategic deployment of the Bible to justify such theft and plunder. "That was already Hitler speaking!" Césaire says. And he prefers to let the colonizers confess in their own words to their sickness, put them on trial by history, one by one.

Defendant 1: Either it is Colonel de Montagnac in Algeria, a French lieutenant colonel sent to Africa to colonize territory for the French in 1845, massacred innumerable Algerians, before being killed at the Battle of Sidi-Brahim that same year, saying: "In order to banish the thoughts that sometimes besiege me, I have some heads cut off, not the heads of artichokes but the heads of men."

Or it is Count d'Hérisson, secretary and interpreter to General Montauban, commander of the French forces in the Anglo-French expedition to China during the Second Opium or Arrow War, 1860: "It is true that we are bringing back a whole barrelful of ears collected, pair by pair, from prisoners, friendly."

Defendant 2, Armand-Jacques Leroy de Saint-Arnaud, Marshal of France, Minister of War until the Crimean War (1853–1856): "We lay waste, we burn, we plunder, we destroy the houses and the trees."

Defendant 3: Thomas Robert Bugeaud, Marshal of France and Governor-General of Algeria (1843–1846): "We must have a great invasion of Africa, like the invasions of the Franks and the Goths."

Defendant 4: Augustin Gérard, Chief of Staff of General Joseph Gallieni in Madagascar, in 1897, when capturing Ambike, the city of the Sakalava, a people that had lived in peace and never imagined creating any defense: "The native riflemen had orders to kill only the men, but no one restrained them; intoxicated by the smell of blood, they spared not one woman, not one child" (Césaire 1955/2000, 40).

Skulls.—Ears.—Burnt-out shells of homes.—Rivers of blood.—Destroyed cities.—Civilizations destroyed.—Cosmologies murdered.—Thriving, peaceful nations macerated. Nothing left. All gone. GONE! Not coming back. . . . All stubbornly showing how colonization "dehumanizes even the most civilized man . . . , inevitably tends to change him who undertakes it; that the colonizer, who in order to ease his conscience gets into the habit of seeing the other man as an animal, accustoms himself to treating him like an animal, and tends objectively to transform himself into an animal" (Césaire 1955/2000, 41). The ruining of other peoples' civilizations.—Force.—Brutality.—Cruelty.—Sadism.—Conflict.—Forced labor.—Intimidation.—Pressure.—Police.—Taxation.—Theft.—Rape.—Degraded masses.—Indigenous peoples turned into colonizing and colonializing instruments.—Into "boys" (the black never grow up, while even the white baby is *baas*, *nkosi*, or *massa*).—Into artisans.—Office clerks.—Interpreters.—Native sergeants.—Prison warders.—Slave drivers.—All tools of force and production (Césaire 1955/2000, 43).

The old racism does not bother or make Césaire indignant: "I merely examine it. I note it, and that is all. I am almost grateful to it for expressing itself openly and appearing in broad daylight, as a sign." It is for him "a sign that the intrepid class which once stormed the Bastilles is now hamstrung, A sign that it "feels mortal, . . . feels itself to be a corpse," which manifests itself in Joseph de Maistre's defense of Europeans who refused to acknowledge the humanity of the peoples of the new world (Césaire 1955/2000, 49). In the "scientist" Lapouge, who hoped that yellow and black populations would never grow, that future society be made of "a ruling class of dolichocephalic blonds and a class of inferior race confined to the roughest labor," and who thus regarded slavery as "no more abnormal than the domestication of the horse or the ox" (Césaire 1955/2000, 50). And from literature, "Psichari soldier of Africa," who took pride in his superior skin color and blood, and had to believe it always, because "when a superior man ceases to believe himself to be superior, he actually ceases to be superior. . . . When a superior race ceases to believe itself a chosen race, it actually ceases to be a chosen race." And so be it. There! "If Europe becomes yellow, there will certainly be a regression, a new period of darkness and confusion, that is, another Middle Ages." But there is more, from Jules Romains, of the Académie Française and the Revue des Deux Mondes: "The black race has not yet produced, will never produce, an Einstein, a Stravinsky, a Gershwin" (Césaire 1955/2000, 51). *Always, the reference is to a white figure of science.*

Césaire mounts a strong rebuttal of Western-and-white writing's claims to objectivity, starting with Gourou's book *Les Pays Tropicaux*, which rejected the existence of any great tropical civilization. Only temperate climates, Gourou had said, were capable of civilization. He was not the only one—the historians, novelists, psychologists, sociologists, journalists, the whole lot, had a "false objectivity," a "chauvinism," a "sly racism." But Césaire hovers on Gourou, who said indigenous populations had "'all taken no part' in the development of modern science" and that any solutions to the problems of the tropics would have to come from outside. Gourou had rightly pointed out how indigenous peoples had suffered from "the introduction of techniques that are ill adapted to them," like slavery, resulting in economic stagnation and regression in their general conditions of living. But then Gourou backs off. What sticks, still, to us, is his comment on science, which Roger Caillois elaborates. Namely "that the West invented science. That the West alone knows how to think." That beyond the West only primitive thinking exists. Caillois was following on Lévy-Bruhl's speculative lies on the "primitive mind," which he later repented, perhaps too late, toward the end of his life: "These minds do not differ from ours at all from the point of view of logic" (Césaire 1955/2000, 69). Where his compatriot Arthur Gobineau had declared that "the only history is white," Caillois could say "the only ethnography is white." Césaire sarcastically simplified what both meant: "It is the West that studies the ethnography of the others, not the others who study the ethnography of the West. A cause for the greatest jubilation, is it not?" (Césaire 1955/2000, 71). Caillois concludes thus: "For me, the question of the equality of the races, peoples, or cultures has meaning only if we are talking about an equality in law, not an equality in fact" (Césaire 1955/2000, 72–73).

My hopes are dashed just as I was beginning to agree with Caillois's critique of anthropology as an instrument of systemic racism! And his closing sentiment is so French—the law is nothing but a piggy bank in which the French hide their systemic racism—not to end it, but to save and even grow it.

Ironically, the priest Placede Tempels called on his European audience to respect Bantu philosophy, while shutting his eyes to Belgium's material colonial plunder of Congo's human and natural resources to develop itself at home. The black man's "own, particular human spirit is the only reality that prevents us from considering him as an inferior being," Tempels said, and it would be a "crime against humanity" if the colonizer destroyed it in the process of freeing the blacks from their primitive state. Rather "what they desire first and above all is not the improvement of their economic or material situation, but the white man's recognition of and respect for their dignity as men, their full human value." What was more, Tempels continued, preserving Bantu philosophy was good for the colonizer, since these people had "integrated us into their hierarchy of life forces at a very high level" (cited by Césaire 1955/2000, 59). That allowed Leopold to assume his place at the top of the Bantu's hierarchy of life forces; to question the colonial state was to question Bantu philosophy itself and Leopold's authority and the loopholes that allowed Belgium to sustain its rule over BaKongo.

To be fair, some honest white men could not ignore what they had witnessed, like the German, Leo Frobenius, who described the Africans he encountered as "civilized to the marrow of their bones!" and concluded that the notion of a "barbaric Negro is a European invention" (Césaire 1955/2000, 53). But such white men were few. The Bible was now invoked, specifically in the example of Mannoni in Madagascar, to portray the African as eternally a child. Whereas the white man had heeded the commandment "Thou shalt leave thy father and thy mother," to become a father, the blacks remain "big-children" who "desire neither personal autonomy nor free responsibility." When they do, it is only because the foolish white agitators have put them to it! Or "that is purely neurotic behavior, a collective madness, a running amok" (Césaire 1955/2000, 61). Whatever they know is simply a detail that fits into a whole. It is negligible, likely nothing, or superstitious myths that impede human progress—the human being obviously white.

And all this is a distraction to prevent us from seeing the material workings and experiencing of the colonial: the blood-stained money accumulating in the Banque d'Indochine and Banque de Madagascar.—The bullwhip.—The taxes.—The martyrs.—Innocent people murdered. The kinds of things Walter Rodney returns to in *How Europe Underdeveloped Africa*:

> The dignity of the non-white peoples.
> The freedom from extermination (Indians and Australians).
> From being lynched (enslaved Africans).
> From being sent to the gas chamber or bonfires (Jewish people).
>
> All these did not owe to their own merits but the colonizer's magnanimity.—
> Conscience.—
> Mercy.—
> Charity.
>
> And mercy was
> To dominate others
> Take their land
> Enslave them
> Send them to the gas chamber
> All the while claiming it was all for their own good!

> They whom the white man and woman had for centuries turned into beasts of burden
> Whose human heart stopped beating right before entering the barracoon.
> Turned into living compost
> Whose body was flesh flour to bake cane and cotton
> Branded with crimson irons
> Whose body was thrown into excrement
> As—not like—cows, goats, and swine.
>
> Auctioned at the town square alongside and for
> Bales of English cloth
> Salted meat from Ireland
> Europe's public said nothing
> Cited verses to bless and thank the lord for a successful market-day
> Yes, Praise the Lord Indeed.
>
> Until Hitler happened to its people.
>
> (paraphrasing Césaire 1939/2001, 33)

HOW SYSTEMIC RACISM FEELS

No wonder why, from a felt experience, the black is a self-mortifying mess, who psychologically relegates himself to being incapable of invention:

> And these tadpoles hatched in me by my prodigious ancestry!
> Those who invented neither powder nor compass
> those who could harness neither steam nor electricity
> those who explored neither the seas nor the sky
> but who know in its most minute corners the land of suffering
> those who have known voyages only through uprootings
> those who have been lulled to sleep by so much kneeling
> those whom they domesticated and Christianized
> those whom they inoculated with degeneracy
> tom-toms of empty hands
> inane tom-toms of resounding sores
> burlesque tom-toms of tabetic treason.
>
> (Césaire 1939/2001, 38)

The blacks are very "mulish" or passive before the self-awakening. They can't see themselves as technological, scientific, adventurous, risk taking, and mobile. They negotiate slavery, a life uncertain, rising and falling like sea waves—as "the paddles vigorously plow the water," the enslaved submit "totally, without reservation," accept the propaganda that their race is unclean, afflicted with leprosy, and only good for perennial servitude to white people.

"I accept. I accept." The black people embrace their curse and inferiority complex.

"Forgive me massa," they yelp in painful submission, as the "twenty-nine legal blows of the whip" dance on their four hundred years of unhealed black flesh.

> Where one expects the blacks to stand up to their tormentor, the blacks become meek, like sheep: Look, am I humble enough? Have I enough calluses on my knees? Muscles on my loins? Grovel in the mud. . . . Dead of the mud.

(Césaire 1939/2001, 46–50) And every day their cowardice grows.

Frantz Fanon's *Black Skin, White Masks* helps us to understand why the black is such a pulverized mess of flesh. The black in Fanon is the native, the vermin being; he is not a person, but occupies "a zone of nonbeing," what his mentor Aimé Césaire had called "thingification" (Fanon 1952/1967, 8). When Fanon says "the black man is not a man," he is referring to that sense of being made to feel as if one does not at all belong to the human species because of a skin color one does not choose at birth. But for the color, the black man tried everything and checked all the boxes required for assimilation, only to find at last that the white wo/man was "sealed in his whiteness, the black in his blackness." Because the French policy of assimilation is (in Fanon's time) based on the blacks leaving their ways of blackness and becoming white, "for the black man there is only one destiny. And it is white" (Fanon 1952/1967, 8, 11, 17). The alienation of the black, the installation through law of an inferiority complex, leads to the internalization of this inferiority complex; it begins to feel epidemic, physically and socially genetic, what one is born to become, what one's children, grandchildren, and their children are destined to be.

Feeling like a nonbeing elicits in the black both aggressiveness and passivity—the passive-aggressive complex. This state of being, this enforced existential uncertainty, leaves the educated black feeling that his own race no longer understands him. Nor does he understand his own people. The educated black—promised assimilation into French citizenship and equality through education, denied equality with even the most illiterate white because of race—lives in a state of limbo, one part identifying with fellow blacks, the other with the white man, sandwiched between his base and his ceiling. He adopts a strange language that aids his own dislocation and separation, while increasing further his own inferiority complex among those whose (white) status he aspires toward. Instead of being a giant because of his high and big education, they ensure that he feels low like an earthworm, they add "native" to his "doctor" to remind him that true doctors are white. That he is not *like* nothing. He *is* nothing itself. Not even the European clothes he wears, the French furniture he sits on, or the English language that hugs his tongue—none of that is a vehicle to transport him to this new social destination. He will always be their Kunta Kinte (Fanon 1952/1967, 25–27).

In one sense the white man contributes in the black man an outrage—through racist profiling.—Imprisonment.—Whip.—Gun.—Thankless toil.—The sheer paternalism!—Condescension.—Pretending to understand.—Indifference. The violence of erasure, the demotion from thing to nothing leaves the black man without culture, civilization.—No historical and knowing past except through Europe. With that move, the white man creates this whirlpool of rage inside the black man "to prove the existence of a black civilization to the white world at all costs. Willy-nilly, the Negro has to wear the livery that the white man has sewed for him." He is in a constant hurry to prove his good behavior, to flee his own shadow populated by the white man's stereotype. The black man is fleeing a shadow bigger and denser than the contents of his own flesh and mind, his body. "Yes, the black man is supposed to be a good nigger" (Fanon 1952/1967, 35).

Fanon also tackles the question: Who must carry the burden of educating white people not to be racist? A question that reduces racism to ignorance. Let's put this into perspective: "Outside the university circles there is an army of fools," Fanon admits. "What is important is not to educate them, but to teach the Negro not to be the slave of their archetypes . . . these imbeciles [who] are the product of a psychological-economic system" (Fanon 1952/1967, 35). Translation: it is not black people's job to educate racists out of their prejudice, as if being racist to us, violating our with dehumanization, was not already burden enough. When the government, company, or the university asks black folk to educate white society out of its bigotry, it is a sign of indifference to the humanity of black people. Like, as Fanon says, "you people" must be grateful we gave you a chance to be here. Like I must be "a good nigger" or (in South Africa or Britain) "a good *kaffir*" or "native" and accept my place: "We have brought you up to our level and now you turn against your benefactors. Ingrates!" (Fanon 1952/1967, 35).

Except nobody gave me anything.
I created it!
I worked for it!
Nobody brought me here.
I came here!
I dared to be here!
My success is not your act of charity.

I analyzed my problems
Carved them up into opportunities!
Took enormous risks
Many sleepless nights
Painful hard work
Against systemically racist thorns and poisons
Impossible odds!
Even against the obstacles you imposed
I simply took what was rightfully mine!

Maker, not artifact.

That's what you don't see when you don't see me.

Fanon observes the white man's "fixed concept of the Negro," which expresses itself in awkward questioning of place and capabilities: "How long have you been in France? You speak French so well" (Fanon 1952/1967, 35). And that's just the start.—"Where do you come from? You speak with an accent."

A French society where, still today, one can only claim rights as a citizen, not (as part of) a racial group, assimilation through merit was on the face of it open to black people. To turn white through attaining a certain level of education and finance might be seen as a rite of purification (white=pure, black=polluted). As we saw earlier, the supposed promise of education and wealth was a movement from darkness to enlightenment, from night to day, to "the color of daylight" and "sunlight under the earth," from insignificant to significant. For the black there was only one way forward, and it led into the world of the white man (Fanon 1952/1967, 51).

To make that move one had to deny their blackness.—Dress white.—Live white.—Befriend white.—Straighten their hair white.—Music white.—Heroes white.—Food white.—Beverages white.—Materialistic white.—Individualistic white.—Accent white.—Think white.—See white.—What then? Bleach the skin, straighten the hair, hate yourself and your parents for giving you this wretched skin. *Cursed be the night you were conceived.* As if the slavery of the ancestor was not enough, here comes the enslavement by inferiority, a "neurotic orientation, . . . a constant effort to run away from his own individuality, to annihilate his own presence. . . . The Negro, having been inferior, proceeds from humiliating insecurity through strongly voiced self-accusation to despair." The black sees in himself only "a negating activity," and it fills him with "third-person consciousness [and] certain uncertainty" (Fanon 1952/1967, 60, 110–111).

Without any recognition, unable to escape his "inborn complex," and having been both hated and hurt, the black man resolves to "assert [him]self as a BLACK MAN."

That resolve comes out of thingification. Here is a situation brought upon the Negro as described by Fanon:

> In my case everything takes a new guise. I am given no chance. I am overdetermined from without. I am the slave not of the "idea" that others have of me but of my own appearance.
>
> I move slowly in the world, accustomed now to seek no longer for upheaval. I progress by crawling. *And already I am being dissected under white eyes, the only real eyes. I am fixed. Having adjusted their microtomes, they objectively cut away slices of my reality. I am laid bare.* I feel, I see in those white faces that it is not a new man who has come in, but a new kind of man, a new genus. Why, it's a Negro!
>
> I slip into corners, and my long antennae pick up the catch-phrases strewn over the surface of things—nigger underwear smells of nigger—nigger teeth are white—nigger feet are big—the nigger's barrel chest—I slip into corners, I remain silent, I strive for anonymity, for invisibility. . . .
>
> Shame. Shame and self-contempt. Nausea. When people like me, they tell me it is in spite of my color. When they dislike me, they point out that it is not because of my color. Either way, I am locked into the infernal circle. (Fanon 1952/1967, 115–116; my own emphasis)

2 HISTORY AS A TOOL FOR ENGINEERING AN AFRICAN PERSONALITY

> As those who have suffered affliction in a foreign land, we have no antecedents from which to gather inspiration.
>
> All our traditions and experiences are connected with a foreign race. We have no poetry or philosophy but that of our taskmasters. The songs that live in our ears and are often on our lips are the songs which we heard sung by those who shouted while we groaned and lamented. They recited their triumphs, which contained the record of our humiliation. They sang of their history, which was the history of our degradation.
>
> —Edward Wilmot Blyden, *Christianity, Islam and the Negro Race* (1888), 105–106

Edward Wilmot Blyden (August 3, 1832–February 7, 1912) was a Liberian educationist, one of the most prolific black and conscious authors of all time, a diplomat and politician in West Africa (Liberia and Sierra Leone).[1] He was born in the Caribbean (what was then called the Danish West Indies), and after being refused admission to theological seminaries (including Rutgers) in the United States because he was black, he traveled to Liberia to work initially as a journalist, heading *Liberia Herald* from 1855 to 1856, and in Nigeria and Sierra Leone for the first newspapers to emerge there. He arrived in Liberia just as the board of foreign missions of the Presbyterian Church in the United States was setting up a high school under Princeton graduate, the Reverend Dr. David A. Wilson of Missouri, and then became its headmaster. In 1862, he was appointed as a professor in the new College of Liberia; two years later, he was appointed as Secretary of State of Liberia alongside his professorship. In both roles he traveled extensively across the world. He left academia in 1871, visiting Europe briefly before spending two years as a diplomat in Sierra Leone and Britain and then presidential candidate in the years before 1888 (Blyden 1888, xiv). Like Garvey later, his "Back to Africa" method alienated

FIGURE 2.1
Edward Wilmot Blyden (1832–1912). *Source*: Edward Wilmot Blyden Museum, Columbia University.

those Africans who felt their ancestors had built the United States and were determined to seek full civil rights in the land of their birth rather than go back to Africa, where they were strangers.

This chapter is a very selective reading of Blyden the critical thinker and builder of future Négritude and Pan-Africanism, which was needed for the mind surgery prerequisite to black self-emancipation, and not his personal (social) life, which is mostly good for argumentation in elite academic circles but is worthless material for self-rehumanization and self-reintellectualization. Thus I will not spend time—beyond this cursory mention—on his problematic comparison of Theodore Roosevelt to Christ, eulogizing the Royal African Society, and his exclusionary perspectives on race, specifically mixed-race people. What I hold dear about Blyden is his steadfast insistence that African societies have a right to futures contrived by themselves, which inspired the likes of Caseley Hayford, Mary Kingsley, Kwame Nkrumah, Nnandi Azikiwe, and many others.

THE AFRICAN PERSONALITY

Blyden was the first exponent of "Pan-Africanism" and the "African Personality . . . with its own identity, values, capabilities, [and] accomplishments, distinguished," one that had emerged out of favorable geographic conditions, communion with nature, and conformity with the environment (Blyden 1862, 15–57). Blyden contrasted the African's communalism and communion with nature to the European's individualism and destructiveness that emerged out of the harsher geographic conditions of Europe. Therefore Blyden argued that busy cities should not be built in Africa, because Africans preferred to live in the countryside where they engaged in farming. Europeans would supply industrial goods, "while the African, in the simplicity and purity of rural enterprises, will be able to cultivate those spiritual elements in humanity which are suppressed, silent and inactive under the pressure and exigencies of material progress" in Europe (Blyden 1887, 126).

Blyden set the tone for the project of self-rehumanization, inverting the blackness that was the white man's object of pathology into the very cornerstone of self-emancipation. In short, spirit and mind engineering first, everything else possible thereafter. And in his travels and the travel accounts of unbiased European observers, Blyden saw the spirit of hospitality that embodied the African personality:

> There is nothing in Africa resembling the poverty which one sees in Europe. The natives in some regions plant a portion of their land especially for the stranger and wayfarer, so that they can indulge in a hospitality unknown in civilized countries—a genuine and unpremeditated hospitality. [Verney Lovett] Cameron, the English traveller, author of *Across Africa*, told me that on one occasion when in the heart of the continent, several weeks' journey from the coast, his supplies gave out and he had nothing to offer the natives in exchange for the necessaries of life; but he experienced no inconvenience, much less suffering. He was the object of abundant and assiduous hospitality from people who had never seen him before and who would never see him again. ("In what country of Europe or America," he asked, "would such a thing be possible?"). (Blyden 1895, 334)

Blyden as a Christian black man understood why Islam succeeded in Africa where Christianity had met with indifference and outright failure: Arabs and Africans became brothers and sisters in flesh and spirit rather than colonizer and colonized, masking greed and domination as a white civilizing and enlightenment mission to the primitive and dark races. Muslim teachers inspired self-respect, while Christian missionaries struck in the convert fear of their god and racial inferiority. In Christianity, Africans existed under Noah's curse that condemned them to eternal inferiority; no such "racial heresy" was to be found in the Qur'an (Blyden 1905, 161). Blyden identifies three key practical objections raised by African Muslims of his time that made the task of conversion to Christianity onerous, objections that can readily be extended beyond non-Muslim African societies. Such objections, I argue, call attention to Africans as intellectual agents, engaging with inbound forces from their own self-invented idioms with a view to strategically and, perhaps, pragmatically deploying them to order and live their lives.

The first objection was to the desocializing influence of the missionary method, which split family ties and disintegrated communities, whereas Islam consolidated and built up on families, increased the population, and made the community stronger. This was also a hallmark of African faiths, incoming Islamic influences notwithstanding. Second, the Christian missionary separated himself from the people, "erecting special and strange buildings for him and furnishing him with comforts and luxuries which the people do not enjoy," usually on top of the mountain to look down upon the meek flock below, whereas the *imam* (Muslim teacher) lived with the people, ate what they ate and gave, and slept in the quarters they provided, "as Christ commanded His apostles to do," Blyden added (1905, 169).

Blyden extended his analysis to the place of the African in the Qur'an, especially in the ancient western Sudanic states, which Arabs called *bilād as-sūdān* (بلاد السودان), or "the lands of the Blacks," in what is today West and northern Central Africa. Here the black person is exalted and dignified equal in the Qur'an, whereas in the Bible the black person is the cursed castaway of Ham, one of the sons of Noah, while the white people have allocated themselves a divinely beautiful place as the blessed descendants of Noah's other son, Shem. One narrative takes the believer toward the equality of races, the other toward racism and hubris. We have seen the play of the Original Sin in the writing of Sylvia Wynter, but writing a century earlier, Blyden cites one outstanding example:

> It is said that this recognition of the African in the Koran was natural, because the Prophet of Islam was descended in part from an African woman; and throughout the history of the religion the exploits of great Africans have been celebrated. In the military history of Islam a Negro slave distinguished himself at two turning points in the early struggles of the faith: at the battle of Ohud [Uhud], when Medina, the refuge of the exiled Prophet, was threatened, in a desperate combat he killed Hamzah, Mohammed's uncle, but one of his bitterest enemies and leader of the opposition. Afterwards, with the same lance with which Wahshi (for that was his name) had killed Hamzah, he slew in a sanguinary battle the impostor "Museilama, the liar," who claimed to be the equal and rival of Muhammad. Among the poets of pre-Islamic days, the Negro Antarah held a high position. (Blyden 1905, 163; also Blyden 1902)

Felix Du Bois, speaking of literature among the blacks of the Sudan, mentions, among others, "the very learned and pious" Sheik Abu Abdallah, who owned no property, spent all his wealth upon the poor and unhappy, "bought slaves that he might give them their liberty," and lived in a house with no door, where "everyone entered unannounced" a man much loved by everybody. And Abu Abdallah was not alone; Du Bois talks of other blacks at the universities of Fez, Tunis, and Cairo, "the most learned men of Islam by their erudition . . . installed as professors in Morocco and Egypt," even while the Arabs were "not always equal to the requirements of Sudanic scholarship. A cerebral refinement was thus produced among certain of the Negro population . . . which gives the categoric lie to the theorists who insist upon the inferiority of the black races" (Du Bois, *Timbuktu the Mysterious*, chapter 13, cited in Blyden 1905, 164).

Blyden uses Timbuktu as a backdrop to the West African technological of his own milieu. He wants the reader to know first that Timbuktu was by design (not some happenstance) a place of transshipment and "the meeting-place of all who travel by camel or canoe," a commercial center and transportation hub, where "the camels transfer their burdens to the canoes, and the vessels confide their cargoes to the camels" and second that the city was an important literary and religious center—the home to many mosques (specifically Sankoré mosque), libraries, and the University of Sankoré, whose founding dates to the tenth century.

To underscore—not simply learning and prayer, but also production born of invention and engineering, of which the buildings are the most iconic and visible. Blyden also undertook several visits along the shores and into the interior to learn about the economic activities African societies were engaged in, at a time of European partition. Referring to the two thousand miles of coastline in British-occupied West Africa, he observed the following:

> They cultivate their lands; they wear cotton dresses of their own manufacture, dyed with native dyes; and they work in iron and gold. The native loom is very primitive, but the native cotton is excellent. The native cotton dresses are much thicker and better than any produced in Manchester, whose manufacturers try hard to imitate them. The African dyes are far brighter and more enduring than the foreign. The African indigo is said to resist the action of light and acids better than any other. . . . *The introduction of foreign cloth into the interior instead of diminishing the manufacture of the native article has increased it, and it more than holds its own side by side with the foreign product, the natives decidedly preferring the African original to the European imitation, and paying much higher prices for it.* They sometimes buy English "bafts"—the trade term for the pieces of cotton of which their dresses are made—which are a clever imitation of their own make, but only because they are very much cheaper. As long as the Africans retain their superiority in manufacturing cotton goods, foreign competition will not interfere with

> the work produced by their primitive appliances. *They also manufacture their own agricultural implements from iron taken from the soil. They make beautiful gold trinkets and their workmanship in that metal is not only curious, but often really beautiful.* The gold mines of Bouré, in the interior of Sierra Leone, and others in the interior of Liberia, yield abundantly with the application of very little labor or capital. (Blyden 1895, 333–334; my own emphasis)

The public debt was manageable. Gold was abundant in Liberia, the government being content with African miners mining sustainably in the interior, bringing their gold to the coast. "We prefer the slow but sure, though less dazzling process, of becoming a great nation by lapse of time, and by the steady growth of internal prosperity—by agriculture, by trade, by proper domestic economy," said Blyden (1888, 119). Commercially, three steam liners from England and Germany and US trading vessels were docking at Liberian ports. Palm oil, camwood, ivory, rubber, gold dust, hides, beeswax, and gum copal were produced in unlimited quantities, with other natural resources untapped and waiting for enterprising investors (Blyden 1888, 119). Liberia was already exporting coffee seed, considered one of the best in the world, to planters in Ceylon (Sri Lanka) and Brazil, and immense opportunities beckoned for carving out extensive farms in the forests (Blyden 1888, 119).

It is in light of this history of the African personality that was humane and industrious that Blyden interpreted what he now defined as "the African problem," subdivided into two. One was the African problem in Africa, which culminated in the European partition of Africa into colonies at the Berlin Conference of 1884–1885 that redefined the Africa-Europe relation as one of coloniality, facilitated by the Christian Church, and occupied by military force, rupturing communities, families, and territories. The other was the African problem in America that began with the start of the trade in Africans as slaves, also ripping up families and communities, the sea separating Africa the homeland from the Africans in bondage in the New World. Europe's eventual occupation started with exploration for natural transport infrastructure, specifically navigable rivers, which, once established, encouraged the search for minerals, and, eventually occupation. France deployed its military, Britain conducted her occupation through charters to companies, and Germany a combination of the two. The European's "intelligence, energy and science" would cleanse the swamps of tsetse and mosquito, open the interior for the extraction and exportation of tropical goods, free the country of "cruel savagery," superstitious, and bloodthirsty tendencies, and expose it to "the regenerating influence of enlightened nations," turning them into "centers of a stable rule, of educational and industrial impulse" (Blyden 1895, 328). But amidst this European self-portraiture, Blyden despaired that "the unfailing signs of approach to a European settlement or to so-called civilization are empty gin bottles and demijohns" (Blyden 1895, 332).

Which is why he saw the Qur'an as "an important educator" exerting "a wonderful influence" among the Hausa, Fulani, Mandinga, Chinese, Hindu, Persian, Turks, and other Muslims, who spoke in their own tongues and read the Qur'an in Arabic (Blyden 1888, 8). In Christian lands, the Negro was portrayed as docile, servile, slow, primitive, struggling for existence, and living at the pleasure of the "civilized" powers (in Haiti and Liberia); in Islamic lands, the African was portrayed as having extraordinary intelligence, enterprise, self-reliance, independence, and dominance, supporting Arabia "without the countenance or patronage to [a] parent country whence their political, literary, and ecclesiastical institutions came. The Muslims of Sierra Leone had no want of aid or instruction from Britain or local government, built their own mosques and schools, sustained their own services, and still supported visiting imams from Arabia, Morocco, or Futah Jalon, where their Christian counterparts bemoaned their lack of independence and initiative" (Blyden 1888, 12).

How then to explain these differences? As Blyden points out, the Muslims found the locals in freedom and independence; they did not claim the high mountain of teaching or giving self-determination, or force conversion as a condition for co-existence:

> The native missionaries—Mandingoes and Foulahs—unite with the propagation of their faith active trading. Wherever they go, they produce the impression that they are not preachers only, but traders; but, on the other hand, that they are not traders merely, but preachers. And, in this way, silently and almost unobtrusively, they are causing princes to become obedient disciples and zealous propagators of Islam. Their converts, as a general thing, become Muslims from choice and conviction. . . . They received [Islam] as giving them additional power to exert an influence in the world. It sent them forth as the guides and instructors of their less favoured neighbours, and endowed them with the self-respect which men feel who acknowledge no superior. *While it brought them a great deal that was absolutely new, and inspired them with spiritual feelings to which they had before been utter strangers, it strengthened and hastened certain tendencies to independence and self-reliance which were already at work. Their local institutions were not destroyed by the Arab influence introduced. They only assumed new forms, and adapted themselves to the new teachings.* In all thriving Mohammedan communities, in West and Central Africa, it may be noticed that *the Arab superstructure has been superimposed on a permanent indigenous substructure; so that what really took place, when the Arab met the Negro in his own home, was a healthy amalgamation, and not an absorption or an undue repression.* (Blyden 1888, 13; my own emphases)

The Oriental aspect of Islam was largely modified, "not, as is too generally supposed, by a degrading compromise with the Pagan superstitions, but by shaping many of its traditional customs to suit the milder and more conciliatory disposition of the Negro." It was also largely affected, Blyden observed, "by the geographical and racial influences to which it has been exposed. The absence of political pressure has permitted native peculiarities to manifest themselves, and to take an effective part in the work of assimilating the new elements" (Blyden 1888, 14).

By contrast, Western Christianity introduced itself to Africans as the religion of the master and the blacks as slaves, shipped across the Atlantic to be worked on the plantation—as a business, to scale up hectarage and production on land that white people had seized from indigenous people; that is how that land moved from mere forest to *plant*ation. The priest designated the black person as a sinner and outcast whose whole life must be dedicated to asking for forgiveness and toil, whose reward was in heaven, even as the white man enjoyed his right here on Earth, at the black person's expense and much toil. The white priest "put new songs in their mouths" declaring they were now spiritually free even while the *chicote*, *sjambok*, and *palmatoria* lacerated their bare flesh:

> The standard of all physical and intellectual excellencies in the present civilisation being the white complexion, whatever deviates from that favoured colour is proportionally depreciated, until the black which is the opposite, becomes not only the most unpopular but the most unprofitable colour. Black men, and especially black women, in such communities, . . . never feel home. They always feel themselves strangers. . . . And [because of] this feeling of self-depreciation intimated above, by the books they read, women, especially, are fond of reading novels and light literature; and these writings that make flippant and eulogistic reference to the superior physical and mental characteristics of race, which, by contrast, suggests the inferiority constantly made [by] the Caucasian the dominant race. . . . Still, we are held in bondage. The African must advance by methods of his own, must possess power distinct from that of the European. (Blyden 1888, 89)

In European Christianity, theirs was a "necessarily partial and one-sided, cramped and abnormal" development, their "independent individuality . . . repressed and destroyed," ideas and aspirations "expressed only in conformity with the views and tastes of those who held rule over them. All avenues to intellectual improvement were closed against them, and they were doomed to perpetual ignorance" (Blyden 1888, 15).

Whereas in Islam, faith and education were coeval; reading lessons happened right after conversion, and no knowledge was withheld or simply stolen. The new convert to Islam chose to wield Qur'an and sword simultaneously as an equal to another Muslim, where the black Christian remained slave, ape, or puppet; in fact, all three (Blyden 1888, 16). Christianity made whiteness closer to godliness, and the illiterate black could pray even in New York thus: "Brethren, imagine a beautiful white man with blue eyes, rosy cheeks, and flaxen hair, and we shall be like him," whereas the black Muslim saw the creator's greatest in the land as designed by his hands. Moral, intellectual, and physical characteristics of Europeans were godly, and the illiterate black considered it "an honour if he can approximate—by a mixture of his blood, however irregularly achieved—in outward appearance." And with that, all sense of human dignity was lost. All written texts were antiblack, whereas in Islam color or race was no barrier to accessing the highest privileges entitled a Muslim. "The slave who becomes a Mohammedan is free," Blyden remarked. And distinguished examples of blacks populated Islamic texts and everyday life. It was a black man, Bilal ibn Rabah, who first made the Azanor (Call to Prayer), that continues to summon all Muslims of the world at the same times to devotion daily. And the Prophet Mohammed responded by appointing Bilal the *muezzin*, or crier. The ninth-century Caliph of Baghdad, Ibn Khallikan, a black man, traveled with Alexander the Great in his Asia Campaigns, and he described him as a brilliant scholar, the subject of praises from the poets of his time: "Blackness of skin cannot degrade an ingenious mind, or lessen the worth of the scholar or the wit. Let blackness claim the colour of your body; I claim as mine your fair and candid soul" (Blyden 1888, 19). Of the poets, two passages in particular illuminate the triumph of brotherhood in faith over color in racist prejudice. One was Abu Ishak Assabi, who says of his black slave, Yumna, who he has the most affection for, "The dark-skinned Yumna said to one whose colour equals the whiteness of the eye? Why should your face boast its white complexion? Do you think that by so clear a tint it gains additional merit? Were a mole of my colour on that face it would adorn it; but one of your colour on my cheek would disfigure me." In the second poem, Assabi says, "Black misbecomes you not; by it you are increased in beauty; black is the only colour princes wear. Were you not mine, I should purchase you with all my wealth. Did I not possess you, I should give my life to obtain you" (Blyden 1888, 19).

Blyden emphasizes that Christian literature has nothing comparable, nor does Islamic literature possess any epithets like "nigger." The Muslim missionary's conversion of Africans was not through preaching, but intermarriages, trade, and deep roots over time, whereas the Christian convert "acquires very low opinion of himself, learns to depreciate and deprecate his own personal characteristics" (Blyden 1888, 23–24). He might not have up-to-date information of the scientific experiments of the day, or move around with apparatuses or manuscripts, but the *imam* (Muslim leader) speaks the language of the people and is deeply immersed in the idioms of local everyday life. "West Africa has been in contact with Christianity for three hundred years, and not one single tribe, as a tribe, has yet become Christian," Blyden could say (1888, 25). In the chapter "Christianity and the Negro Race," Blyden explains why:

> Everybody knows how it happened that the Africans were carried in such large numbers from Africa to America; how one continent was made to furnish the labourers to build up another; *how the humanity of a Romish priest, while anxious to dry up tears in America, was indifferent to unsealing their fountains in Africa*. It was out of deep pity for the delicate Caribs, whom he saw groaning under the arduous physical toil of the Western hemisphere that Las Casas strove to replace them by robust and indefatigable Africans. Hence the innumerable woes which have

> attended the African race for the last three hundred years in Christian lands. In justice, however, to the memory of Bartolome de las Casas, it should be stated that, before he died, he changed his mind on the subject, and declared that the captivity of the Negroes was as unjust as that of the Indians . . . and even expressed fear that, though he had fallen into the error favouring the importation of black slaves into America from ignorance and goodwill, he might, after all, fail to stand excused for it before the Divine Justice. (Blyden 1888, 32; my own emphasis)

An apology too late. In 1620 the first ship carrying enslaved Africans landed on North American shores. English writer and religious thinker William Penn bought some. The Reverend George Whitefield did. Even university presidents owned Africans as slaves—like Jonathan Edwards (Princeton); Samuel Johnson, Benjamin Moore, William Samuel Johnson, William Alexander Duer, and Frederick A. P. Barnard (Columbia University); Benjamin Wadsworth, Nathaniel Eaton, Increase Mather, Joseph Willard, and Edward Holyoke (Harvard); and MIT's founding president, William Barton Rogers, among others. Universities traded in and/or owned enslaved Africans, among them Brown University and the Brown family, Columbia and Barnard College, Dartmouth, Georgetown, Hamilton College, Harvard and Harvard Law School, Johns Hopkins University, University of Pennsylvania, Princeton, Rutgers, University of Virginia, College of William and Mary, and Yale—to mention a few.

Time has not afforded us the opportunity to know which ones among the United States' past ten presidents, or its senators and congressmen and women, would have owned slaves had they had a chance or governed before 1865. They would, of course, not be limited for company. George Washington, Thomas Jefferson, James Madison, James Monroe, Andrew Jackson, Martin van Buren, William Henry Harrison, John Tyler, James K. Polk, Zachary Taylor, Andrew Johnson, and Ulysses S. Grant owned some. At least 1,800 members of Congress owned them. To say nothing of the governors. The New York Stock Exchange and New York City, and countless companies, made their money out of trading in and enslaving black people.

John Wesley would call this long macabre moment the "sum of all villanies" (Blyden 1888, 34). What was so nonhuman (not just human) was how generations of Huguenots and Puritans could train their children on the idea of a God-given right to enslave the African forever, to deploy the bible to convince converts to Christianity that their sole purpose on Earth was complete submission to the white master, and to prepare special books on the "oral instruction" of the enslaved, who were banned from reading for themselves lest they discover the truth and start questioning. The Right Rev. William Meade, Bishop of the diocese of Virginia, stressed in his book of sermons, tracts, and dialogues, for masters and slaves, that: "Some He hath made masters and mistresses for taking care of their children And others that belong to them. . . . Some He hath made servants and slaves, to assist and work for their masters and mistresses, that provide for them." The next two pages addresses the enslaved: "Almighty God hath been pleased to make you slaves here, and to give you nothing but labour and poverty in this world, which you are obliged to submit to, as it is His will that it should be so. Your bodies, you know, are not your own; they are at the disposal of those you belong to, &c." Then again on page 132: Violence is "correction," and when it comes your way, you deserve it, and "Almighty God requires that you bear it patiently." To disobey the master, therefore, is to "yield to the temptation of the devil" and disobey God (Blyden 1888, 35).

South Carolina Governor General James Henry Hammond, writing to the English abolitionist Thomas Clarkson, would say on January 28, 1845:

> I firmly believe that American slavery is not only not a sin, but especially commanded by God himself through Moses, and approved by Christ through His Apostles. . . . I endorse without reserve the much-abused sentiment of Governor McDuffie, that "slavery is the corner-stone of our Republican edifice;" while I repudiate as ridiculously absurd that much-lauded but nowhere accredited dogma of Mr. Jefferson, that "all men are born equal." Slavery is truly the "corner-stone" and foundation of every well-designed and durable Republican edifice. (Blyden 1888, 37)

Wesley was a protestant, and Blyden was quick to commit to posterity that the Protestant Church "proclaims the doctrine of the brotherhood of man, and then tramples upon that which prophesies revere" (Blyden 1888, 45). Only black Roman Catholics had up until that date successfully overthrown oppression—Haitians under Toussaint L'Ouverture. In Rome itself, the names of distinguished black males and females had attained sainthood. Some, like "el Negro Juan Latino" in Cervantes's *Don Quixote* poem, were published poets and had white wives. Benjamin Banneker—naturalist, mathematician, astronomer, almanac author, landowner, surveyor, farmer—was up until then the most distinguished black professor in any Protestant country. Brazil had the decorated general, Henry Diaz (Blyden 1888, 48).

Blyden goes biblical:

> If we come down to New Testament times, we find, again, Africans and their country appearing in honourable connections. When the Saviour of mankind, born in lowly circumstances, was the persecuted babe of Bethlehem, Africa furnished the refuge for his threatened and helpless infancy. African hands ministered to the comfort of Mary and Joseph while they sojourned as homeless and hunted strangers in that land. In the final hours of the Man of Sorrows, when His disciples had forsaken Him and fled, and only the tears of sympathising women, following in the distance, showed that His sorrows touched any human heart; when Asia, in the person of the Jew, clamoured for His blood, and Europe, in the Roman soldier, was dragging Him to execution, and afterwards nailed those sinless hands to the cross, and pierced that sacred side—what

> was the part that Africa took then? She furnished the man to share the burden of the cross with the suffering Redeemer. Simon, the Cyrenian, bore the cross after Jesus. "Fleecy locks and dark complexion" thus enjoyed a privilege and an honour, and was invested with a glory in which kings and potentates, martyrs and confessors in the long role of ages, would have been proud to participate. (Blyden 1888, 177)

The moral of the story is simple: those things that the white man claimed to be bringing, to be the marker of white civilization, were in fact originally made in Africa. That meant that Africans must not come to them as visitors, but inventors. Blyden was unwavering in defense of African values, in the face of an onslaught of white character assassination, depicting Africans as unsophisticated, cannibalistic, to him "a purely fictitious being, constructed out of the traditions of slave-traders and slave-holders, who have circulated all sorts of absurd stories, and also out of prejudices inherited from ancestors, who were taught to regard the Negro as a legitimate object of traffic" (Blyden 1888, 68). They attacked African polygamy as primitive and sinful. Yet in Prussia everyone was allowed three wives (literally slaves), and when their husbands died, they were required to inter themselves on the pier with their husbands or commit suicide so that they join them in the afterlife. Dagobert: three wives and a "multitude of concubines." Charlemagne: two wives, many concubines. By the tenth century, England and Ireland had started trading in and working other humans as slaves. The Teutons, Swedes, Prussians, Goths, and Saxons engaged in human sacrifices to their gods and to win wars. So too the Danes, who every nine years congregated at Lederun, their capital, and "sacrificed to their gods, 99 horses, 99 dogs, 99 cocks, 99 hawks, and 99 men. . . . And what shall we say of those human hecatombs offered during a period of three hundred years by Christians to the god of the slave trade?" (Blyden 1888, 69–70). The King of Dahomey explained the practice of sacrifices this way: "You have seen that only a few are sacrificed, and not the thousands that wicked men have told the world. If I were to give up this custom at once, my head would be taken off to-morrow. These institutions cannot be stopped in the way you propose. By-and-bye, little by little, much may be done; softly, softly, not by threats" (Blyden 1888, 70).

Negro Muslims were not "benighted Africans" or *tabula rasa* upon which Europe could simply inscribe its values or "turn them into Romans," Blyden concluded. Only by "drawing forth the native energies of these nations, while it left them free to develop their own national peculiarities in their own way," would Europe's relationship with Africa be cordial (Blyden 1888, 76–77). A century later, Thabo Mbeki called it "African solutions to African problems," a key tenet of his quiet diplomacy.

AN EDUCATION TO ENGINEER THE AFRICAN PERSONALITY

Intellectually, Blyden became a key figure, especially following his searing critique of European anthropological assertions in the 1860s, specifically the racist writing of the British anthropologist James Hunt, who had stated that "The Negro is inferior intellectually to the European [and] can only be humanized and civilized by Europeans" (Hunt 1864, 51–52). In his later writings, therefore, Blyden blasted the African who subjected himself to European education, read European texts, and imbibed European ideas and values that served to emphasize his blackness as sign of inferiority. He described the Africans of the European academy as "caricatures of alien manners, who copy the most obvious peculiarities of their teachers, with all their drawbacks and defects." The educated African became "alienated from himself and from his countrymen . . . neither African in feeling nor in aim." For him, "everything is Europe and European" (Blyden 1887, 95–105).

Blyden viewed as "only temporary and transitional" Liberia College's offering of an education like any in Europe or the United States with universalistic humanities and social sciences. But he also was clear that "the College is only a machine, an instrument to assist in carrying forward our regular work—devised not only for intellectual ends, but for social purposes, for religious duty, for patriotic aims, for racial development" (Blyden 1888, 82–83). So Liberia College designed

> a system of instruction more suited to the necessities of the country and the race—that is to say, more suited to the development of the individuality and manhood of the African—to bring the institution more within the scope of the co-operation and enthusiasm of the people; where *the students may devote a portion of their time to manual labour in the cultivation of the fertile lands which will be accessible, and thus assist in procuring from the soil the means for meeting a large part of the necessary expenses*; and where access to the institution will be convenient to the aborigines. The work immediately before us, then, is one of reconstruction, and the usual difficulties which attend reconstruction, of any sort. (Blyden 1888, 84; my own emphasis)

The goal of this reconstruction of not just the curriculum, but also the African personality, was to nurture a class of minds with health and a wealth of ideas to dare to invent a new society, wherein the skills learned would have deep roots in African culture, "imbued with public spirit, who will know how to live and work and prosper in this country, how to use all favouring outward conditions, how to triumph by intelligence, by tact, by industry, by perseverance . . . men who will be determined to make this nation honourable among the nations of the earth" (Blyden 1888, 84).

After all, Blyden emphasized, the object of all education was to "secure growth and efficiency, . . . to produce self-respect; a correct education." *The West African Reporter*

FIGURE 2.2
Liberia College (now University of Liberia) in Monrovia, 1893. *Source*: University of Liberia (https://ul.edu.lr/about/history-of-the-ul/).

(Sierra Leone) rightfully lamented the lack of relevance of the Western-and-white education for African students:

> We find our children, as a result of their foreign culture (we do not say, in spite of their foreign culture, but as a result of their foreign culture), aimless and purposeless for the race—crammed with European formulas of thought and expression so as to astonish their bewildered relatives. Their friends wonder at the words of their mouth; but they wonder at other things besides their words. They are the Polyphemus of civilisation, huge but sightless—*cui lumen ademptum*. (Blyden 1888, 86)

To which Blyden added:

> There are many men of book-learning, but few, very few, of any *capability*—even few who have that amount, or that sort, of culture, which produces self-respect, confidence in one's self, and efficiency in work. Of a different race, different susceptibility, different bent of character from that of the European, they have been trained under influences in many respects adapted only to the Caucasian race. Nearly all the books they read, the very instruments of their culture, have been such as to force them from the groove which is natural to them, where they would be strong and effective, without furnishing them with any avenue through which they may move naturally and free from obstruction. Christian and so-called civilized Negroes live, for the most part in foreign countries, where they are only passive spectators of the deeds of a foreign race; and where, with other impressions which they receive from without, an element

> of doubt as to their own capacity and their own destiny is fastened upon them, and inheres in their intellectual and social constitution. They deprecate their own individuality, and would escape from it if they could. And in countries like this, where they are free from the hampering surroundings of an alien race, they still read and study the books of foreigners, and form their idea of everything that man may do, or ought to do, according to the standard held up in those teachings. Hence, without the physical or mental aptitude for the enterprise which they are taught to admire and revere, they attempt to copy and imitate them, and share the fate of all copyists and imitators. Bound to move on a lower level, they acquire and retain a practical inferiority, transcribing, very often, the faults rather than the virtues of their models.
>
> Besides this result of involuntary impressions, they often receive direct teachings which are not only incompatible with, but destructive of, their self-respect. (Blyden 1888, 86–88)

Yet despite being intelligent enough to revolt against the depictions of black people in geography, travel, and history books and their own experiences of such misrepresentations, the youth were obligated to study such "pernicious" teachings. After leaving school, those same images of the African were amplified *ad nauseum* in newspapers, reviews, novels, and scientific works, and it made the youth begin to feel it could in fact be true after all that they were inferior to white people. "Such is the effect of repetition," the power of multimedia (Blyden 1888, 88). Blyden drives home the point of dehumanization and deintellectualization as a psychological engineering process:

> Having embraced, or least assented, to these errors and falsehoods about himself, he concludes that his only hope of rising on the scale of respectable manhood is to strive after whatever is most unlike himself and most alien to his peculiar tastes. And whatever his literary attainments or acquired ability, he fancies that he must grind the mill which provided for him, putting the material furnished to his hands, bringing no contribution from his own field; and of course nothing comes out but what was put in. . . . *Therefore, he never acquires the self-respect or self-reliance of an independent contributor. He is not an independent help, only subordinate help*; so that the European feels that he owes him no debt, and moves on in contemptuous indifference of the Negro, teaching him to contemn himself. (Blyden 1888, 88; my own emphasis)

Africans had to show they were able to "carve out our own way," refusing to take satisfaction with European influences shaping politics, laws, and social atmosphere. Anglo-Saxon methods were not "final"; Africans had something to contribute. "We look too much to foreigners, and are dazzled almost to blindness by their exploits," Blyden says (1888, 90).

The power of traveling in foreign countries lay in witnessing "the general results of European influence," to have misgivings and anxieties about Western-and-white civilization and universalism, to know what to take home and what to leave alone. What was beneficial to Europeans could be ruinous to Africans; what was beneficial on the surface or short term could be hurtful deep down and over time. Indiscriminate appropriation was "often imbibing overdoses" and destroying balance, to become experts in the geographies, histories, and cultures of foreign countries, yet ignorant of our own (Blyden 1888, 91, 101–102).

Blyden preferred Classics (Greek and Latin) and Mathematics as part of the Liberia College curriculum because "there are not . . . a sentence, a word, or a syllable disparaging to the Negro. He may get nourishment from them without taking in any *race-poison*. They will perform no sinister work upon his consciousness" (Blyden 1888, 97; my own emphasis). And, but for the influence of Greece and Rome and their "so-called dead languages and the treasures they contain," Europe would have remained uncivilized for a lot longer (Blyden 1888, 97). The Islamic civilization in thirteenth-century Spain and the Crusades came much later and scaffolded upon Rome and Greece. Then, through the medium of Arabic translations and the Chinese invention of print, the West now gained access to the writings of these philosophers and poets, and the extensive literature of antiquity. Classical learning and knowledge became to Europe "the highest culture then within the reach of mankind" (Blyden 1888, 97). Somewhat misguidedly, Blyden believed that "the study of the Classics also lays the foundation for the successful pursuit of scientific knowledge," stimulates the mind and arouses interest in scientific problems (Blyden 1888, 100), and did not imagine the African capable of also being scientific. But he was right that the Classics and Mathematics qualified the African for "the practical work of life." The curriculum at Liberia College would also include Arabic and "some of the principal native languages—by means of which we may have intelligent intercourse with the millions accessible to us in the interior, and learn more of our own country" (Blyden 1888, 101).

To accelerate Liberia's "future progress," and give this "advance" permanence, the College would have to produce "men of ability," or "capability," which Blyden defined as "the power to use with effect the instruments in our hands." It did not help to have "the learning of Solomon," yet be unable to bring that learning to perform any useful purpose: "A man without common sense, without tact, as a mechanic or agriculturist or trader, can do far less harm to the public than the man without common sense who has had the opportunity of becoming, and has had the reputation of being, a scholar" (Blyden 1888, 102). The bible would be the College's textbook, but "without note or comment," particularly because it would be studied in its original language, alongside Arabic and the Qur'an, to establish connections.

Where then does a people that have suffered affliction in a foreign land get inspiration? Or those born under colonial rule and who have never known freedom. Blyden answers this question as follows:

As those who have suffered affliction in a foreign land, we have no antecedents from which to gather inspiration.

All our traditions and experiences are connected with a foreign race. We have no poetry or philosophy but that of our taskmasters. The songs that live in our ears and are often on our lips are the songs which we heard sung by those who shouted while we groaned and lamented. They recited their triumphs, which contained the record of our humiliation. They sang of their history, which was the history of our degradation. To our great misfortune, we learned their prejudices and their passions, and thought we had their aspirations and their power. Now, if we are to make an independent nation—a strong nation—we must listen to the songs of our unsophisticated brethren as they sing of their history, as they sang the songs of their traditions, the wonderful and mysterious events of their tribal or national life, of the achievements of what we call their superstitions; we must lend a ready ear to the ditties of the Kroomen who pull our boats, of the Pesseh and Golah men, who till our farms; we must read the compositions, as rude as we may think them, of the Mandingoes and the Veys. We shall in this way get back the strength of the race, like the giant of the ancients, who always gained strength, for his conflict with Hercules, whenever he touched his Mother Earth. (Blyden 1888, 105–106; my own emphasis)

And he ends by imploring us to gird our loins:

We have a great work before us, a work unique in the history of the world, which others who appreciate its vastness and importance, envy us the privilege of doing. . . . Let us show ourselves equal to the task. *The time is past when we can be content with putting forth elaborate arguments to prove our equality with foreign races. Those who doubt our capacity are more likely to be convinced of their error by the exhibition, on our part, of those qualities of energy and enterprise* which will enable us to occupy the extensive field before us for our own advantage and the advantage of humanity—for the purposes of civilisation, of science, of good government, and of progress generally—*than by any mere abstract argument about the equality of races.* (Blyden 1888, 107; my own emphasis)

The education of girls was already in Blyden's plans for Liberia College by 1888, ideas way ahead of their time, and one and a half decades prior to the start of the women's suffrage movement in the United States. Already, a female department had opened at Liberia College, headed by a female principal. Blyden explained the moral and human imperative for ensuring equal opportunities for boys and girls in the creation of a self-determined society:

I cannot see why our sisters should not receive exactly the same general culture as we do. I think that the progress of the country will be more rapid and permanent when the girls receive the same general training as the boys; and our women, besides being able to appreciate the intellectual labours of their husbands and brothers, will be able also to share in the pleasures of intellectual pursuits. We need not fear that they will be less graceful, less natural, or less womanly; but we may be sure that they will make wiser mothers, more appreciative wives, and more affectionate sisters. (Blyden 1888, 102–103; my own emphasis)

IN CLOSING: FOREIGNERS, THE DIASPORA, AND AFRICAN DEVELOPMENT

For Blyden, the responsibility for opening up and developing Africa fell on Africans themselves, not outsiders. His enthusiasm for the African colonization, as expressed in the American Colonization Society (ACS), was different from among others abolitionists like Frederick Douglass, William Lloyd Garrison, and William Wells Brown, who saw this as an attempt by those slave-owning ACS members to dispose of free blacks to avoid them inciting to freedom the still-enslaved ones. Many members did not believe free blacks capable of integrating into white-dominated US society and values, which could be better served if they were evacuated to Africa to be among other blacks.

Blyden was convinced that the enslaved African should return to Africa to civilize his fellow Africans who were not yet exposed to Europe. In the wake of the US Civil War and the emancipation in 1865, and a full two decades before the European partition of Africa began, Blyden floated the idea of mobilizing African Americans, "despairing of a redress of their grievances in the United States," to undertake a colonizing mission of sorts and carve out "progressive new empires" in Africa, to show that "a Black Man can run a country" (Henriksen 1975, 281). And Liberia, established in 1847 to accommodate African Americans who wished to return to the native land, was the perfect place to begin (Shepperson 1960, 301). The "Back to Africa" project rested on massive mobilization of skilled African Americans and the promotion of the US constitution (Lynch 1965). Blyden first approached the British government, then the United States, and finally France, to supply the men and materiel to colonize a large swath of territory stretching from Liberia through the Sudan to the Indian Ocean. After these white men had done their job, the tsetse fly, mosquitoes and other tropical maladies would soon afterward either finish them off or chase them out, leaving Africans to govern themselves and run the resultant technological, political, and economic infrastructure.

The Berlin Colonial Conference of November 15, 1884, to February 26, 1885, put paid to Blyden's idea of a black colonization of Africa by black people designed to create pan-African cross-fertilization of ideas and practices. The partition that followed was a culmination of several organizations in nearly all European countries formed for the purpose of exploring and occupying Africa, all with royal patronage. They included the International African Association (formed in Brussels in 1876) with King Leopold II as patron; the Italian National Association for the Exploration and Civilisation of Africa; Espanola para Esploracion Africa; the German Society for the Exploration of Africa (established by the German Geographical Associations in 1872 and receiving funding from government); the Afrikanische Gesellschaft (formed in Vienna in 1876); the

Hungarian African Association (set up in Budapest in 1877); the National Swiss Committee for the Exploration of Central Africa; and so forth (Blyden 1888, 110).

These organizations were no longer interested in speculative and sentimental accounts of Africa, but accurate knowledge driven by demand for new markets and raw materials for the depressed factories of Lancashire "waiting to be inspired with new life and energy . . . so that Africa, as frequently in the past [slavery], will have again to come to the rescue and contribute to the needs of Europe" (Blyden 1888, 111).

For Blyden, European plans for "opening up" Africa, alone without Africans, were inadequate and contrary to the continent's sustainable development. Commerce, science, and philanthropy might set up stations and thoroughfares, but they were "helpless to cope fully with the thousand questions which arise in dealing with the people" (Blyden 1888, 111). "Mere trade" and missionary work alone were not enough; neither was bad. Trade was a good thing, and missionaries were welcome. Sort of.

> Place your trading factories at every prominent point along the coast, and even let them be planted on the banks of the rivers. Let them draw the rich products from remote districts. We say, also, send the missionary to every tribe and every village. Multiply throughout the country the evangelising agencies. Line the banks of the rivers with the preachers of righteousness—penetrate the jungles with those holy pioneers—crown the mountain-tops with your churches, and fill the valleys with your schools. No single agency is sufficient to cope with the multifarious needs of the mighty work. *But . . . the results of your work [must be] beneficial and enduring [to Africa].* (Blyden 1888, 111–112; my own emphasis)

Pro-colonization agitators in Britain under the lead of Sir Thomas Fowell Buxton loved the Blyden language when describing "a scheme for the regeneration Africa by means of her civilised sons, gathered from . . . exile, and, at great expense, sent out an expedition to the Niger, for the purpose of securing on that river, hundred square miles of territory on which to settle the returning exiles." Earlier Captain William Allen, British commander of the first Niger expedition, had noted in 1834 that such an exalted settler community "would excite among its neighbours desire to participate in those blessings," and offer a "normal or model society" that he could see "gradually spreading to the most remote regions," with rich dividends for commerce. Buxton and his colleagues basically copied the language of the American Colonization Society in their "Friend of Africa" agenda, stressing "native agency, or the agency of the black people themselves to forward their own cause." The candidates to labor in this "African vineyard" would come from emancipated slaves, who would be "people with religious views and with intellectual capacity equal to the whites," not two or three only, but more, "enough to form a society" (Blyden 1888, 113–114).

To Blyden these were "sound" views. One writer in the *London Times* (May 31, 1882) had particularly impressed him when giving the example of Zanzibar in the wake of the abolition of slavery: "It is the formation of self-sustaining communities of released slaves in the countries whence they were originally brought by the slave dealers, in order that, by their example and influence, they may teach to the surrounding people the advantages of civilisation. The sight of a body of men of the same race as themselves, living in their midst, but raised to a higher level by the influence of Christianity and civilisation, has naturally produced in them a desire of raising themselves also" (Blyden 1888, 114).

Liberia was an equally compelling example, especially in matters of education, with the college at the apex of a public school system of education in literary, religious, industrial, mechanical, and commercial courses. Influential chiefs were especially targeted, the idea being to convince them of the virtues of Western-and-white-derived education first, so that they could then decree it for their subjects. This would become the colonial settler state's *modus operandi*, but the key difference is that a black man was driving it (Blyden 1888, 118). Liberia was not a reproduction of the United States: "The restoration of the Negro is the restoration of the race to its original integrity, itself; and working itself, for itself and from itself, it will discover the methods for its own development, and they will not be the same as the Anglo-Saxon methods," Blyden insisted (1888, 126).

3 NÉGRITUDE: CULTURE AS A TOOL FOR RE-MEMBERING BLACK HUMANITY

Négritude was a cultural-intellectual movement whose origins are attributed to two key literary developments. One was the work of the Haitian journalist, anthropologist, and politician Anténor Firmin (1850–1911) reacting with a sense of outrage at the racist writing of Joseph Arthur de Gobineau; unlike Edward Wilmot Blyden, however, Firmin set the tone for the nonterritorializable quest for an African personality that would come to be called "Négritude." He appropriately titled his book *De l'égalité des races humaines* (1885), translated to *On the Equality of Human Races* (2000), as a direct response to Gobineau's *Essai sur l'inégalité des races humaines* (1853), translated as *Essay on the Inequality of Human Races* (1915). Often regarded as one of the fathers of Pan-Africanism, Firmin inverted Gobineau's "black as pathology" into "black is beautiful," thereby *performing restorative surgery upon the dehumanized black personality*. The echoes of Blyden in Firmin are no coincidence. While Blyden developed the idea, Firmin, alongside fellow Haitian Bénito Sylvain and Henry Sylvester Williams, a Trinidadian lawyer, turned it into a community; they convened the First Pan-African Conference in London in 1900 that cemented Pan-Africanism as a movement. They invited W. E. B. Du Bois and bestowed upon him the task of drafting the conference's general report.

The second artistic cultural development was the Harlem Renaissance because it was active in Harlem, New York, in the 1920s and 1930s, led by, among others, Alain Locke, Langston Hughes, and later Richard Wright, and was seized with the *noireism* (race and blackness) of its times. Like Blyden, it condemned a violently racist, imperialistic, and enslaving Europe that celebrated and rewarded selfishness and mistook arrogance for toughness and predatory competition and materialistic individualism for progress. One can see the influence of Blyden throughout.

FIGURE 3.1
Antenor Firmin. *Source*: Wikicommons.

In the Paris of the 1920s and 1930s too, the black diaspora was preoccupied with *noireism*, frustrated by the French policy of assimilation, on white terms, into a metropolitan white society. These black men and women of letters were saying Yes to association with France and French culture, but NO to assimilation into it. Even association was on the terms of the white man, and for these young scholars from the French colonies and territories, it too was unacceptable. They included women like Paulette Nardal and her sister Jane Nardal, proud owners of Clamart Salon, the teashop where all the black intellectuals in Paris met. Clamart Salon was the birthplace of the Négritude movement. The catalyst was the founding, by Paulette Nardal and the Haitian Leo Sajou, of the journal *La revue du Monde Noir*, publishing in French and English, and bringing together its Caribbean, African American, and African scholarly readers in Paris.

FIGURE 3.2
Paulette Nardal. *Source*: Wikicommons.

It was the Nardal sisters who introduced the Harlem Renaissance's inspirational writings to black intellectuals in Paris fed up with French policies of assimilation. This is not an insignificant moment: it formed an umbilical link between the Harlem Renaissance and its Anglophone writers (mostly poets) like Locke, the unofficial "dean" of the Harlem Renaissance, James Waldon Johnson, Hughes, W. E. B. Du Bois, Claude McKay, Jean Toomer, Countee Cullen, later Sterling Brown, and women like Zora Neale Hurston, author, anthropologist, and one of the publishers of the journal *Fire*! The African Americans, Césaire admits, were "the first to teach us the rudiments" of what would become Négritude. "They were the first to say 'Black is beautiful.' . . . It was the beginning of a cultural revolution, a kind of revolution of values. It was in no way a refusal of the outside world, it was . . . a desperate quest for the Negro 'Self'" (Césaire 1989, 51).

In effect, the Nardal sisters joined Pan-Africanism and Négritude together, into what George Shepperson coined the "African diaspora," which is increasingly being called Global Africa. In 1931 Paulette and Leo Sajou (a Haitian) founded the English and French journal *La revue du Monde Noir* (*The Black World Review*) to provide vent for

FIGURE 3.3
From left: Langston Hughes, Charles S. Johnson, E. Franklin Frazier, Rudolph Fisher, and Hubert T. Delany in 1924. *Source*: Schomburg Center for Research in Black Culture, Photographs and Prints Division, New York Public Library.

Caribbean and African scholars in Paris. Most of them were poets, not least "the three fathers" (*les trois pères*) of Négritude: the Martinican Césaire, the French Guyanese Leon Damas, and the Senegalese Léopold Sédar Senghor. Through traveling they discovered a black commonality across geographies that they might not otherwise have had in the localized experiences of colonialism in their own individual countries. It is this black pride, this Global Africa, and its inspirational value for those who dare to invent the future, that anchors the current project, and deserves further exploration.

Présence Africaine (*African Presence*) was to Négritude what *The Negro World* was to the Harlem Renaissance and Garveyism: a platform and a vehicle for cultural self-rehumanization. Alioune Diop had founded it in 1947, adding a publishing house and a

FIGURE 3.4
0. Léopold Sedar Senghor. 1. Aimé Césaire. 2. Léon Damas. *Sources*: 0 and 2, Wikicommons; 1, AZMartinique.com.

bookstore in Paris. But it was when Césaire and Senghor joined him that *Présence* became the voice of the Négritude movement. Nine years after its formation, *Présence* hosted the first International Congress of Black Writers and Artists in Paris. Among its speakers were Césaire, Senghor, Jacques Rabemananjara, Cheikh Anta Diop, Richard Wright, Franz Fanon, and Jean Price-Mars. Pablo Picasso designed the conference poster; in fact, among others, it was Picasso, Maurice Vlaminck, and Georges Braque who made the Negro fashionable in France (Césaire 2000).

FIGURE 3.5
Wifredo Lam. *Source*: Wikicommons.

Much later, in 2000, Césaire elaborates on the African influences on Pablo Picasso via the Cuban virtuoso Wifredo Lam and, by extension (in ways that are also autobiographical for me, for the influence of my mother), African mothers:

> Picasso admired Lam very much. When you look at Wifredo Lam's paintings, you can find the jungle, Voodoo, Macumba and Santeria in them. You find in them the fundamental Gods, the fundamental paganism of the African. He used to tell me, in the same way Leopold [Senghor] tells me about the village poetess who has influenced his aesthetic so much . . . well, Wifredo used to tell me about Ma'Antonica. She was a mambo who influenced him very much, who initiated him. (Césaire 1989, 67)

Those Afro-Cuban elements filtered into Lam's art when he returned to Havana from France in 1941, and came face to face with the reality of the oppression the descendants of the enslaved were still facing, their art simply relegated to caricature for touristic display. Later he would recall:

> I decided that my painting would never be the equivalent of that pseudo-Cuban music for nightclubs. I refused to paint cha-cha-cha. I wanted with all my heart to paint the drama of my country, but by thoroughly expressing the negro spirit, the beauty of the . . . art of the blacks. In this way I could act as a Trojan horse that would spew forth hallucinating figures with the power to surprise, to disturb the dreams of the exploiters. . . . What has really broadened the range of my painting is the presence of African poetry. (Richards 1988, 91–92)

From this biographical note, Césaire's poem dedicated to and entitled "Wifredo Lam" begins to make sense as a tribute to the painter and sculptor's contribution to Négritude, an art that is not simply about making artifacts and selling, but sculpting the self-contrived, necessary self in the plural of black community, the "we":

> Nothing except the shivering spawn of forms liberating themselves
> from facile bondages
> and escaping from too premature combinings . . .
> in the combats of justice I recognized
> the rare laughter of your magical weapons
> the vertigo of your blood
>   and the law of your name.
> (Césaire 1983/2001, 712)

Another pivotal moment in the growth of Négritude came with Césaire's "discovery" of Haiti, which itself followed his awakening to the Harlem Renaissance, as he sought to "explore the totality of the black world" (Césaire 2000, 90). Haiti was where, from August 14, 1791, to January 1, 1804, Toussaint L'Ouverture led the first successful revolution of enslaved black people to liberate themselves. And Césaire exclaimed, "Haiti, where Négritude stood up for the first time" (Césaire 1955/2000, 90). In a later conversation he elaborates, "The first Negro epic of the New World was written by Haitians, people like Toussaint L'Ouverture, Henri Christophe, Jean-Jacques Dessalines, etc." (Césaire 2000, 90). *And Haiti traveled to Africans through its example, a confirmation and demonstration of Négritude in action*, an affirmation of "determination to shape a new world, a free world." And the black nation's authors Hannibal Price, Louis-Joseph Janvier, Antenor Firmin, Justin Lhérisson, Frédéric Marcelin, Fernand Hibbert, and Antoine Innocent spoke the need to reclaim African cultural and aesthetic values to counter "the total and colorless assimilation" that the French had whitewashed the earlier generation of Haitian writers with.

And it was Senghor who introduced Africa in depth to Césaire, to its modes of knowledge production, its artistic productions, and its storytelling and dances and trances. Despite mastering Greek and Latin culture, Senghor was to Césaire the *Dyâli*:

> A man perfectly anchored in his identity, a Senegalese identity, an African one taking in all that the world could bring him, in a "Self" not assimilated but assimilating. . . . What you have in Senghor comes from the Serere, from the court poetry of the Wolof, from the society that valued honor, from these gymnic poems, from this eulogistic poetry when one salutes the athlete on entering the stadium. . . . I wrote him a poem and the word that immediately came to my mind was Dyâli. Literally, the one who says words, the poet. I sent him the poem. He was very moved by it. The Dyâli is also the one who shows the way. (Césaire 1989, 51)

Among other things, Dyâli the poem lauds Senghor as "inventor of the people and of its budding . . . master of its word . . . teller of the essential . . . what must always be told again" (Césaire 1989, 53–55).

NÉGRITUDE

When asked what they were trying to do when they started Négritude, Césaire's answer reveals the spirit of daring to invent the future as simultaneously a "quest for ourselves." And he captures it beautifully, admitting to inauspicious, unstructured beginnings of what then becomes a big idea and attitude of being:

> I think you should not look for motivations too much. Our initial actions were not completely rational; what we did was not planned at all. Incidentally, that is what made it pure. It was really an intuition; it was a rebellion, but it had not been thought out. We were really "possessed" . . . by an . . . idea that was the ferment of a great enthusiasm and that was life-giving. . . . This idea of Négritude, this desperate quest for ourselves, this determination to rehabilitate a history, this feeling of a solidarity to develop, this feeling of a faithfulness towards our "ancestors"; even though I don't like this word very much. Anyway, the word we abhorred was renouncement; we would never renounce anything. We belonged with those people who accept history, who try to understand it, and who try to make things advance. (Césaire 1989, 62)

The spirit of Négritude as a strategic and necessary invention can best be summarized in the Maryland Pan-Africanist John Bruce's declaration: *"I am a negro and all negro. I am black all over, and proud of my beautiful skin"* (Shepperson 1960, 310; my own emphasis). Later, in a 1989 interview, Césaire would declare, *"Nothing that is negro can be alien to me. My heart bleeds in Soweto, but it also bleeds in Harlem"* (Césaire 1977, 53; my own emphasis). Négritude was a strategic and reconstitutive appropriation, turning the term *"nègre"* or "nigger," which the white man had used in racist, dehumanizing terms, into the foundation of black self-rehumanization, the ultimate symbol of black pride.

FIGURE 3.6
John Edward Bruce. *Source*: Wikicommons.

For Césaire, Négritude is a concrete, not abstract, "coming to consciousness" born in the experience and atmosphere of rejection and thingification that makes one feel and develop "an inferiority complex." It can grind you under or it prompts a fightback, a re-/search for your identity, to "establish" your identity, "a concrete consciousness of what we are—that is, of the first fact of our lives: that we are black; that we were black and have a history, a history that contains certain cultural elements of great value . . . beautiful and important black civilizations" (Césaire 2000, 91). These are the civilizations Césaire talks about in *Discourse*, at the start of chapter 1, completely erased from white-centric histories of the world that devoted not "a single chapter to Africa, as if Africa had made no contributions to the world. Therefore," Césaire concludes emphatically:

> We affirmed that we were Negroes and that we were proud of it, and that we thought that Africa was not some sort of blank page in the history of humanity; in sum, we asserted that our Negro heritage was worthy of respect, and that this heritage was not relegated to the past, that its values were values that could still make an important contribution to the world. . . . Universalizing, living values that had not been exhausted. *The field was not dried up: it could still bear fruit if we made the effort to irrigate it with our sweat and plant new seeds.* So this was the situation: there were things to tell the world. We were not dazzled by European civilization. We bore the imprint of European civilization but we thought that Africa could make a contribution to Europe. It was also an affirmation of our solidarity. That's the way it was: I have always recognized that what was happening to my brothers in Algeria and the United States had its repercussions in me. I understood that I could not be indifferent to what was happening in Haiti or Africa. Then, in a way, we slowly came to the idea of a sort of black civilization spread throughout the world. (Césaire 2000, 92; my own emphasis)

And for Césaire—and Senghor—there was no contradiction between the universal and particular; rather, that one reaches the universal through the exploration and understanding of the particular: "So we told ourselves: the blacker we are going to be, the more *universal* we'll become. I don't think in terms of antagonism. I am myself wherever men stand and struggle. Hence my way of relating (this is paradoxical) to this land, the tiniest township in the universe, this speck of an island that is, for me, *the world*" (Césaire 1989, 67; my own emphasis).

In a 1977 interview, Césaire is clear that in trying to address the "particular," that which hegemonic structures have designated "universal" must be pragmatically, strategically, and selectively researched, understood, and deployed as a tool of self-emancipation. This in no shape or form is to be confused with assimilation to that which hegemonic cultures have designated universal. *In and of itself the designated universal is not a tool a priori; it only becomes so in view of those struggling assigning it purpose within a scheme conceived and executed by them*:

> No-one can ignore marxism, but it must be used as a tool. There is no question of making of it an ideology or a new dogma. Marxism is an extraordinary analytical tool; quite exceptional. I said just now that all mystifications must be destroyed: we must recognize that marxism has conspicuously helped us to demystify or demythify colonialism. This having been said, it is clear that marxism remains one instrument among others which are complementary to it. . . . But it could be objected that ethnography, linguistics and psycho analysis are matters for the élite and that the people need to eat: these famous animal needs, to remain within Marx's vocabulary. (Césaire 1977, 46)

Having acquired access to wider academic discourses, the role of the black intellectual, Césaire says, is to be "the conscience of a community":

> And this conscience must not be passive but militant. Among the values at the heart of his fight there are first justice, man and truth. This is very important particularly in our case. We are living at a time when it is essential to be lucid: to dissipate the myths, to destroy the mystifications, to see and make things seen, not to lie to oneself and not to lie to others . . . , *select what we need and follow our own road. . . . To seek a particular African path, at the same time taking advantage of the contributions of the other worlds, but well knowing, fully realizing that in reality nobody has thought for us or can think for us*. (Césaire 1977, 45; my own emphasis)

In two interviews, Césaire explains why it had to be Négritude and not some other attitude. The first was in 1989, when he confirms that Négritude is not a philosophical concept that they set out to write for academic purposes. The times demanded it:

> If Senghor and I spoke of Négritude, it was because we were in a century of exacerbated Euro-centrism, a fantastic ethnocentrism, that enjoyed a guiltless conscience. No one questioned all that—the superiority of European civilization, its universal vocation—no one was ashamed of being a colony. . . . They had interiorized the colonizer's vision of themselves. In other words, we were in a century dominated by the theory of assimilation. . . . So Négritude was for us a way of asserting ourselves. First, the affirmation of ourselves, of the return to our own identity, of the discovery of our own selves. . . . *I think that as long as you will have Negroes a little everywhere, Négritude will be there as a matter of course. . . . And, believe me, this fight against alienation is never totally over.* (Césaire 1989, 55–57; my own emphasis)

The second interview was the epilogue in the 2000 edition of *Discourse*, in which he elaborates on Négritude as creative disobedience in action. Césaire is at pains to show that Négritude was carefully chosen for its inversive politics, one that no other word could express. It had to be the word that stipulates the malign, the pathos of being black, black that is not *noir* but *négre* or the nigger, racist when spoken by those that fashioned it to hurt and discriminate, and, when used as nigga, at once inversive appropriation to the black who references brotherhood in struggle borne of historical thingification when we say it to each other as black people:

> Our struggle was a struggle against alienation. That struggle gave birth to Négritude. Because Antilleans were ashamed of being Negroes, they searched for all sorts of euphemisms for Negro: they would say a *man of color*, a dark-complexioned man, and other idiocies like that. That's when we adopted the word *négre*, as a term of defiance. It was a defiant name. To some extent it was a reaction of enraged youth. Since there was shame about the word *négre*, we chose the word *négre*. I must say that when we founded *L'Etudiant noir*, I really wanted to call it *L'Etudiant négre*, but there was a great resistance to that among the Antilleans. . . . There was in us a defiant will, and we found a violent affirmation in the words *négre*, and négritude. (Césaire, interview Depestre 2000, 89; my own emphasis)

When one black person calls another "My Nigga" or "Nigga please," that is endearment, a sense of community, that "you're my people," and "I gut yu!" If a nonblack person calls me that word, I will interpret it as "nigger" (the racist version) not "nigga" (as called me by a fellow black person in endearment). When Césaire and Senghor settled on *négre,* they were upstaging the othered, racist meaning and reinvesting it with positive, self-defined values.

As a concept for mental surgery, Négritude had two major tensions. First, "to admire or to hate" European civilization? Movement in any one direction compromised the balance Négritude sought to inhabit. Césaire's poetry sought both to "redeem and embrace," to "forgive and revenge," to "resist and yield" to the European system that colonized and oppressed him (Moore and Beier 1963, 14–26). And Senghor refused to reject either European or traditional African values in total, seeking instead to draw selectively from both, "a synthesis of civilisations, retaining only the fecund elements of each," he insisted. For him, a "straight choice between isolation and assimilation" was a false choice. Here he repeated Blyden: blacks were full of the "warmth of human feeling—some joy and much pain" and "superior without invention and conquest" (Le Baron 1966, 268), while the whites were "hard and cold," with all their technology and science. And in the poem, *Prayer to Masks,* Senghor would add, the black man "would teach rhythm to the world that has died of machines and cannons . . . [and] return the memory of life to men with a torn hope" (Senghor 1963, 49).

It is easy to read Césaire as recusing Africans from being capable of the scientific and the technological; instead, his is an eloquent rejection of the tethering of African identity to the "colon-," to a mere victim of someone else: the colonizer. I read *Prayer to Masks* as a satire against the white man's "former childish fantasies," which erase our descent "from the Amazons of the king of Dahomey," "from the princes of Ghana with eight hundred camels," the "wise men in Timbuktu under Askia the Great," "the architects of Djenne, nor Madhis, nor warriors," from yesteryear's carriers of lances. I, descendent of the blackest of black, the proudest of proud African greatness, am rendered just "a sheep grazing his own afternoon shadow, . . . pretty mediocre dishwashers, shoeblacks without ambition, at best conscientious sorcerers and the only unquestionable record that we broke was that of endurance under the *chicote*" (Césaire 1939/2001, 33).

Césaire now mounts a proud defense of non-European civilizations and their admirable ethic of communality, societies, he says, that were "ante-" and "anti-" capitalist, democratic, "always," cooperative and fraternal—destroyed by imperialism. They had their faults like any other human society, but were neither hated nor condemned (in apparent reference to Noah's curse to Ham's descendants, and before then, the Adamic Original Sin dismissed earlier by Wynter). They were "content to be" (Césaire 1955/2000, 44). They were inventors of science and technology, innovators, and entrepreneurs. The Egyptians invented arithmetic and geometry, the Assyrians' astronomy, the Arabs' chemistry. Rationalism was not new to Islam by the time Europeans started talking about it

(Césaire 1955/2000, 70). Here we see continuities from Blyden in the way Césaire rereads Islam.

As their Négritude awakens, the blacks begin to appreciate values derived from his African heritage and cosmologies—as if to say, also that the science and technology as defined by Europe ought not to be the yardstick for measuring civilization, that in fact, the world would be all the poorer if it only had machines, without the essence of things. Never invented anything, yes, never explored anything, yes, and never conquered anything, yes, because that is not what excites the black. Hence, "ignorant of surfaces but captivated by the motion of all things"; "indifferent to conquering, but playing the game of the world." The black gives the world wisdom (hence "eldest sons"), channels and enables the life and life-giving elements to nourish the world (as opposed to monopolizing them for selfish ends); Africa is the energy of, gives energy to, the world. The blacks are "the eldest sons of the world, porous to all the breathing of the world, fraternal locus for all the breathing of the world, drainless channel for all the water of the world, spark of the sacred fire of the world, flesh of the world's flesh pulsating with the very motion of the world!" They who humanize and socialize what the European deems natural objects of dry science, renders them spiritual, even fragile beings in need of care and love, what!—the African cosmology is to be human without discrimination based on race or species. *Things that the white man has made dry, filled with concrete, or hard as steel, and made unexciting, the black imbues with humane values*; hence the dawn is "tepid" like the temperature of water, the black's blood is "aroused by the male heart of the sun," the black knows "the femininity of the moon's oily body," exalts the antelopes and the stars, and his survival "travels in the germination of grass" (Césaire 1939/2001, 41–42). Humanity is inclusive, it is "enclosed concordance," it is humanizing, it cares, it derives and imbues life and humanity from that which the white man has adjudged nonhuman, unbreathing, feelingless, spiritless, static, and mindless. Hard. Cold. Fixed.

And the white man is antithetical to this black cosmovision, where everything is imagined and related with humanely. Césaire pities and mocks "our omniscient and naïve conquerors," "horribly weary from [their] immense efforts," "stiff joints crack under the hard stars," "blue steel rigidities pierce the mystic flesh," touting defeats as "deceptive victories," making "grandiose alibis" out of "pitiful stumbling" (Césaire 1939/2001, 42).

The second tension said of Négritude was that it was too elitist, alienated from ordinary people, and declaring its "blackness" from exile without any purchase on the everyday realities of Africans in Africa. Hence Wole Soyinka's indictment: "*'A tiger does not proclaim his tigritude, he pounces.' In other words: a tiger does not stand in the forest and say: 'I am a tiger.' When you pass where the tiger has walked before, you see the skeleton of the duiker, you know that some tigritude has been emanated there*" (Jahn 1969, 265–266; my own emphasis).

Much later, in a 1977 interview, Césaire replied to Soyinka without referring to him personally. He found the condemnation of Négritude "vociferous" and "brilliant," but, he said, when he hears it, he just smiles. Négritude, he said, made it possible for these scholars to find an object of condemnation, whereas Césaire and his peers were pioneers who have "the merit of having searched, of course with the risk of making mistakes":

> It is very easy to stay in one's ivory tower and to pronounce excommunications, maledictions etc. So we wanted to clear a path. . . . We all thought at the time, Senghor, Damas and I, that while there are contingencies, secondary matters, there is nevertheless one value which must be maintained and which is fundamental: the virtue of identification. To be oneself. To remain oneself. I also believe in a feeling of fidelity and solidarity. The need to pursue the quest. I do not think there is any renewal without the maintenance of these values and without a perpetual return to the source for inspiration. (Césaire 1977, 52)

In a later interview, Césaire emphasizes why the work of mind reengineering is critical for Africa and why he writes:

> Africa is hungry for its own being. And that's why I don't want to be a Messiah or a prophet. My only weapon is my tongue; I speak, I awaken, I am not a redresser of wrongs. I don't perform miracles, I'm a redresser of life. I speak, I give Africa back to herself. I speak and I give Africa to the world. (Césaire 1989, 61)

Césaire, at last having resolved on a return to his native land, marvels at leaving a Europe "utterly twisted with screams," a Europe that "proudly overrates itself." Where in *Discourse* we see the black trammeled by the colonizer, now he wears "this egotism beautiful and bold," plowing confidently forward like "an implacable cutwater" (Césaire 1939/2001, 29).

In other words, having tired of the life of feeling sorry for his black self, the black man wears his blackness as, not like, a cloak of honor, and declares all his blackness. It comes in the form of prayer, wherein the black asks Providence for several weaponries. The first is that of leadership (at the time male-centric), and the power to make—to "mold"—the power of sorcery (engineering):

> Grant me the savage faith of the sorcerer
> grant my hands the power to mold
> grant my soul the sword's temper
> I won't flinch. Make my head into a figurehead
> and as for me, my heart, do not make me into a father nor a brother
> nor a son, but into the father, the brother, the son, nor a husband, but the lover of this unique people.
>
> (Césaire 1939/2001, 43)

In a 1989 interview, Césaire explains the essence and intent of *Notebook* as a return:

> If all I can do is speak at least I will speak for you. . . . My tongue shall serve those miseries which have no tongue, my voice the liberty of those who founder in the dungeons of despair. This *Notebook*, in my mind is, in spite of its being short, the fundamental book. It is from this book that all the rest came. In all things there is a fundamental intuition. This intuition, this fundamental vision, is in the *Notebook*. People do not have a very keen sense of history; you must go back to the period of this *Notebook*. You must try to imagine what the life of an eighteen-year-old man of color, a young Negro isolated in Paris, was like. So I arrive in Paris. What do I know of the vast world? Not much. (Césaire 1989, 49)

Second, he asked for capabilities to be a change agent, a man of genius not vanity, the extended arm of good, a beginner and an ender, a medium, and a germinator. In short, a creative black:

> Make me resist any vanity, but espouse its genius
> as the fist the extended arm!
> make me a steward of its blood
> make me a trustee of its resentment
> make me into a man for the ending
> make me into a man for the beginning
> make me into a man of mediation
> but also make me into a man of germination.
> (Césaire 1939/2001, 44)

Above all, an executor—a doer of work, "a hoer for this unique race," who must be brave (gird his loins) and love, for the confrontation he was embarking on engendered hatred (Césaire 1939/2001, 44). And not just an executor but a generator of ideas.

Négritude man becomes bullish; he starts to reimagine the world from and through himself, shouting, "Mulish reason you will not stop me from casting on the waters." Currents of thirst inside him smash the white man's "form" (order) into "deformed islands." His defiance disrupts the white man's end (intent). The black man turns "Annulose islands" (the Caribbean, where black humanity is reduced to feel like tiny worms) into "one single beautiful hull" (Césaire 1939/2001, 48). He turns Europe's maps of empire into "the world map made for my own use, not tinted with the arbitrary colors of scholars, but with the geometry of my spilled blood, . . . and the determination of my biology, not a prisoner to a facial angle, to a type of hair, to a well-flattened nose, to a clearly Melanian coloring, and Négritude, no longer a cephalic index, or plasma, or soma, but measured by the compass of suffering" (Césaire 1939/2001, 49). A volcano of defiant energy has erupted in the black man:

> Suddenly now strength and life assail me like a bull and
> the water of life overwhelms the papilla of the morne,
> now all the veins and veinlets are bustling with new blood
> and the enormous breathing lung of cyclones and the fire
> horded in volcanoes and the gigantic seismic pulse
> which now beats the measure of a living body
> in my firm conflagration.
>
> (Césaire 1939/2001, 50)

This is no mere mortal versus mortal combat; the ancestral spirits are leading the war; they are the commanders; their four hundred years of bones arisen, breathing energy and direction to their heirs-in-flesh. And Césaire ups the ante: "We are standing now, my country and I, hair in the wind, my hand puny in its enormous fist and now the strength is not in us but above us, in a voice that drills the night and the hearing like the penetrance of an apocalyptic wasp" (Césaire 1939/2001, 50).

Now the black man starts challenging Europe's narrative of Africa; "Europe has force-fed us with lies and bloated us with pestilence," the voice says. It rejects that "the work of man is done that we [the blacks] have no business being on earth, that we parasite the world, that it is enough for us to heel to the world whereas the work of man has only begun . . . No race has a monopoly on beauty, on intelligence, on strength and there is room for everyone at the convocation of conquest" (Césaire 1939/2001, 50).

The "trial by the sword" has begun; the "lance of night" of the Bambara ancestors demands the blood, fat, liver, and heart of man, not chicken. Only the brave "men's hearts which . . . beat with warrior blood." There will always be those resigned to remaining where the white man has sat them, the epicenter of whose brains the colonizer sits on—the colony within. They remain convinced of their "not being made in the likeness of God but of the devil, those who believe that being a nigger is like being a second-class clerk." Them house niggers! Loyally sitting in their pews, the good boys and girls, even snitches, deluded by material comfort into thinking they now fit in, that they are part of the place. The word is "sellout," "informant," "snitch," or "coconut." All they care about is their "upward mobility," and once they have arrived or made it, they never look back or around them. They become numb, feel absolutely nothing. They start to "beat the drum of compromise in front of themselves, those who live in their own oubliette, [and] who say to Europe: 'You see, I can bow and scrape, like you I pay my respects, . . . I am not different from you; pay no attention to my black skin: the sun did it'" (Césaire 1939/2001, 51). Césaire calls this type "the nigger pimp, the nigger Makarios, and all the zebras shaking themselves in various ways to get rid of their stripes in a dew of fresh milk" (Césaire 1939/2001, 52).

THAT VOLCANO of energy erupts into revolution in the poem, "Knives of Noon":

When the Blacks do the Revolution they start on the Champs de Alars
by uprooting giant trees flinging them at the sky's face like dog bayings
that lay to rest a bird-cool current in the warm air where it draws fire
and blood from the white. Draws blood from the white? Yes food for
white is the power and the right fought by the black roots at their
hearts. They won't quit conspiring in the little, well-made hexagons
of their pores. . . .

Gentle Lord
I spit harsh in the faces of those starving us, insulters'
and embalmers' faces, dismemberers, tearers. Harsh Lord.

(Césaire 1982c, 7)

This rebel is also the subject of Césaire's *Une Tempete* (*A Tempest*) adaptation for black theater of William Shakespeare's play, *The Tempest*, except that he strategically deploys it toward understanding the colonial project and black struggle against it. Prospero is recast as the colonizer and two slaves, Caliban, who is the rebel slave, and Ariel, who accepts to be a slave until his master, Prospero, grants him his freedom. It has been suspected that Caliban is Malcolm X, especially as he tells Prospero, "Call me X. That would be best. Like a man without a name. Or, to be more precise, a man whose name has been stolen" (Césaire 1985, 18). Ariel, by contrast, says he has had constantly a "wonderful dream that one day Prospero, you and I, we would all three set out, like brothers, to build a great world" (Césaire 1985, 26). He believes in nonviolence and conciliation, and as such has been interpreted as Martin Luther King Jr. When interviewed in 1989, Césaire draws attention to his choice of "Une" ("a") instead of "La" ("the"), and to write the former as experienced and felt by black people: "Yes, it is *Une Tempete*. No one can write *La Tempete* again. *Une Tempete* is the point of view of the loser (Caliban) not that of Prospero: the viewpoint of the colonized, not that of the colonizer" (Césaire 1989, 63).

"The old Négritude progressively cadavers itself," Césaire at last declares, asserting now the initiative, adventure, risk taking, creativity, inventiveness, and the technological and scientific nature of the black. But first, Césaire needs to finish with the old Négritude. The black's grandfather dies—the "good nigger. The Whites say he was a good nigger, a really good nigger, Lassa, good ole darkly." Césaire will not moan this grandfather, unless mockery is mourning: "He was a good nigger indeed, poverty had wounded his chest and back and they had stuffed into his poor brain that a fatality impossible to trap weighed on him; that he had no control over his own fate; that an evil Lord had for all eternity inscribed Thou Shall Not in his pelvic constitution; that he must be a good nigger; must sincerely believe in his worthlessness, without any perverse curiosity to check out the fatidic hieroglyphs. He was a very good nigger" (Césaire 1939/2001, 53). So good a nigger that it "never occurred to him that he could hoe, burrow, cut anything else really than insipid cane. He was a very good nigger!" (Césaire 1939/2001, 53).

The slave ship "cracks everywhere. . . . Its belly convulses and resounds." The black is in revolt; the captain hangs the "biggest loud-mouth nigger," throws him overboard or "feeds him to his mastiffs." It's pandemonium: the nigger who must always be seated, is standing up, and standing up everywhere—in the hold, cabins, wind, sun, blood. Standing, free. In the rigging, at the tiller, at the compass, at the map, under the stars, standing. Free.

This, black we, is not the time to be sitting down. It is time to dance, to dance a self-determined dance, to music composed by us, our own rhythm, invented by us, who dare to invent the future. To unchain the future:

Rally to my side my dances
you bad nigger dances
to my side my dances
the carcan-cracker dance
the prison-break dance
the it-is-beautiful-good-and-legitimate-to-be-a nigger-dance
Rally to my side my dances and let the sun bounce on the
racket of my hands
but no the unequal sun is not enough for me
coil, wind, around my new growth
light on my cadenced fingers
to you I surrender my conscience and its fleshy rhythm
to you I surrender the fire in which my weakness smolders
to you I surrender the "chain-gang"
to you the swamps
to you the non-tourist of the triangular circuit
devour wind
to you I surrender my abrupt words
devour and encoil yourself
and self-encoiling embrace me with a more ample shudder
embrace me unto furious us
embrace, embrace us
but have also bitten us

to the blood of our blood bitten us!
embrace, my purity mingles only with yours
so then embrace
like a field of even *filaos*
at dusk
our multicolored purities
and bind, bind me without remorse
bind my black vibration to the very navel of the world
bind, bind me, bitter brotherhood
then, strangling me with your lasso of stars
rise, Dove
rise
rise
rise
I follow you who are imprinted on my ancestral white cornea.
Rise sky licker
And the great black hole where a moon ago I wanted to drown
It is there I will now fish the malevolent tongue of the night in its motionless veneration!

(Césaire 1939/2001, 56–57)

Grandfather is dead. Where he had never imagined bondage as a choice, or rebellion as an option, grandson never imagines freedom as a choice, or bondage as an option. Freedom is the only future. "To Freedom or Death!," as Thomas Sankara will vow in later pages. True freedom can't be given; it has to be taken, made, innovated. It is work. Weapons. Methods. Movement. The colonizer cannot volunteer out of power; he has to be forced out. Force is the work of militancy. As Malcolm X would say, By Any Means Necessary! Necessity justifies the means. The situation defines fit for purpose. Means and ways. The technology-to-whom thing! Whose? As Fanon would say later in his seminal chapter in *The Wretched of the Earth*, entitled "Concerning Violence," "The colonized man finds his freedom in and through violence" (Fanon 1961/2004, 86). For that he defers to and elaborates on his mentor Césaire's poetry:

THE REBEL (harshly):

My name—an offense; my Christian name—humiliation; my status—a rebel; my age—the stone age.

THE MOTHER:

My race—the human race. My religion—brotherhood.

THE REBEL:

My race: that of the fallen. My religion . . . but it's not you that will show it to me with your disarmament. . . .

'tis I myself, with my rebellion and my poor fists clenched and my woolly head. . . .

(Very calm): I remember one November day; it was hardly six months ago. . . . The master came into the cabin in a cloud of smoke like an April moon. He was flexing his short muscular arms—he was a very good master—and he was rubbing his little dimpled face with his fat fingers. His blue eyes were smiling and he couldn't get the honeyed words out of his mouth quick enough. "The kid will be a decent fellow," he said looking at me, and he said other pleasant things too, the master—that you had to start very early, that twenty years was not too much to make a good Christian and a good slave, a steady, devoted boy, a good commander's chain-gang captain, sharp-eyed and strong-armed. And all that man saw of my son's cradle was that it was the cradle of a chain-gang captain.

We crept in knife in hand . . .

THE MOTHER:

Alas, you'll die for it.

THE REBEL:

Killed. . . . I killed him with my own hands. . . .

Yes, 'twas a fruitful death, a copious death. . . .

It was night. We crept among the sugar canes.

The knives sang to the stars, but we did not heed the stars.

The sugar canes scarred our faces with streams of green blades.

THE MOTHER:

And I had dreamed of a son to close his mother's eyes.

THE REBEL:

But I chosen to open my son's eyes upon another sun.

THE MOTHER:

O my son, son of evil and unlucky death—

THE REBEL:

Mother of living and splendid death,

THE MOTHER:

Because he has hated too much,

THE REBEL:

Because he has too much loved.

THE MOTHER:

Spare me, I am choking in your bonds. I bleed from your wounds.

THE REBEL:

And the world does not spare me. . . . *There is not anywhere in the world a poor creature who's been lynched or tortured in whom I am not murdered and humiliated.* . . .

THE MOTHER:

God of Heaven, deliver him!

THE REBEL:

My heart, thou wilt no deliver me from all that I remember. . . .

It was an evening in November. . . .

And suddenly shouts lit up the silence;

We had attacked, we the slaves; we, the dung underfoot, we the animals with patient hooves. . . .

Then was the assault made on the master's house.

They were firing from the windows.

We broke in the doors.

The master's bedroom was wide open. The master's room was brilliantly lighted, and the master was there, very calm . . . and our people stopped dead . . . it was the master . . . I went in. "It's you," he said, very calm.

It was I, even I, and I told him so, the good slave, the faithful slave, the slave of slaves, and suddenly his eyes were like two cockroaches, frightened in the rainy season. . . . I struck, and the blood spurted; that is the only baptism that I remember today. (Césaire, *Les Armes Miraculeuses, Et les chiens se tai saient,* 133–137, in Fanon 1959/1965, 87–88; my own emphasis)

II FREEDOM STRUGGLES AS INNOVATION

Everywhere I go in Africa, when I engage young people, they are generally dismissive about struggles for independence.

"What is there to show for it?"

"We graduate into unemployment, so what is the purpose of going to school . . . getting a degree?"

"Even graduates are now hustling at streetlights selling sweets and airtime."

"It would appear that our leaders liberated power from the white man so that they can become the new white man. They are corrupt. They don't care for the common person. All they do is loot and cling onto power."

"It is hard to reclaim our histories of struggle as inspirations when those who led them went on to become thieves and tyrants, and mass unemployment reigns supreme."

"Look how those who became presidents turned out—power-hungry, cruel dictators who turned thriving economies into ruins!"

"Well, Africa's first born (Ghana) is sixty-four years old, the last born (South Africa) is twenty-seven. Surely, Africa is no longer a child, even an adolescent. If it was a person, it would be married with kids by now!"

Some have gone as far as saying: "The colonizer was far better. At least he could oppress us because he was white but the economy thrived. Now we are oppressed by fellow blacks, the economy has collapsed, and unemployment reigns supreme."

To those who reply that they fought a long difficult war to free the country the reply is: "Go and tie the country back where you untied it and see if we cannot liberate it. Liberating us is no justification for the new slavery."

Which reminds me of the late African reggae icon Lucky Dube, when drawing the stark choice between colonial rule and independence: "Do you want to be a well-fed slave, or a hungry free man?"

It is hard to argue with these statements. Elsewhere I have written against the appropriation of the war contribution in Zimbabwe by a few political elites as if all of us were enjoying the sun on the beaches of Hawaii while they were busy liberating us. I have repositioned liberation struggles in general as communal efforts involving not guns and guerrillas alone, but many objects, knowledges, infrastructures, and combatants-without-guns.

In my conversations with young Africans, however, I always turn the tables on them: "Well, it could be said that struggling against colonial rule was the generational mandate of our fathers' and mothers' generation. Each generation's mandate is defined by its own realities. What is your generational mandate?"

"The past is not a place of residence, but of reflection, learning, and inspiration. What are we to learn from it? We cannot relitigate and remake or live in the past, but history can be an ingredient to build better futures."

I have always said that *death is kind to those political leaders who die during struggles for freedom or fighting for good causes, when they are still exceedingly popular, for it leaves us—and history—with positive memories and prevents us from seeing the monsters they might later become*. The reason for young people's disillusionment with the founding fathers and what became of that generation of nationalists is a generational one and is justified. However, there is a risk of assuming that these men fought alone, that their power was self-generated absent the people, and that, because the outcome was disappointing, we also dismiss and disown the struggle.

I see where the despondency is coming from. The presidency has become a monarchy, a gravy train, and a looting machine. In Kenya when a new government is coming into power, they call the process IOTTE (it's our turn to eat). At the general level in Africa, the language of patriotism becomes a cover for looting, and to bludgeon critics both rhetorically and physically. The same machinery of state the colonizer put in place was retained at independence, perfected, and extended to stay in power. Meanwhile the economies are, barring a few countries like Rwanda, plagued by massive youth unemployment, electricity blackouts, and deteriorating and collapsing rail, road, and water infrastructure.

Those who died in the moment of struggle, when freedom was an aspiration still to be attained, were probably not dreaming of looting the country, betraying the struggle, and disappointing us all. Would they have become just as corrupt? Most probably, going by their comrades who saw independence and went on to put the country in their own pocket. I look up for inspiration to those who died young and never lived to test power and what they were saying and doing at the time. They constitute a beacon of inspiration for the way they deployed their knowledge in the service of turning problems into spaces of solving with the people. They are not the only ones, but the idea is that those I have left out will inspire others to write about them. The intention is to change the narrative of the liberation struggle from a political one that elephantizes the elites, to reclaim the struggle as something we did, all of us and not just a few people we put in power. The best starting point is Fanon and Cabral, culminating in chapter on Guerrilla Healthcare Innovation in Zimbabwe[1], as examples of how to reclaim these struggles to show that the people are not bystanders while academicians, tech companies, and politicians invent the future. They are co-authors.

4 STRATEGIC DEPLOYMENTS

In the end, Négritude as mind reengineering is not necessarily an end in itself or a means to an end, but a prerequisite toward the physical engineering of freedom.[1] Frantz Fanon's *A Dying Colonialism* (1959/1965) acknowledges the effects of white racism on black people but moves on beyond not only victimhood, but also Négritude itself. In it, Fanon showcases the creative resilience that renders the death of colonialism inevitable and self-rehumanization possible. With Fanon, we see the convergence of Négritude and Pan-Africanism.

Whereas Fanon wrote *Black Skin, White Masks* while he was still somewhat numbed and yet to awaken from the trauma of racist dehumanization, he penned *A Dying Colonialism* while he was in Algeria and then Tunisia, putting his military experience and academic knowledge to the service of the Algerian Revolution. In 1953 Fanon had left France for Algeria, where he had served the Free French forces during the war, and took up the position of psychiatrist at the Blida-Joinville Psychiatric Hospital. It is often browsed over that Fanon was a dedicated physician who radically transformed his treatment of mental illness by adopting a psychotherapeutic method that approached patients from their culture and dedicated a copious amount of time to training and mentoring nurses and interns in this new practice. One year later, November 1, 1954, the Front de Libération Nationale (FLN) began the Algerian Revolution.

It so happened that a French doctor, Pierre Chaulet, used to visit Blida Hospital and had conversations with Fanon. And Chaulet was not just any physician; he was secretly working for the FLN, performing secret operations on injured fighters as well as providing secret shelter to its leader, Ramdane Abane. In time his cover was blown and he was deported, only for him to travel to Tunisia to link up again with the FLN, which had

FIGURE 4.1
Frantz Ibrahim Fanon. *Source*: Wikicommons.

headquarters and bases there. By the time he was deported, Chaulet had already introduced Fanon to the FLN, right there at Blida, in 1955.

Fanon's work changed with the start of the war. He found himself treating French soldiers suffering from traumatic stress arising from their torture activities against Algerians as well as the victims of that torture. He undertook countless missions across Algeria, in Kabylie more than other places, to investigate the psychological and cultural lives of Algerians, using his identity as a scholar-psychiatrist, meanwhile collecting vital intelligence. His conscience weighed upon him, and he resigned his post in 1956 to devote his energies full time to the struggle for Algerian liberation. The following year he was deported to France, only to squirrel off to Tunis to join the editorial collective of *El*

Moudjahid, serve as ambassador to Ghana for the Provisional Government of the Algerian Republic (GPRA), and attend many an Organization of African Unity (OAU; now African Union [AU]) meeting.

It was also during this time (1953–1960) that he published two books that are my favorite Fanons, written while he was fighting, working for change, struggling and not just sitting in the armchair: the collection of essays later published as *Pour la Revolution Africaine* (*Toward the African Revolution*, (Fanon 1964) and *L'An V de la Révolution Algérienne* (*Year Five of the Algerian Revolution*, which was then published by Monthly Review as *Studies in a Dying Colonialism* (Fanon 1959/1965) before being shortened to *A Dying Colonialism* by Grove Press for marketing purposes in 1965. While the book is generally read as a history of the Algerian Revolution, I reread it as *a text written from experience as analytical location, in one sense, and knowledge in the service of and through problem-solving, with the venue of liberation struggle as a space generative of theory, in another. The problem to be solved was that of colonial occupation* not just of Algeria, the geographic territory that Algerians lived in, but the occupation of the Algerians themselves.

Yet, as Fanon warned, an occupied people is never disarmed; and it never awakens because it has never slept, nor does it remember because it has never forgotten. Why occupied? Because, says Fanon, a colonized people is an occupied people. What is occupied is not just the soil, ports, airports—physical space. Colonialism has "settled itself in the very center of [us]" and undertaken "a sustained work of cleanup work, of expulsion of self, of rationally pursued mutilation" that is neither "occupation of territory" nor independence. It renders our history, our way of life, how we think "contested, disfigured, in the hope of a final destruction." And, Fanon adds, "under these conditions, the individual's breathing is an observed, . . . occupied breathing. It is a combat breathing." The moment the colonizer occupies, "the real values of the occupied quickly tend to acquire a clandestine form of existence" (Fanon 1959/1965, 65). An occupied people are never just victims. Occupied, yes, but never disarmed, always leaving the possibility of rising up against the occupation. Enslaved, yes, but never slaves, always leaving the possibility of self-liberation. The process is creative resilience.

Don't let's make the mistake the French made. They only considered how many guns the Algerian had, and completely ignored "the holy and colossal energy that keeps a whole people at the boiling point" (Fanon 1959/1965, 29). The revolution's true strength was not in men or weapons, but that spirit of creative resilience. This measurement of power, not by how big and how many heavy machine guns (as the colonizer deluded himself), but by human energy, exposed a key difference in the comparative cosmologies of weaponry.

I would like to demonstrate this story of African creative resilience and strategic deployments through a close reading of three chapters in *A Dying Colonialism*. Namely, the veil, radio, and medicine as felt, experienced, and strategically deployed by Algerians under and fighting for freedom against French colonial rule. First, some keywords.

By strategic deployments I mean notions of technology derived not always from human-made, human-making, and substantial human-modification, but from human knowing, but positionings and assignments and learnings vis-à-vis/from animals, plants, atmospheres, places, and so on informed by detailed observation, critical thinking, and abstraction and that offer benefits without always disadvantaging, doing harm, or touching. As experienced in Africa, the Western/colonizer has exercised strategic deployments to exploit, destroy, and dominate, whereas Africans have typically deployed it *to be technological without necessarily touching or hurting.* As a mode of thinking about technology, "strategic deployments" invests more attention to how Africans assign meaning and purpose to things. Nothing arrives to them as technology *a priori*, but only becomes so in view of their strategic deployments, which assign the thing purpose contrived by them. This is one of the most understated attribute of Africans.

Creative resilience is the phenomenon whereby Africans, even when death is certain, when their backs are against the wall, do not just roll over and die, but fight to not die or, if nothing else, die fighting. That is the second equally underestimated strength sustainable African futures must be built upon. It can also be extended to other peoples who have suffered adversity, but my intent here is modest.

VEIL, ENVEILING, UNVEILING

In this first section, I illustrate two forms of strategic deployments of the veil, one by the colonial government, the other by the FLN. Both illustrate a version of the technological, not in the banal forms we are accustomed to when we talk about technology and innovation, but something more complex, emerging only within the crucible of the experiential location. And that experiential location produces and feeds off its own archives, vocabularies, and theories.

To a visitor to 1950s' North Africa, the veil was the distinctive feature of Arab society and women in Tunisia, Algeria, Morocco, and Libya. For men, the *fez* was worn in the cities, *turbans* and *djellabas* (long hooded cloaks) in rural areas. Especially for women, the veil was "a uniform which tolerate[d] no modification, no variant" (Fanon 1959/1965, 36). The veil in question is specifically the *haïk* (the big one covering the whole face and body), so that the Algerian woman is encountered as "she who hides behind a veil" (Fanon 1959/1965, 36).

The reason why family is so strong in Africa is because it is home. It is the institutional memory, the server, of the nation. The home, Fanon says, is "the basis of the truth of society, but society authenticates and legitimizes the family" (Fanon 1962/19 And in Algerian society, home could not exist without the veil, because it defined the dignified female and made her a woman. It is what made a male a man, and the male partner a husband, the sequel to mother and father, respectively. Each African society has the adage: home is woman/woman is home (*musha mukadzi* in *chidzimbahwe*).

The occupier from the onset determined to destroy the family to break the backbone of African society by controlling African birthrates, as we saw during slavery, and then in Africa during occupation, with family planning disguised as public health. The occupier imposed taxes and passed laws to force men to work on white farms, mines, and cities. Meanwhile, he criminalized women's stay with their husbands, turning the countryside into a space for women and children, and the elderly, having creamed out all the able-bodied males (muscles) into forced or cheap labor camps called "hostels." The colonial city developed at the expense of the rural. A few women worked as domestics, but their presence in the city was criminalized; inspections to sniff them out were rigidly enforced.

The veil became a battleground between the French colonizer and Algerians. In the early 1930s, French colonial officials "committed to destroying the people's originality" (read creativity, innovation) and preempt a nationalist reawakening, focused on the veil. European sociologists, ethnologists, and native commissioners (usually one and the same thing) had concluded the veil was "a symbol of the status of the Algerian woman," that it was men's weapon to oppress them.

The purpose of Europe's anthropological project in Algeria, as exemplar of its practice Africa-wide, was unambiguously clarified: "If we want to destroy the structure of Algerian society, its capacity for resistance, we must first of all conquer the women; we must go and find them behind the veil where they hide themselves and in the houses where the men keep them out of sight" (Fanon 1959/1965, 37–38). The occupier started with the hungry and economically vulnerable women, and those against the veil. Good money was invested in these projects, quite similar to current "impact investment" geared toward women in Africa. Fanon takes us way back to the 1930s:

> After it had been posited that the woman constituted the pivot of Algerian society, all efforts were made to obtain control over her. . . . In the colonialist program, it was the woman who was given the historic mission of *shaking up the Algerian man*. Converting the woman, winning her over to the foreign values, wrenching her free from her status, was at the same time

FIGURE 4.2
"Aren't You Pretty? Unveil Yourself!" Colonial poster, Algeria, 1958. *Source*: Lambert 2016 (*The Funambulist Magazine*).

achieving a real power over the man and attaining a practical, effective means of destructuring Algerian culture. (Fanon 1959/1965, 39; my own emphasis)

Each veil removed, the more "the flesh of Algeria laid bare," "open and breached," the happier the occupier, the more Algerian women's faces "offered . . . to the bold and impatient glance of the occupier." The Algerian woman was, in so doing, "accepting the rape of the colonizer," to him "a romantic exoticism, strongly tinged with sensuality." Cue the remarks of a European lawyer visiting Algeria upon seeing some unveiled Algerian women: "These [Algerian] men . . . are guilty of concealing so many strange beauties." Such a "cache of . . . prizes, of such perfections of nature, owes it to itself to show them, to exhibit them. If worst came to worst, he added, it ought to be possible to force them to do so" (Fanon 1959/1965, 43). To unveil her body was to reveal her beauty, to bare her secret, thus "breaking her resistance, making her available for adventure," to bring her "within his reach, to make her a possible object of possession." The veiled woman can see without being seen; it "frustrates the colonizer. . . . She does not yield herself" (Fanon 1959/1965, 44).

For Fanon, the European woman of his time is already revealed; she embellishes in order to further reveal her beauty and conceal her blemishes—through hair styling, fashion, and makeup; the Algerian woman veils, covers, "to cultivate the man's doubt and desire." Either way, the rapacious eye of the occupier has monopoly to see. French conquest of Algeria—of villages overrun, property seized, women raped, the country pillaged—evokes a sadistic freedom to penetrate or create "faults, fertile gaps" to do so. Thus, Fanon concludes, "the rape of the Algerian woman in the dream of the European is always preceded by a rending of the veil. We witness a double deflowering" (Fanon 1959/1965, 46).

Here, the purpose of European colonial education was a strategic deployment of the cynical sort: to make you our instrument for extending our civilized Western-and-white habits to your people! You people! (Those of us subjected to racism know what that means). You must play your role as an agent of cultural imperialism, now that you have tasted the good life. Even after being unveiled—set free—or educated, you were still a native. The Algerian could never be an expert; he was always a "native" first. Like some local flora and fauna, infrahuman, human game, and if resisting those labels, vermin beings.

But three decades later, the French's elaborate dream of turning Algerian society to Western-and-white values "by means of 'unveiled women aiding and sheltering the occupier'" not only remained elusive, but had backfired spectacularly.

What is at stake in such projects is never the emancipation of Africans but to control and bend them to the will of Western-and-white values; it is about turning Africans to submit to the hegemony of a culture in which they will always be phonies, imitators, almost-but-never-assimilated-into-Frenchman, subordinate to those who invented and own that culture, having discarded and now unable to return to their own, to forge their own self-determined culture. The occupier calls the colonized person's refusal of acculturation the primitive mindset. What he misses is "the organic impossibility of a culture to modify any one of its customs without at the same time re-evaluating its deepest values, its most stable models" (Fanon 1959/1965, 42).

During their revolution, Algerians strategically deployed the veil as a technology for innovating self-liberation through and with their own cultural resources and artifacts, innovations that "were at no time included in the program of the struggle. The doctrine of the Revolution, the strategy of combat, never postulated the necessity for a revision of forms of behavior with respect to the veil" (Fanon 1959/1965, 47).

In fact, FLN had no place for women before 1955: it feared they would compromise the total secrecy needed; they must be kept completely in the dark. Even when the decision was made to include them, the sentiment for or against it was strong: it meant changing "the very conception of the combat." And with that, the entire idea and practice of struggling, of war, better yet, revolutionary war, was changed.

It meant the demasculinization of who could be a revolutionary, what could be revolutionary, and the obligations not only of the participant, but the recognition of their role and obligations to them. Not just using women to subvert the entire nation to a political program—the occupier had done that with unveiling and flopped. Same sacrifices as men? Same confidence? Strength of character? The militants hesitated. Inclusion of women into the "machine" of revolution must be done "without affecting its efficiency" (Fanon 1959/1965, 49).

No, the enemy was too ferocious. Too criminal. Brutal. The leaders had done time in the occupier's jails, talked to survivors, or had intelligence: where only men had been tortured, now the women too became legitimate targets of torture. No, it was too dangerous.

The debates raged within the FLN . . .

The woman had kept the revolutionary spirit alive in the home, the family. Revolutionary war was a "total war" that might come at the cost of breaking up family—and Algerian society itself. Or, come to think of it: wasn't that the meaning of revolution—creating a new society on the ashes of the old? What would be the consequence to family and society when the woman was no longer just a mother or daughter "knitting for" or "mourning" the male whose marker of being a man was now a freedom fighter, when the woman was now at the heart of combat, enduring arrest, rape, torture, and gunfire? (Fanon 1959/1965, 66).

Ultimately the decision had to be taken, a new layer of being a woman was added (or removed); the revolutionary woman, fighter for freedom. The veiled woman who had never gone to a formal school was now tasked with memorizing and carrying in her head and verbally delivering messages in the combat zones. From this experiential space, one sees a wireless communication that is fully human; to capture a woman was to capture information.—To capture an intelligence report (document).—A container of information.—A mobile carrier of information.—A courier.—A wealth of information on two legs.—*A server.*

The military commanders are now passing through the cities. Every one of them is wanted by the cops. The revolutionary women scouts "lead the way," one hundred to two hundred meters ahead, carrying the commanders' and troops' automatic pistols, revolvers, and grenades, sometimes all three. The commanders follow close behind, ready to move in immediately and recover the weapons if things turn ugly. She wears the veil as they maneuver the countryside. That is—veil as camouflage, material, visual, and psychological. The Algerian woman discards her veil on the edge of the city, walks

FIGURE 4.3
Algerian women under training during the revolution. *Source*: Wikicommons.

a hundred meters ahead of ordinary-looking men, carrying a suitcase full of weapons or rifles, pistols, and grenades. She is, Fanon says, "the group's lighthouse and barometer," warning when there is impending danger, as the file navigates roads and alleys infested by police cars and patrols (Fanon 1959/1965, 51).

The unveiled female combatant "walks the street"; young men get attracted to her. Why? She looks North African, wears no veil! She has freed herself from Muslim custom! And immediately they see her as an object of their sexual fantasies! Perfect camouflage for her actual mission. She wants them to inhabit that fantasyland; she strategically deploys herself, unveils herself to turn this lascivious mind of the occupier's culture into a weapon of self-destruction, presenting an attractant that draws and fixes the eye and mind to the face and quickly to the "relevant chapters" of the woman's body, seeing an object of desire moving, meanwhile numbed blind to the guerrilla inside her. She smiles beautifully; she swaggers her hips; the occupier's mind gets even more twisted. OR the young men shout obscenities at her: "When such things happen, she must grit her teeth, walk away a few steps, elude the passers-by who draw attention to her" (Fanon 1959/1965, 53). And the venue is Paris.

From 1956 onward, as the French intensified the use of terror tactics, massacring Algerian civilians, the FLN resolved to fight terror with terror, a tactic it had abhorred until

FIGURE 4.4
FLN female bombers. From left to right: Samia Lakhdari, Zohra Drif, Djamila Bouhired, and Hassiba Bent-Bouali. *Source*: Wikicommons.

then. The French had used it against the Nazis during the German occupation, derailing trains and setting off bombs. The use of the *fidaï* (meaning "death volunteer" in Islam) risked increasing civilian casualties, distorting the purpose of the revolution, and upsetting progressive allies that agreed on the object but not such means. Some even wondered: was independence from French rule worth that method? Others felt the enemy must be confronted "individually and by name," in reply to the enemy torture and murder of freedom fighters "shot and killed while trying to escape." Places where colonizers met and specific members of the police became targets (Fanon 1959/1965, 57).

In this new phase of confrontation, the Algerian woman's role intensified. In her bag, she now carried the grenades and revolvers the *fidaï* would plant in the bar. An operation involving two people, giving the public appearance of being strangers to one another. The one a courier and camouflage, a decoy, the unveiling transforming her into "a European woman, poised and unconstrained," engendering no suspicion. The other "stranger" going about his own journey (Fanon 1959/1965, 55–56). One (the woman) was the weapon delivery system, albeit one guided by human intelligence, her own and that of the operational orders given; the other the weapon-by-self-destruction (Fanon 1959/1965, 57). (Our analysis must not cloud us; human lives were lost here.

Innovation, strategic deployment, is never just about nourishing life; it is also about creativity that goes into mutually assured (self-)destruction.)

Unveiling as a performance of camouflage requires powers of invention; the Algerian woman does not just get orders and execute effortlessly or intuitively. In traditional Algerian society, the woman's body (womanhood) reveals itself to a girl "by its coming to maturity and by the veil," which covers the body and "disciplines it, tempers it" at its most effervescent stage. Hence the veil "protects, reassures, isolates," whereas the unveiled body "seems to escape, to dissolve," makes the woman feel naked, incomplete, anxious, unfinished, disintegrating, distorted.

Unveiled, she must "invent new dimensions for her body, new means of muscular control . . . create for herself an attitude of unveiled-woman-outside," conquer timidity and awkwardness to pass off as a European, yet without overdoing it lest she exposes instead of camouflaging herself. She walks like this into the European city, feeling revealed, "stark naked," relearning her body. She "re-establishes it in a totally revolutionary fashion," sets up a "new dialectic of the body and of the world." This is not ANY Algerian woman; this is a "revolutionary woman" (Fanon 1959/1965, 59).

The colonizer discovers he is being outwitted. Now all Algerian women are suspects; by increasing *surviolence* (surveillance as violence), the colonizer does the freedom fighters' work of mobilizing the whole nation for them. Colonizer puts on a brave face, drags servants, poor women, and prostitutes into public spaces, and symbolically unveils them. But they refuse this gift of liberation by Charles de Gaulle. *The Algerian woman does not take freedom as a gift; she conquers freedom.* Algerian women who have long put away their *haïk* bring them back on—to reject the occupier! "Colonialism wants everything to come from it. . . . Colonialism must accept the fact that things happen without its control, without its direction," Fanon says (Fanon 1959/1965, 63). But colonialism never learns.

RADIO

The story of radio in African hands during the national liberation moment (from 1951 in Tunisia until 1994 in South Africa) is the story of how Africans made the radio African and technological. I have said earlier that things do not arrive in Africa as technology *a priori*, that "the African technological" is about African initiative and in the first person ("We import . . ." or "We make . . ."), not the passive language of receiving and not inventing or manufacturing anything ("Technology arrives in Africa," "Technology is transferred to Africa," "Technology adoption," "Accelerating technology transfer and diffusion to Africa," "Sustainable sources of Critical Raw Materials [CRMs] from Africa"), and so on. When imposed on Africans, things incoming do not acquire traction or meaning unless Africans themselves say or show so. The start of the liberation wars in Tunisia (1951–1952) and Morocco (1952–1953) and then Algeria (November 1, 1954) piqued Algerians' interest in owning or accessing radio as a "specific technique for the dissemination of news" (Fanon 1959/1965, 75). The radio in the Algerian Revolution is one such case of strategic deployments.

Like the gun and Bible, in the mind and hand of the occupier, radio was a colonializing instrument. The colonizer strategically deployed Radio-Algier as an extension of the French National Broadcasting System to advance colonial society and its values (Fanon 1959/1965, 69).

The European settler of Algeria saw owning a radio—along with owning a villa, car, and fridge—as a sign of participating in Western-and-white modernity, a sound confirmation of being a "civilized man," and of the "living and palpitating reality" of colonial society and power. It was a constant reminder to never "go native." As an "instrumental technique in the limited sense," Fanon says:

> The radio receiving set develops the sensorial, intellectual, and muscular powers of man in a given society. . . . As a system of information, as a bearer of language, hence of message, the radio may be apprehended within the colonial situation in a special way. Radiophonic technique, the press, and in a general way the systems, messages, sign transmitters, exist in colonial society in accordance with a well-defined statute. Algerian society, the dominated society, never participates in this world of signs. The messages broadcast by Radio-Algier are listened to solely by the representatives of power in Algeria; the non-acquisition of receiver sets by this society was precisely the effect of strengthening this impression of a closed and privileged world that characterizes colonialist news. (Fanon 1959/1965, 73)

Most Europeans in Algeria owned a receiver; that is 95 percent of all radio sets prior to 1945, with Frenchmen speaking to other Frenchmen. From 1947, the numbers increased marginally, as the station started playing more national, Algerian music and recruited "native" broadcasters to appeal to Algerians with a view to Western-and-white-washing them. The white man in Algeria adopted radio at a rate similar to Europe and North America.

Only "developed bourgeois" Algerians (the *évolué* class) had them. The obscene and clownish content came across as offensive to Algerians; listened to as family, the radio created unease. Hence the slow uptake by the people. They saw radio as a technique by which the occupier sought to destabilize their society and cultural values through the imposition of Western-and-white values and norms of reason and sociability (Fanon 1959/1965, 71–74). They rejected the "colonial" radio, not the artifact.

The European settler farmers' response to outbreak of the revolution was to round up their workers and tell them spiritedly that the entire "gang of rebels" had been macerated

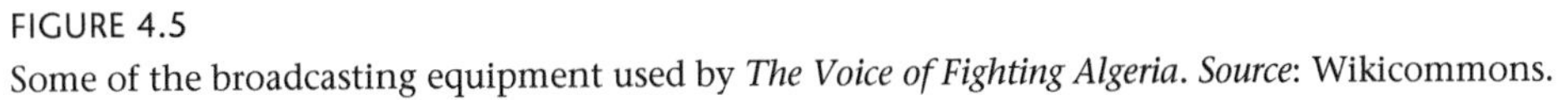

FIGURE 4.5
Some of the broadcasting equipment used by *The Voice of Fighting Algeria*. *Source*: Wikicommons.

in the Aures Mountains or in Kabylie. House servants were now shocked to be invited to a lemonade or cake in the aftermath of suspected "terrorists" being executed a mile or two from the farm. These were inducements to betray, to sell out, to inform, to be informants, against their own kind.

Tired of the occupier's propaganda, the Algerian sought his own source of information, replaced "the enemy news with his own news," proved the oppressor's "truth" as "an absolute lie," a "menaced lie," and countered it with "an acted truth." The effectiveness of Algerian creative resilience put the occupier on the defensive; his resistances, reactions, and brutality betrayed "the effectiveness of national action and made that action participate in a world of truth. . . . Because it avowed its own uneasiness, the occupier's lie became a positive aspect of the nation's new truth" (Fanon 1959/1965, 76). No longer pained and desperate refusal but *daring to invent a different information system, a network, a world wide web of anticolonial uprising in search of self-conceived, self-designed, and self-accomplished freedom.*

Let me emphasize this.

The project of creativity in service of self-liberation meant designing and building a news distribution system—the Arab telephone—as a "balance-restoring element" to reveal to the Algerian "whole sections of truth" hitherto hidden from them. What had once been a pejorative—the Arab telephone—or word of mouth was now a safe, wireless revolutionary tool, in contrast to hi-tech. Here, then, is *one of those instances when slow-tech or no-tech becomes the highest form of the technological.* A different form of real-time communication based on physical meeting and conversation (which mobile technology is killing in the West), not broadcast like radio, which the colonizer could jam or intercept, but discretely communicated in person by word of mouth. Starts with one or a few, who tell others, the numbers of those who now know continues to multiply, and, being members of family, village, district, once they know, it is not just them who know. Everybody.

Meanwhile, the urban-based Algerians who could afford it bought newspapers considered democratic that came from France—*Libération*, *L'Express*, *France-Observateur*, *Le Monde*, and *L'Humanité*. These European newspapers were shipped in not for Algerians, but for the Frenchmen. They were sold at kiosks only, for whites only; the Algerian was expected, even required, to read only the Western-and-white-washing local press, sold by public criers. To ask for a copy of *Le Monde* or *L'Express* was virtually an act of treason, a confession to supporting the FLN. To buy or hold it was deemed "a nationalist act . . . a dangerous act . . . equivalent to an act of war" (Fanon 1959/1965, 81). From 1955 on, the adults started sending kids to buy the papers for them. They were soon found out. The kiosks started refusing to sell to minors. The revolution's political directorate responded with orders for Algerians to boycott the local press, which the majority did not read anyway because "the people's generalized illiteracy left it indifferent to things written" (Fanon 1959/1965, 82).

It was the guerrilla who made radio a wireless Algerian technology, not the French. By 1955, it had become the only non-French source of information about the revolution. As an addition into a household or *douar* (village), the wireless—or radio set—was a gradual and belated introduction. Nobody was stampeding to get a wireless until the end of 1956 when the FLN distributed flyers announcing *The Voice of Fighting Algeria*'s airing schedule and wavelength. What had been a disinterested nonaudience turned into an avalanche to listen to "this voice 'that speaks from the *djebels* [mountains]'":

> In less than twenty days the entire stock of radio sets was bought up. In the *souks* (markets or shops) trade in used receiver sets began. Algerians who had served their apprenticeship with European radio-electricians opened small shops. . . . The absence of electrification in immense regions in Algeria naturally created special problems for the consumer. For this reason battery-operated receivers, from 1956 on, were in great demand on Algerian territory. In a few weeks several thousand sets were sold to Algerians, who bought them as individuals, families, groups of houses, *douars*, *mechtas*. (Fanon 1959/1965, 83)

One could think the hunger was the fault of the French occupier, who had strategically deployed wireless as a technology of lying and misinformation, a technical instrument to kill Algerian culture through French cultural imperialism. But that is giving the colonizer credit for what was in fact a strategic choice by Algerians armed with their own meanings of truth, news, information, and fact, who simply felt what they were getting was false. As Fanon notes, people did not "adopt" the radio because it was "a modern technique for getting news," but as "obtaining . . . access to the only means of entering into communication with [and living with] the Revolution" (Fanon 1959/1965, 83).

In the hands of the FLN, through the mind and voice of its cadres, broadcasting from their base, the radio becomes a means of revolutionary communication, a force for societal transformation, and a catalyst in the self-engineering of freedom out of slavery. The technical instrument of the radio receiver loses its identity as "an enemy object"—as "colonial radio"—and became a "new signaling system brought into being by the revolution . . . consolidating and unifying the people" (Fanon 1959/1965, 83–84). In a word, a "revolutionary radio."

The occupier reacts.

All radio and battery sales—banned. All spare batteries—withdrawn from the market. The Algerian dealers see this problem as an "opportunity to . . . supply the people with spare batteries with exemplary regularity by resorting to various subterfuges" (Fanon 1959/1965, 85). The French soon detect the wavelengths of the broadcasting stations and systematically jam them. Soon, *The Voice of Fighting Algeria* is barely audible.

Now, listen to Fanon, talking about African creative resilience on display, the listener as innovator, caught between a signal broadcast by the FLN, jammed by the French. Listen to listening as guerrilla warfare with the wave:

> Tracts were distributed telling Algerians to keep tuned in for a period of two or three hours. In the course of a single broadcast a second station, broadcasting over a different wave-length, would relay the first jammed station. The listener, enrolled in the battle of the waves, had to figure out the tactics of the enemy, and in an almost physical way circumvent the strategy of the adversary. Very often only the operator, his ear glued to the receiver, had the unhoped-for opportunity of hearing The Voice. The other Algerians present in the room would receive the echo of this voice through the privileged interpreter who, at the end of the broadcast, was literally besieged. . . . The whole nation would snatch fragments of sentences in the course of a broadcast and attach to them a decisive meaning. Imperfectly heard, obscured by an incessant jamming, forced to change wave lengths two or three times in the course of a broadcast, *The Voice of Fighting Algeria* could hardly ever be heard from beginning to end. It was a choppy, broken voice. (Fanon 1959/1965, 85)

And yet *The Voice*'s presence was "felt," its reality "sensed," a "phantom-like character." The act of listening was itself one's own "clandestine participation in the essence of the Revolution." Every night, nine till midnight, "the voice of the combatants was here. Behind each modulation, each active crackling, the Algerian would imagine not only words, but concrete battles" (Fanon 1959/1965, 86–89). There was no separating listening to *The Voice of Fighting Algeria* from being "at one with the nation in its struggle"; the act of tuning in and listening to radio became a process of new national formulation, to be in the micromobilities of struggle with the fighter among the rocks, on the *djebels*. Ducking. Diving. Escaping. Attacking. Winning. Dying for us! *For me!* For Algeria.

Even the cultural tools and informatics the occupier had intended for imperialism were extended to, inverted into, instruments of self-liberation. To speak in or understand French was no longer treasonous to revolution or evidence of being culturally occupied; under the purposive and strategic deployments of *The Voice of Fighting Algeria*, the French language became an instrument of self-liberation. The Algerian "occupied" the language of the occupier and owned it, no longer for purposes of assimilation or to impress the occupier for purposes of obtaining the status of *évolué*, but to eject the colonizer. No longer looking to the French for approval!

The French looked for fear.

Saw none.

Self-consumed in their own vanity, the French misread the inversion to mean "Aha, at last the natives have realized the folly of resisting civilization and now want to assimilate. Well then. What a great opportunity to advance a 'French Algeria'! Let's make French the only realistic means of communication across all tribes—Kabyles, Arabs, Chaouias, Mozabites, everyone." Unaware that while they had invented French, the Algerian had extended it—strategically deployed it—to Algeria to invent revolutionary code (Fanon 1959/1965, 91). The daydreaming occupier was unaware of an unfolding reinvention of "an attribute of the occupier" through rendering it open to Algerian cosmologies, signs, symbols, and influences.

And Fanon's conclusion is neither a celebration nor condemnation—merely an observation of the bind that is still with us as we deal with technologies of our own time:

> In August 1956, the reality of combat and the confusion of the occupier stripped the Arabic language of its sacred character, and the French language of its negative connotations. The new language of the nation could then make itself known through multiple meaningful channels. The radio receiver as a technique of disseminating news and the French language as a basis for a possible communication became almost simultaneously accepted by the fighting nation. (Fanon 1959/1965, 92)

SYRINGE

Because it came with race and humiliation and was a tool of racist oppression, Western-and-white medical practice was not innocent, objective, or science—it could only become scientific (or useful knowledge) only after, not before, the African placed a hand, mind, and view on it, within an order of knowledge familiar to and defined by the African. To be science, information has to be knowledge, and to be knowable, the rationale for knowing self-determined, never simply imposed. At best, this incoming thing is knowable through the way Africans feel and experience its strategic deployments in the hands of its bringer, who in this case was a racist oppressor. As a descriptor of top-down meaning, we can call it colonial medicine.

That is why it elicited from Africans an "ambivalent attitude," as a sign and felt hand of "the occupier's modes of presence," the colonial doctor's compulsory visit always preceded by police forcibly rounding up entire villages to attend (Fanon 1959/1965, 121–122). It is here that we see the naivety of Western-centric science and technology studies circumscribed by world-minority cosmologies of science as knowledge of scientists driving national or global technological advances. Accompanying it is a shockingly narrow view of "ethics" or "bioethics" reduced to institutional review board criteria of Western-and-white practice, often naïvely and dangerously extended to govern research outside the Western-and-white geographic and the Western-and-white cosmological.

This is what it feels like from the experiential location of the African rounded up to meet the colonial doctor:

> [The colonized] perceives the doctor, the engineer, the schoolteacher, the policeman, the rural constable, through the haze of an almost organic confusion. The compulsory visit by the doctor to the *douar* is proceeded by the assembling of the population through the agency of the police authorities. The doctor who arrives in this atmosphere of general constraint is never a native doctor but always a doctor belonging to the dominant society and very often to the army. (Fanon 1959/1965, 121)

The occupied saw statistics on sanitary improvements as yet more proof of the occupier's firm grip, not progress against diseases. The occupier bragged about what he was doing for the natives, who would be blighted by ignorance without him. Whereupon Fanon quips, "Nobody asked you for anything; who invited you to come? Take your hospitals and your port facilities and go home" (Fanon 1959/1965, 122). The equation? French medical service=French colonialism: the Algerian "rejected doctors, schoolteachers, engineers, parachutists, all in one lump."

A doctor-patient relationship is impossible in a situation governed by coloniality. In the West or among the colonial settlers, a sick white person approaches a doctor confidently; he trusts the doctor, "puts himself in his hands. He yields his body to him." The doctor can awaken or worsen pain during examination, but the patient trusts that it will result in "peace in his body." Trust in the physician is guaranteed, even where the technique is doubtful (Fanon 1959/1965, 123). Medicine is medical because it is deployed to heal, as it should be. It is felt so, as it should be.

Not under coloniality. The patient never trusts the doctor—much like a German prisoner operated by a French surgeon in World War II (who begged him not to kill him moments before he was given an anesthetic). So too the French prisoner to a German doctor.

Worse under colonial occupation, Fanon says: the Algerian refused hospitalization because of his "lingering doubt as to the colonial doctor's essential humanity," and a feeling of being experimented upon. Which was true: in some hospitals, the French "doctors" recalled Algerian patients to the hospital under the scientific pretext of "further examinations" and experimented upon them, including inducing epileptic fits to determine the pain threshold of each "tribe" (Fanon 1959/1965, 124). Resistance to hospitalization was based not on "the native's" fear of cities, distance, being away from the protective shield of family, what people would say, or attachment to traditional medicines, sorcerers, or healers.

"The colonized not only refused to send the patient to the hospital," Fanon says, "but he refused to send him to the hospital of the whites, of strangers, of the conqueror," where he was told to abandon all his ways of therapy, was "literally insulted and told [he was] a savage" for making scratches on his son's forehead to insert medicines direct into the vein. The issue was not that the white doctor diagnosing a condition as meningitis was wrong, but that "the colonial constellation is such that what should be the brotherly and tender insistence of one who wants only to help me is interpreted as a manifestation of the conqueror's arrogance and desire to humiliate" (Fanon 1959/1965, 126).

(Fanon of course was rather ambivalent: simply because the white man said it was meningitis and that his treatment worked did not mean other modes of therapy were false. Hence Fanon is useful to me as a resource for self-rehumanization after colonial dehumanization, what Césaire called "thingification," but not self-reintellectualization after being rendered unthinking and savage, my cosmologies dismissed as fable and primitive superstition, as seen through the violently imposed monopoly of Western-and-white cosmology).

Therefore, the celebrated role of science in the West is not a universal, in view of the impossibility of the occupied-seeking-self-liberation and the colonizing society "to agree to pay tribute, at the same time and in the same place, to a single value" (Fanon 1959/1965, 126). Where cosmologies converged, the occupier always interpreted that as the natives finally seeing the light and assimilating to Western-and-white civilization. Thus for Fanon, "going to see the doctor, the administrator, the constable or the mayor are identical moves." The colonized, enslaved, and discriminated against see not a doctor, but "both a technician and a colonizer" and "a spokesman for the occupying power" (Fanon 1959/1965, 127, 131).

Contact between the colonizer and colonized was not possible at a social, uncoerced level; hence the colonial doctor sees pain as "protopathic, poorly differentiated, diffuse as in an animal, . . . a general malaise rather than a localized pain." He seeks answers from the clinical examination, from the body, which proves "equally rigid," its "muscles . . . contracted." The Algerian wonders why this man calling himself doctor asks him what is wrong instead of diagnosing him like the healer does. The Algerian is skeptical: "*I know how to get into their hospital, but I don't know how I'll get out—if I get out.*" The colonial doctor concludes: "*With these people you couldn't practice medicine, you had to be a veterinarian*" (Fanon 1959/1965, 127–128; my own emphasis). If medical at all, colonial medicine is veterinary practice—animal medicine.

At which point Fanon concludes: "*When the colonized escapes the doctor, and the integrity of his body is preserved, he considers himself the victor by a handsome margin.* For the native the visit is always an ordeal. . . . The medicines, the advice, are but the sequels of the ordeal" at the hands of "*the colonizing technician.*" The colonizing doctor is faced with an uncooperative, even defiant patient who "cannot be depended upon to take his medicine regularly, . . . takes the wrong doses," does not honor appointments, and does not conform to prescribed diet (Fanon 1959/1965, 127–128; my own emphasis).

Here then, is a patient who defies being patient and being a patient, who flees, disengages. His is "not a systematic opposition, but a 'vanishing'," "a whole complex of resentful behavior," a "refusal." Coerced into Western-and-white medicine, yet equally believing that "two remedies are better than one," penicillin and the healer simultaneously. To "gradually adopt an almost obsessional respect for the colonizer's prescription often proves difficult. The other power, in fact, intervenes and breaks the unifying circle of the western therapy" (Fanon 1959/1965, 129–131).

The Algerian's ambivalence persists event when the doctor is a "native doctor," so long as he uses the occupier's techniques and norms. Here, rewind to *Black Skin, White Masks*, to the black who quests for, and ultimately gives up, assimilation, to be white, at last embracing his blackness, black pride as attitude, hence Négritude, as defined by Césaire, Senghor, and Damas. Even though "the native" questing for assimilation distinguishes himself as a doctor or engineer, in the eyes and words of the occupier, he will always be a "native" first and a doctor or engineer second. Still, for himself and to curry favor with the occupier, he must separate himself from the primitive herd called the tribespeople, show civility (being civilized, that is to say, embracing Western-and-white values), as "a Europeanized, Westernized doctor" or engineer, no longer a part of the occupied, tacitly now in the oppressor's camp, "having acquired the habits of a master" (Fanon 1959/1965, 132). He attacks his ancestral medicines "to demonstrate firmly his new admission to a rational universe."

Meanwhile, the patient who is persuaded by the veracity of Western-and-white technique soon feels the native technician to be not good enough; he wishes to be examined and treated by "the true possessors of the technique." A subject upon which Fanon concludes: "It is common to see European doctors receiving both Algerian and European patients, whereas Algerian doctors generally receive only Algerians. . . . We are dealing here with the drama of the colonized intellectuals before the fight for liberation" (Fanon 1959/1965, 132).

In a long footnote, Fanon calls medical practice in the colonies "systematized piracy." Algerians were given injections filled with distilled water but told it was penicillin or vitamin B12. Chest X-rays and radiotherapy sessions "to stabilize a cancer" were "given" without any radiological equipment, the doctor placing the patient on a sheet for fifteen to twenty minutes or "taking X-rays" with vacuum cleaners. It was about making money, said one doctor: "I fill three syringes of unequal size with salt serum and I say to the patient, 'Which injection do you want, the 500, the 1000 or the 1500 francs one?' The patient," so the doctor explains, "almost always chooses the most expensive injection" (Fanon 1959/1965, 133).

Besides being almost always a landowner or settler farmer, the doctor was also a militia officer organizing local "counterterrorist" operations. In cases of torture, he did not find evidence incriminating French troops; he always issued a certificate of natural death where torture resulted in death. He administered the "truth serum," a chemical substance having hypnotic properties that was injected into a vein, inducing loss of control or consciousness, forcing the captive to disclose without knowing. It left a seriously impaired personality. Says Fanon: "Everything—heart stimulants, massive doses of vitamins—is used before, during, and after the sessions to keep the Algerian hovering between life and death. Ten times the doctor intervenes, ten times he gives the prisoner back to the pack of torturers" (Fanon 1959/1965, 138).

When at the start of the war the French imposed an embargo on antibiotics, ether, alcohol, and antitetanus vaccine, the Algerian people had already "decided no longer to wait for others to treat them," since the occupier could give or take medications and surgical instruments as he pleased. People had watched relatives die of tetanus, the sale of whose antidote to Africans was prohibited. On occasion, they sent a European to buy it for them—without being arrested. And Fanon says: "Science depoliticized, science in the service of man, is often non-existent in the colonies." Medications became weapons, strategically deployed through denial to Algerians (Fanon 1959/1965, 139–140). When doctors wield such weapons of war, they become enemy combatants, not physicians.

Like the Algerian tradesman innovating ways to supply radios to the people, so too the Algerian pharmacist, nurse, and doctor clandestinely making antibiotics and dressings available to the wounded. A "Ho Chi Min Trail" of sorts stretched from Tunisia and

FIGURE 4.6
Doctor Frantz Fanon and his medical team, Blida-Joinville Psychiatric Hospital, Algeria. *Source*: Frantz Fanon Archives/IMEC.

Morocco, streaming medical supplies into every part of Algeria. This guerrilla public health system now replaced the visit to and of the *occupying doctor*, the traditional role of the pharmacy, and the colonial idea of a hospital. Bombs and raids were killing ten civilians for every single French soldier. The FLN started aggressively recruiting medical students, nurses, and doctors.

Before the war, the Algerian doctor is not a doctor; he is a native first and doctor second. Native doctor. His role: to play his part in spreading Western-and-white values, to cause upheaval within his societal norms and values (Fanon 1959/1965, 140). The native doctor is an occupied territory. Now, in the revolution, the Algerian doctor is reintegrated, no longer 'the' doctor, but 'our' doctor, 'our' technician. "Populations accustomed to the monthly or biennial visits of European doctors now see Algerian doctors settling permanently in their villages. The Revolution and medicine manifest their presence simultaneously" (Fanon 1959/1965, 142).

WHAT WE (MUST!) LEARN FROM THIS

Every struggle requires a reengineering of the mind first and foremost. It has to be self-reengineering—it has to come from us, its recipe must be from us, its ends defined by us. It cannot simply be "decolonizing the mind." It's important, yes, but absolutely not enough. Too many thinkers. Too many charlatans shouting "Decolonizing and decoloniality" from the rooftops just to draw a cushy salary and carve a niche in academia, while doing nothing for the people they write about. The people they exploit. No different from the colonial they critique. Vampires feeding on the blood of the everyday person. It could be data, crude oil, or minerals. Makes no difference: all extractive practices.

The mind—critical thinking, empirical data, publishing—must be placed in doer mode, so that everything we do must always be addressing a specific real-life problem facing Africa or black diasporic communities. What we research or write must not imitate Western-and-white privilege because we are still recovering from slavery and colonialism against a deck heavily stacked against us because of systemic racism. Hence my interest in a narrative useful for building things and aiding action, not just comment on discourse. I insist in my work on theory from the Global South, in this case theory from Africa (Mavhunga 2014, 2017, 2018). Fanon is one of the anchors of that, recognizing that he succeeds in taking me from a necessary critique of colonialism and expands beyond Césaire (who takes us from an understanding and critique of colonial racism to reconstructing a confident black personhood) to self-rehumanize that personhood building on Négritude. But Fanon does not help me much in thinking as an African and having the necessary African cosmologies and vocabularies to do it. Cheikh Anta Diop sets a model, but his vocabularies were not followed up with a critical thinker-doer mode. In that mode, writing such as this serves as a mind-engineering tool, to go well beyond decolonial and other colon-based options, beyond critique to the project of self-determined thought in service of problem-solving. My ambitions are circumscribed by the emergencies I am confronting at the village level in Africa. Good theory, theory-qua-theory, however wonderful, is irrelevant at the grassroots unless it is directly applied to render problems legible, and opens the way for new directions toward a solution. That is why *A Dying Colonialism* is important.

5 THE WEAPON OF THEORY

One contemporary once called Amílcar Lopes Cabral "the most original and significant African revolutionary thinker to appear since the death of Frantz Fanon." A man whose political thought was characterized by "a combination of careful, painstaking, theoretical analysis and reflection with a patient, step-by-step application of theory to the practical questions of winning a war and constructing a new society. The avoidance of rhetoric and the emphasis on analysis founded on practice" meant his thought was both method and theory (McCollester 1973).

African liberation struggles illustrate the role of the African diaspora as a force for daring to invent the future. Algerian nationalism began almost exclusively among immigrants in France. So too with Guiné Bissau and Cape Verde. Amílcar Lopes Cabral studied agricultural engineering at the Instituto Superior de Agronomia in Lisbon, Portugal, where he met other students from Lusophone Africa, like his countryman Vasco Cabral, Agostinho Neto, and Mário de Andrade (Angola) and Eduardo Mondlane and Marcelino dos Santos (Mozambique). At the second Conference of Nationalist Organizations of the Portuguese Colonies in Dar es Salaam in 1965, Cabral reminisced about that moment:

> I remember very well how some of us, still students, got together in Lisbon, influenced by the currents which were shaking the world, and began to discuss one day what could today be called the re-Africanisation of our minds. Yes, some of those people are here in this hall. . . . You have among you here Agostinho Neto, Mario de Andrade, Marcelino Dos Santos, you have among you Vasco Cabral and Dr. Mondlane. All of us, in Lisbon, some permanently, others temporarily, began this march, this already long march towards the liberation of our peoples. ("The Nationalist Movements" 1965, 62)

They set up student and cultural organizations to write poetry emphasizing the beauty and substance of their African cultures, questioning the basis for Portuguese *assimilado*

FIGURE 5.1
Amilcar Cabral. *Source*: Wikicommons.

(assimilation) policies. Some of the poetry in this "re-Africanization of our minds" movement was published in Margaret Dickinson's *When Bullets Begin to Flower* (1972).

Cabral finished his agronomy training in 1951 and returned to Guiné-Bissau in 1952 to work in the Departamento de Serviços Agrícolas e Florestais. He undertook an agricultural survey in 1953 and 1954, all over the country, talking to people, listening to their problems, their aspirations, and learning about their histories and cultures. He then examined the country's soils and crops, its economic potential, and the effects of colonial oppression on people. Cabral returned to Portugal in 1955 to work as an agricultural consultant on the colonies until 1959, when he teamed up with Aristides Pereira, Julio de Almeida, Elisée Turpin, Fernando Fortes, and his brother Luís Cabral to form the Partido Africano da Independência da Guiné e Cabo Verde (PAIGC). The objective was to secure independence, freedom, economic progress, and socio-cultural advancement ("At the UN" 1962, 31).

Nonviolence was not an option given Portuguese intransigence and brutality; initial attempts at marching and strikes were urban-based, in the classical proletariat-as-vanguard-of-the-revolution mode. The Portuguese banned even African sports and recreational association and kept suspect Africans under surveillance. By 1959, all attempts at lawful action had failed, and the *civilizados*, with several craftsmen and manual workers, resolved to create an underground structure for armed struggle, especially after the dockworkers strike at Pijiguiti on August 3, 1959, was brutally crushed, leaving fifty dead and more than a hundred wounded ("Practical Problems and Tactics" 1968, 109).

The time for peaceful means was over. PAIGC began to organize, train, and mobilize for armed struggle. It consolidated and strengthened its urban organization clandestinely, made the rural its "principal force" for struggling, and sent its recruits for military and technical training and its political-intellectual apparatus across the country to study and plan. The strategy: "To expect the best, but to be prepared for the worst" ("At the UN" 1962, 31). And the worst was a Portuguese force that by 1968 had risen to twenty-five thousand men well equipped with the latest material resources, including US B26 bombers and German Fiat 91 jet fighters ("The Development of the Struggle" 1968, 92).

The party knew nothing of struggles elsewhere when it launched its own. Cabral had not even heard of Mao's works as late as 1961. In September 1959, the cadres left the city for the countryside to mobilize the people. The party also created a political school in Conakry to "prepare political activists . . . in how to mobilize our people for the struggle" ("Towards Final Victory" 1969, 127). From there they began to organize the masses while training cadres militarily, conducting reconnaissance, and planning operations for national liberation. In 1963, the armed struggle began (Davidson 1969, 32). Exactly a decade later, Portugal, fighting three losing wars in Guiné, Mozambique, and Angola, fled. Amílcar Cabral never lived long enough to see Guiné's independence, assassinated by Portuguese agents in 1973 aged just forty-nine.

This chapter is not one of those "radical" Pan-Africanist readings of Cabral. It is not a nostalgic plea on behalf of independent Africa's founding fathers or a rallying cry to defend African nationalism or Pan-Africanism. And it is definitely not to be read—as some among the South African Sociological Association executives erroneously read my 2019 keynote lecture on Fanon in Pretoria—as a historical narrative of the liberation struggles in Africa. To remind each other, the purpose is to look for moments in Africa's story where precedents of critical thought in service of or as by-product of problem-solving are evident, in order to inspire ourselves to use knowledge in service of problem-solving in our own time. In other words, not always or necessarily privileging academic enquiry or epistemology or seeking to know as the objective or primary goal or purpose, but with that as outcome or even by-product of solving problems through making and building.

Where Cabral defined the weapon of theory as a weapon to study and analyze society for the purpose of mobilizing and organizing it to struggle, I extend the remit of this term to something far broader. Namely: as *a potential précis for a new space integrating critical thinking and doing, the critical being simultaneously important and rigorous, hence thinking on topics that are important, and being thoughtful about it, and critical thinker-doer being both doing based on, informed by, generative of, theory, and not the other way round.* First, a very brief orientation.

THE WEAPON OF THEORY I: AS A TOOL FOR PROBLEMATIZING

This section deals with the weapon of theory as a tool for problematizing, that is: to "advance towards the struggle secure in the reality of our land (with our feet planted on the ground)" ("To Start Out from the Reality" 1969, 44). You cannot solve what you do not know, and there is no point in knowing what you cannot solve while ignoring what you can solve and therefore need to know. Such basic research did not acquit with the reality that the people of Guiné, including PAIGC, lived.

The starting point for PAIGC, therefore, was to acknowledge the impossibility of struggle absent knowledge of reality, to recognize that consciousness of reality equips humanity with potential to transform that reality. Those that lead the struggle must "never confuse what they have in their head with reality . . . but must weigh up and make plans which respect reality and not what he has in his head" ("To Start Out from the Reality" 1969, 45). "You have already clearly understood what the people are," Cabral said in a lecture in 1969. But "against whom are our people struggling?" ("Struggle of the People" 1969, 75).

The Europeans were "the human instruments" of Lisbon in Guiné and hostile to national liberation. The number of whites was small, never above three thousand and one thousand, respectively, in Guiné and Cape Verde, principally officials, technicians, businessmen, employees, and so on, all to be seen "in relation to the struggle." Almost all were colonists, with no long-term interest in Guiné, and would "want to go away when they have made their pile" in this "commercial" colony. Then there were a few good whites among them sympathetic to the revolution, including those who helped Luís Cabral escape from Guiné. Thus the mantra was to "let all the forces we can bring together come . . . but not blindly" ("Unity and Struggle" 1969, 35). Initially the attacks avoided the white civilians whom PAIGC understood were in Guiné just for the money and targeted instead the Portuguese military infrastructure and forces. Cities were the last resort in case the Portuguese had not surrendered in four years; that was the position in 1968 ("Practical Problems and Tactics" 1968, 109).

The Guinéan city exposed the Euro-centricity of Marx and Lenin. A general movement of national liberation was unfeasible; there was neither proletariat nor revolutionary peasantry. "When we had made our analysis," Cabral says, "there were still many theoretical and practical problems left in front of us." True, Marxist-Leninist thought helped PAIGC to know that its struggle had to be working class led. "We looked for the working class in Guiné and did not find it." Intellectuals? Nowhere to be found; "the Portuguese did not educate people" ("Brief Analysis" 1964, 54), let alone revolutionary intellectuals, those with profound theoretical knowledge. Here and there one could find the engineers, doctors, bank clerks, and so on, but only with concrete knowledge of Guiné, not the weapon of theory. Only the petty bourgeoisie inspired by "the reality of life" in Guiné.—The suffering endured.—Events in Africa and elsewhere.—Experienced abroad. ("Brief Analysis" 1964, 54). The dockworkers were very conscious of their exploitation but lacked organization; here was a nucleus to rally other wage-earning groups: "If I may put it this way, we thus found our little proletariat" ("Brief Analysis" 1964, 54)

PAIGC divided Guiné's society into urban and rural as a way of understanding where exactly to begin the struggle. Within the city were two main groups. The African petty bourgeoisie was itself composed of Africans with three attitudes. The first were "heavily committed to and compromising" with colonialism (e.g., higher officials, liberal professions), but within them was a revolutionary or nationalist petty bourgeoisie, who were "the source of the idea of the national liberation struggle." Some members of this procolonialist petty bourgeoisie had visited Cabral at his house to try and dissuade him from "spoiling your career as an engineer," and told him, "We are all Portuguese." This group was beyond redemption; the exceptional ones, like Cabral, were tiny outliers. The second substrate of the petty bourgeoisie were simply undecided and sought to get the best of both worlds; whether the Portuguese stayed in power or PAIGC succeeded, they still won. This group could not be trusted, but was not entirely worthless like the first.

The third group was composed of the salaried urban people, who were anti-Portuguese but scared as hell to challenge them. The majority of the wage earners were committed to the struggle but were not easy to mobilize and tended to safeguard whatever little they had ("Brief Analysis" 1964, 48, 51). "What if we lose? We lose our refrigerator, our pay at the end of the month, our radio, our dream of going to Portugal for holidays" ("Unity and Struggle" 1969, 36). This group was easiest to convince about colonial injustice because they earned ten escudos a day while the European got thirty to fifty escudos for the same job. "To take my own case. . . . I was an agronomist working under a European who everybody knew was one of the biggest idiots in Guiné; I could have taught him his job with my eyes shut but he was the boss" ("Brief Analysis" 1964, 52).

It was to the petty bourgeoisie, his own peers, that Cabral gave one choice:

> To strengthen its revolutionary consciousness, to reject the temptations of becoming more bourgeois and the natural concerns of its class mentality, to identify itself with the working classes and not to oppose the normal development of the process of revolution. This means that in order to truly fulfil their role in the national liberation struggle, *the revolutionary petty*

> *bourgeoisie must be capable of committing suicide as a class in order to be reborn as revolutionary workers, completely identified with the deepest aspirations of the people to which they belong* . . . what Fidel Castro recently correctly called the development of revolutionary consciousness. ("The Weapon of Theory" 1966, 89; my own emphasis)

Another group below the salaried workers was the déclassé, made up of the "lumpen-proletariat" (beggars, prostitutes, and so on, relatives of petty bourgeois or workers' families, and newly arrivals from rural areas), and "outrightly against our struggle" ("Brief Analysis" 1964, 48, 51). This group was easy to mobilize as agents of the Polícia Internacional e de Defesa do Estado, the Portuguese secret police ("Unity and Struggle" 1969, 36).

One group defying the rural-urban or petty/bourgeoisie-déclassé categorization was composed of "many lads without work, who can read and write, who work here and there, and who often live at the expense of an uncle in the city." They moved back and forth between rural and urban.—Spoke nearly all the languages of Guiné.—Knew rural customs well.—Had solid urban knowledge.—Were fairly self-confident.—Played football.—Could read and write, "which makes a person an intellectual in our country," Cabral remarked sarcastically. It was among the group that PAIGC found people "with a mentality which could transcend the context of the national liberation struggle . . . because on the one hand they are of the city and on the other they are closely tied to the bush," have little or nothing to lose, have exposure to the city and thirst for fine possessions, but have suffered humiliation from systemic racism ("Unity and Struggle" 1969, 36).

In 1969, after benefitting from many false starts and failures, Cabral could say: "We must struggle without rushing, struggle in stages, develop the struggle progressively, without making great leaps." Unlike many struggles that began with a politburo, general staff, and national liberation army, PAIGC started its own as "one plants a seed, a seedling is born, which grows and grows until it produces flowers and fruit: that is the path of our struggle, stage by stage, step by step, progressively without great leaps, . . . like a growing child who at the start is satisfied with a feeding bottle of milk, or mother's breast milk, but when he is aged three complains if he is given a feeding bottle of milk or the breast, because this is no longer enough for him" ("Our Party and the Struggle" 1969, 67).

"We must consider that we were learning how to wage struggle in step as we were advancing (on the path)," Cabral said. The coastline was different from the interior, Manaco from Oio, Mandinga from Balanta. Guiné and Cape Verde (some 640 km apart) were a tiny 40,000 km^2 in size, the former nine times the size of the Cape Verde and Bijagos islands and islets put together. Senegal and Guiné Conakry flanked the country on either side. How would PAIGC organize a struggle in a country so far apart? ("To Start Out from the Reality" 1969, 46).

Guiné is "in almost its entirety has no mountain, no high point," bar a few hills found around Boé, while Cape Verde is composed of volcanic and mountainous islands. They complemented each other ("To Start Out from the Reality" 1969, 50). Guiné's rivers spread from the Atlantic like "tongues of the sea"—mostly salt water except "the only genuine river in our land"—the Corubal. These tongues of the sea provided many ports for guerrilla activity using boat transport, but also resupply lanes for the Portuguese deep into their barracks in the interior. Geography was a Portuguese ally: "If Bissau were on the mainland, if there were no island of Bissau, if it were not for the Corubal, if the river Mansoa were not on the other side, we should already have been inside Bissau," Cabral says in 1969 ("To Start Out from the Reality" 1969, 51).

Classical guerrilla warfare said mountains were indispensable to a guerrilla war. There were none in Guiné, and the war had to be fought. "If we did not attach importance to our own reality, to put it under analysis and draw conclusions on how to operate, we should have said that it is impossible to wage guerrilla warfare in Guiné because there are no mountains. . . . It is not the mountains which open fire," Cabral noted. On the other hand, Cape Verde had mountains, but these were ten islands, not one: "In which island or islands should we begin mobilizing? . . . We had the difficulties of communications from where we are to the islands, between the islands, etc." ("To Start Out from the Reality" 1969, 51–52).

The mountains were the rural people, which is why theory—research—had to be mobilized to thoroughly understand them. Here Marxist theory, with its classical "peasants," was not a sharp tool. The "peasant" of China or Russia was very different from that of Algeria and that of Algeria was different from the peasant of Guiné ("Towards Final Victory" 1969, 128). The Portuguese had not expropriated the land but allowed Africans in Guiné to cultivate the land. There were no big agricultural companies such as in Angola. The basic structure of land as communal property continued, but exploitation happened at the marketing stage of the value chain—what was the real value of products?

How then to convince the people that the Portuguese were exploiting them? It was not enough to say "the land belongs to those who work on it." There was "more than enough" land. It was a nonstarter to say colonialism must be resisted because it was bad; what was the lived proof? ("Towards Final Victory" 1969, 129). The case for supporting PAIGC had to be drilled down to the individual, using direct language, not the language of class analysis and Marx, "no generalizations, no pat phrases," but starting "from the concrete reality of our people":

> Why are you going to fight? What are you? What is your father? What has happened to your father up to now? What is the situation? Did you pay taxes? Did your father pay taxes? What have you seen from those taxes? How much do you get for your groundnuts? Have you

> thought about how much you will earn with your groundnuts? How much sweat has it cost your family? Which of you have been imprisoned? You are going to work on road-building: who gives you the tools? You bring the tools. Who provides your meals? You provide your meals. But who walks on the road? Who has a car? And your daughter who was raped-are you happy about that? ("Towards Final Victory" 1969, 129).

Any fancy, abstract language would make people nervous. Those assigned to mobilize them had to be genuine insiders, lest the people think that "we were outsiders come to teach them how to do things." PAIGC had to come to people "to learn" from them, not just to teach them ("Towards Final Victory" 1969, 129). To do so, they needed to understand each people's specific history because Guiné was composed of many peoples. The first step involved the strategic assignment of a problem-solving value to Marxist class analysis and then deploying it as an instrument to understand rural society for a purpose: to struggle. *It was not simply an application of Marx but a strategic deployment for the objective of understanding a reality with a view to solve it. Sometimes that route may be the best way to discover the limits of imported concepts and strategies and the beginning of self-driven, innovative making of theory in service of problem-solving.*

The operation of theory as weapon was to analyze these groups, establish the "contradictions between them and within them," and "locate them all vis-à-vis the struggle and the revolution"—before, during, and after the struggle ("Brief Analysis" 1964, 46–61). What Cabral meant was not simply knowledge of the concrete (names, events, places) but a theoretical rendering of such concretes to class analysis. He set about dividing Guinéan history into stages. First, composed of low levels of productive forces, a rudimentary mode of production, no private appropriation of the means of production, no classes, no class struggle, corresponding with a communal agricultural and cattle-raising society, with classical Marxism's horizontal, stateless social structure. Second, increased level of productive forces, private appropriation of the means of production, and classes and class struggles, corresponding to Europe's feudal or assimilated agricultural or agro-industrial bourgeois societies, with a vertical social structure and a state. Finally, the end of private appropriation, class, and class struggle, and the advent of new forces of production, aligning with socialist or communist societies, mainly industrial, stateless, horizontal, and humane. In the same country, stages overlapped, especially under conditions of racialized inequality, where Europeans enjoyed more wealth and freedom, and Africans were poor and oppressed ("The Weapon of Theory" 1966, 78).

Indeed, it was precisely *to exhaust the limits of Marx, who had never waged a revolution, let alone an armed liberation struggle in Africa, and to start from the reality as a problem that needed critical thinking in the service of solving it, not by somebody else, but by those doing this critical thinking and examination of the reality.* It was here that the party mobilized its knowledge of theories of revolution to parse out whether the ingredients were present and, if not, to modify and tailor theory to the specific realities of each society in Guiné, toward designing and executing solutions. In his analysis of Guinéan society in 1964, therefore, Cabral divided the population in the countryside into two distinct groups: semifeudal (Muslim Fula), and stateless groups (animist Balanta). Among the Fula and Manjaco, chiefs were a part of traditional society. The majority Fula and Mandinga had come from outside, conquered, and then assimilated local inhabitants into their Fula and Mandinga identities. Those who refused (to convert to Islam and submit) came to be called *balanta* (to *lanta* is "to refuse," hence *balanta*=the "refusers") ("Unity and Struggle" 1969, 37).

The Fula and Manjaco were "vertical societies" with a chief at the top, followed by religious leaders, then the professions—ironsmiths, goldsmiths, cobblers . . . all recognized in a hierarchy. The artisans like blacksmiths (the lowest occupation) and leather-workers, were critical to Fula socio-economic life and were "the embryo of industry" ("Brief Analysis of the Social Structure in Guiné" 1964, 46–61). Then came the farmers, who tilled the land for the chiefs per custom. The chief, appointed God's representative on earth by custom, blessed all seeds first before the farmers ploughed, then tilled their own, and were then obligated to till the chief's too ("Unity and Struggle" 1969, 38). These "peasants" were "the really exploited group in Fula society" ("Brief Analysis of the Social Structure in Guiné" 1964, 46–61). Cabral found even the term "peasant" vague: "The peasant who fought in Algeria or China is not the peasant of our country." The Portuguese had not expropriated the land but allowed Africans to cultivate it and exploited people's labor ("The Development of the Struggle" 1968, 91–102). Fula women had no rights. They were critical to production but did not own the produce, their polygamous husbands considering them their property to some extent. PAIGC selectively preserved those traditions concerning collective land ownership and chiefly privileges concerning ownership of land and labor utilization, so that the "peasants" who depended on such chiefs met their work obligation for specific periods per year ("Brief Analysis of the Social Structure in Guiné" 1964, 46–61). This was tactical: to avoid antagonism.

Further, Cabral identified several "intermediary positions" (tellingly he did not call them classes), with chiefs, nobles, and religious figures at the top, artisans and Dyulas (itinerant traders) in the middle, and "peasants" at the bottom. The merchants transported merchandise over long distances, selling or bartering in and outside Guiné, including lending money to chiefs ("Unity and Struggle" 1969, 40).

Balantas were a "horizontal society" without any "great chief," each family being autonomous and seeking counsel from a council of elders when needing to settle disputes with neighbors and combining families or the entire community when difficult or much work needed to be done quicker or easier. It was the Portuguese who created chiefs from the Mandinga or former African policemen and imposed them on Balanta. From the outside it looked as though Balanta offered no resistance and accepted this imposition, but they were "only play-acting towards the chief." In practice they continued as they had done before. When one family planted more land and harvested more, it did not hoard the food, which was spent or given away, "for one individual cannot be much more than another. . . . If someone has grown a great deal of rice, he must hold a great feast, to use it up" ("Unity and Struggle" 1969, 38).

This was the principle of Balanta society: that property and land belonged to the village, while the instruments of production were family or individually owned. Monogamous women owned their produce and were "fairly free" except that their children belonged to the head of the family: the husband ("Brief Analysis" 1964, 46–61). As described by Cabral in 1969, a great many Balanta joined the struggle "because of their type of society, a horizontal (level) society . . . of free men, who want to be free, who do not have oppression at the top, except the oppression of the Portuguese. The Balanta is his own man and the Portuguese is over him, because he knows that the chief here, Mamadu, is in no way his chief, but is a creature of the Portuguese" ("Unity and Struggle" 1969, 38).

Theory enabled PAIGC to establish first the extent to which each group depended or not on the colonial regime; their position vis-à-vis the struggle; their nationalist capacity, and their revolutionary capacity. The Fula chiefs were too tied to colonialism. The artisans were extremely dependent on the chiefs, so many simply followed them while others tried to break free and oppose Portuguese colonialism. Dyula were itinerant traders without any real roots anywhere, sought nothing but bigger and bigger profits. Cabral acknowledges their technical role in the war: "It is precisely the fact that they are almost permanently on the move which provided us with a most valuable element in the struggle . . . to mobilize people. . . . All we had to do was give them some reward, as they usually would not do anything without being paid" ("Brief Analysis of the Social Structure in Guiné" 1964, 46–61). The "peasantry" had suffered much exploitation, but tended (especially the Fula) to follow their chiefs. Extra work would be required to politicize and mobilize them against their chiefs toward the struggle. Balantas and other groups without state organization were Islamized but not Islamic, were "thoroughly impregnated with animist conceptions," and resisted Portuguese rule the most ferociously; PAIGC found them the "most ready to accept the idea of national liberation" ("Brief Analysis of the Social Structure in Guiné" 1964, 46–61).

Overall, the "peasantry" was a "physical force" but not a "revolutionary force"; it was "almost the whole of the population, it controls the nation's wealth, it is the peasantry which produces," but it was difficult to convince them to fight. Contrast that to China, whose peasantry had a history of revolt. The PAIGC soldier did not find "a very warm welcome" among all groups; many had to be won over ("Brief Analysis of the Social Structure in Guiné" 1964, 46–61). PAIGC had analyzed the living conditions of the Balanta, Fula, Mandinga, and Pepel, the petty bourgeoisie, salaried workers, shop staff, port workers, and descendants of Cape Verdians and how they would respond to the struggle. But it never took into consideration the role of Fula and Manjaco traditional chiefs and notables. The assumption that people whose forebears had fought the Portuguese would automatically welcome PAIGC proved mistaken.

How deeply racial discrimination, European force, and indeed colonialism itself felt depended on the proximity or remoteness of Africans to the Portuguese, whether it was a settlement, a police post, or patrols. Those living in the "deep bush" could "well die without ever having seen a white man." When Cabral was touring Oio with a Portuguese agronomist, children had come up to them "and rubbed his arm to see why he was white like that." It was their first time to see a white man, whereas an African living in the city saw white people daily ("Unity and Struggle" 1969, 35).

The reality was that Guiné was an economically backward country, without any "real" industry and an agriculture that "belongs to the age of our grandparents" ("To Start Out from the Reality" 1969, 52). The slave trade and colonization had completely destroyed the social and economic structure of African society. "Tribal structure," already disintegrating through African nation building (Samori Touré in West Africa, Shaka in Southern Africa), destroyed the structure but preserved the superstructure, "meaning those who were ruling the tribes, or the groups, so that these would serve as intermediaries to help the colonialists to rule" ("To Start Out from the Reality" 1969, 62). The few chiefs (*regulos* in Guiné, *sobados* in Mozambique and Angola) who remained were puppets deployed to bend their people to the colonizer's will at pain of torture or death. Councilors "representing" Africans (99.7 percent of whom did not vote) were typically Europeans, with a few African appointees serving their masters. Only one out of 120 deputies in the Lisbon "Parliament" (from Sao Tomé) was an African, and they were advertised at the United Nations and other gatherings as the multiracial, equal opportunities and diversity face of Portuguese rule. Except the government propaganda mouthpiece, *Brado Africano*, all African press was banned, as was any organization or demonstration ("The Facts about Portugal's African Colonies" 1960, 27).

Also under Portuguese colonial law, 99.7 percent of all Africans in Angola, Guiné, and Mozambique were classified as "uncivilized" and 0.3 percent as "assimilated" (*assimilados*), who theoretically were Portuguese citizens. To attain *assimilado* status, the "uncivilized" had to show economic stability, higher-than-average living standards, live in a "European manner," pay taxes without fail, do military service, and be literate. Under that standard only one-half the Portuguese population itself would make the cut. Thus rendered uncivilized, the African became property, a vast reservoir of forced and export labor, an object of racial discrimination sanctioned by law. The citizenship that *assimilado* status granted came without rights and privileges, and its beneficiaries saw themselves as too civilized to be natives but never treated as equals by the Europeans. Like every African, they had to carry a pass and observe a 9:00 p.m. curfew or face arrest. The claim to "multiracialism" was a myth ("The Facts about Portugal's African Colonies" 1960, 22).

Almost all "uncivilized" Africans were educated in Catholic missions under an agreement between Portugal and the Vatican to teach the "uncivilized" how to be timid, useful subjects of their local governments. The Catholic monopoly on schools and the elementary nature of the education meant that Africans remained illiterate. Swathes of territory bigger in size than Portugal had no schools. In 1937, only forty thousand out of some four million Angolan children went to school, ten times less than in the Belgian Congo. Few Africans attended secondary school, mostly from *assimilado* or chiefly families—not a single university built, which meant only one hundred out of eleven million Africans could attend university in Portugal. All secondary and most primary school teachers were Europeans except in Cape Verde, dispensing a curriculum that dismissed or mocked African culture, civilization, and being ("The Facts about Portugal's African Colonies" 1960, 25).

The philosophy of education was white supremacist and intended to engineer the African mind to fear the white skin and wither in its presence:

> The white man is always presented as a superior being and the African as an inferior. The colonial "*conquistadores*" are shown as saints and heroes. As soon as African children enter elementary schools, they develop an inferiority complex. They learn to fear the white man and to feel ashamed of being Africans. African geography, history and culture are either ignored or distorted, and children are forced to study Portuguese geography and history. ("The Facts about Portugal's African Colonies" 1960, 26)

Defining technologies of imperialism, Cabral located the strength of Portuguese colonialism to lie in the dehumanization of the African. They manifested as "cannon, . . . the *palmatoria*, the whip, the pistol, the modern rifle, the machine gun, the mortar, bombs of all kinds, including napalm bombs, and torture" lacerating the flesh of the occupied. The human technologies also included the navigators and mariners of former days, the mercenaries, the captains general, the soldiers of the "pacification," the sepoys, the *chefes de posto*, the administrators, the governors, the colonial troops (army, navy, and air force) and the political police ("At the UN" 1962, 26). The Portuguese colonials had always disrespected "*the African personality* and the African culture," which Cabral attributed to the backwardness and ignorance of Portugal. (And here, the connections with Blyden are clear.) Not only did it underdevelop colonies; Portugal itself was underdeveloped and culturally and economically backward. Whereas France, Belgium, and England were full of African works of art, nothing of the sort happened in Portugal. Cabral thinks this contact could have "opened the way to universal knowledge of the abilities of the African, of his culture in general, of his religions and philosophical concepts—in other words the way in which the African confronts the reality of the world with cosmic reality." Cabral says this did not happen in Portugal because the colonials sent to Guiné were ignorant ("The Development of the Struggle" 1968, 91–102; my own emphasis).

Educated in Lisbon or Oxford, and products of Catholic schools, where their nothingness without the white man was drummed into them *ad nauseum*, the African elites that emerged were a compromised lot. Cabral attacks the diaspora intellectuals as political opportunists and tribalists, schooled in European universities, who frequented the cafés of Brussels, Paris, Lisbon, and other capitals and were "completely removed from the problems of their own people [and] at times even look down on their own people" ("Practical Problems and Tactics" 1968, 117). In other African countries, this intelligentsia suddenly sought to be president or minister, to amass diamonds, gold, and women and wear "expensive clothes—a morning coat or even great *bubus* to pretend that they are Africans. All lies, they are not Africans at all. They are lackeys or lapdogs of the whites." Now they had their chance to be the new white man and, when denied the chance, strategically deployed tribalism to tear the country apart after independence ("To Start Out from the Reality" 1969, 62).

When the Portuguese colonized Guiné, all African lands became the property of the Portuguese crown so that Africans stayed on, but their labor and its fruits were heavily taxed or sold at low prices.—A 'sovereign tax' was imposed.—Compulsory crops, forced labor (including exports), and total control were enforced through violence ("The Facts about Portugal's African Colonies" 1960, 17). People were forced to sell their produce—accounting for 70 percent of all production—at artificially low prices ("The Facts about Portugal's African Colonies" 1960, 20). One-fifth of land in Mozambique, four-fifths in Angola, and almost all cultivated land in Sao Tomé were in European hands, seized through violence ("The Facts about Portugal's African Colonies" 1960, 21).

The Portuguese not only seized African land and turned Africans into dehumanized machines in public and corporate projects. They also exported them as forced labor to neighboring colonies. On Sao Tomé, forced laborers from Cape Verde, Angola, and Mozambique worked twelve hours a day. A quarter of a million Angolans and four hundred thousand Mozambicans were forced into labor every year, with one hundred thousand of the latter exported to neighboring Rhodesia and South Africa. Mortality among the laborers was high—as much as 30 percent. Seventy-five percent of their wages were only paid upon return, only after they had paid all taxes and given an extra fifteen days of free labor to the Portuguese government. On paper they were entitled to food and medicine; they got none or little. Most of the workers were adolescents or children. An *assimilado* earned three to four times less than a European worker and was always a "second-class" worker regardless of more or equal skill ("The Facts about Portugal's African Colonies" 1960, 24).

The African health condition under Portuguese rule was pathetic. There were 380 doctors for eleven million people inhabiting a land of 2 million km^2 with the worst communication infrastructure across Portuguese colonies. Most of them were in Cape Verde, with a doctor-to-patient ration of 1:10,000. Angola's hospital-to-person ratio was 1:280,000 people, with a doctor-to-patient ratio of 1:20,000 and nurse-to-patient ratio of 1:10,000 and thirty beds for every ten thousand people. The infant mortality rate was over 40 percent and in some areas 80 percent. Between 1940 and 1950, some forty thousand Africans died of drought and famine in Cape Verde; in 1956, 84 percent of registered deaths were of "obscure or unknown causes" that, curiously, killed only Africans ("The Facts about Portugal's African Colonies" 1960, 25).

The health of the people was grim. When the rains were good, food was abundant, but the norm was drought and famine and death from starvation and malnutrition. Forced labor exports to Sao Tomé and Angola were "transported like animals in the holds. . . . Those who died were thrown into the sea." Malaria was rampant; almost everyone in Guiné had it. Same for intestinal worms. Diseases of all kinds—including leprosy—were the order of the day. Surviving on rice every day—no meat, no milk, no eggs—made the body too weak to resist infection; few made it past thirty years ("To Start Out from the Reality" 1969, 54).

The colonizer drained the wealth and labor of Guiné but "did nothing to develop any resource in our land, absolutely nothing." The ports in Bissau and St. Vicente were merely "worthless . . . mooring quays." The French had built good, big ports in Dakar, Conakry, and Abidjan, the British in Lagos; over twenty ships could easily moor there. "And then we see how much time the Portuguese wasted in teasing us, tinkering and playing with us. They did nothing for our land," just subsistence farming ("To Start Out from the Reality" 1969, 52). Almost five hundred years of economic backwardness, no real industry except "so-called mini-plant for pressing oil from rice shelling. This is not a factory, it is no more than a great 'pestle'" ("To Start Out from the Reality" 1969, 52). Another small plant for treating rubber (tappings).—Another small fishmeal factory in Bijagos.—Three fish-packing stations. When Cabral was in secondary school, his mother worked at one such station to supplement her sewing income. She was paid fifty cents an hour, earning four pesos (escudos) over an eight-hour day when fish catches were good and just one hour a day when fish were scarce ("To Start Out from the Reality" 1969, 53).

The reality of Guiné was that of a people without factories to seize from the colonizer to commence manufacturing. "Today we have vast liberated areas," Cabral explained in 1969. "If there were factories there, it would be useful. Perhaps we could make cloth, perhaps we could make soap on a proper scale, instead of Comrade Vasco's tiny soap bars. We could manufacture other things if we had mines." But there was no bauxite mining or petrol drilling to attract Standard Oil. No threatened vital interests for the United States to pressure the Portuguese to negotiate with PAIGC for a peace settlement: "They think of us as a corridor between the Republics of Guiné and Senegal, a simple passageway." Then again, if the country had important mines and factories, "the imperialists would have come into the war quickly and in greater strength. . . . As it is, at least we have a quieter life, just bush and desert." No experience in iron casting to enable PAIGC to manufacture its own weapons either: "We can only make muskets, but muskets are ineffective." No developed economy.—No inventive culture.—No schools.—No education.—No capacity to operate mortars, artillery, or aircraft. Any method of instruction had to start from that reality ("To Start Out from the Reality" 1969, 53, 61).

The political reality was that Guiné was a Portuguese colony. Whatever black official the Portuguese appointed, the latter alone had power. This is the reality from which all other reality stemmed, that generated the struggle. The purpose of struggling was to take the power back to change the geographical reality, "to make a new human geography, . . . to transform the economic reality of our land." Development self-defined, not as meant by Europe. Hence the liberated zone as a new political reality: "We are ruling ourselves," Cabral declared in 1969 ("To Start Out from the Reality" 1969, 61).

THE WEAPON OF THEORY II: AS TOOL OF PROBLEM-SOLVING AND MAKING

We have seen the way PAIGC mobilized theory as method to study and understand the reality of Guiné in order to change it. In this section, I propose a reading of Cabral's speeches and interviews to be revealing a second dynamic: concerning process, spaces, and moment or atmosphere of national liberation struggling as generative of theory in service of problem-solving. Central to PAIGC's approach was the aggregation and inextricability of the outside and inside—things, spaces, perspective, and actions, the motions of struggling involving testing, questioning, dismissing, affirming, or building beyond received wisdoms, a winnowing process that created its own facts and falsehoods, and sieved out true revolutionaries from phonies.

It is one thing to theorize for the sake of erudition or speculation (academic argumentation), to theorize or suggest something for others to implement, or thought production absent present danger or ultimate sacrifice. It is quite another to engage in theory as strategic deployment—not "good theory" as end in itself but a means to an end, not for others but oneself as the "we" (Africans at large) engaged as a people in a struggle to solve a problem in common: the racist colonial system.

One of Cabral's most spectacular deployments of theory in service of struggle was in reference to physics, casting struggling as engineering, dependent entirely on unity, which he defended spiritedly as akin to "motion, i.e. not like one bottle among the entirety of bottles in the world, or one person as a unity among all the people in the world." Not the unity of one individual but oneness as entirety to achieve a given aim, unity not as standstill or stasis but "in a dynamic sense, in motion" ("Unity and Struggle" 1969, 28; my own emphasis). Like a football team composed of eleven individuals, each with a task different from the other, "different temperaments and personalities, different levels of education, different professions (some doctors, others engineers, others even illiterate), different religions (Moslem, Catholic, traditional), different political parties, and so on. But all must play as a team to win, form a unity. Many fruits, yes, but in one basket ("Unity and Struggle" 1969, 29).

Struggle, Cabral argued, is "a normal condition of all living creatures in the world. All are in struggle, all struggle." It is like one seated on a chair, the body exerting force on the chair and the floor beneath the floor. If the chair has no sufficient force to support the person sitting on the chair, its legs will sink, the force exerted stronger than the soil could withstand. The earth is constantly in motion, rotating, developing a force that repels us off the earth, pushing us away from the center, a centrifugal force. Then another force, the force of gravity, attracts all bodies near it toward it like a magnet: the more the distance and mass, the more the force. "We remain on earth and do not go outwards," Cabral says, "because the force of gravity is much greater than the centrifugal force which draws us outwards." The quest, and capacity, to go to the moon or launch into space was based on seeking to conquer the force of gravity, thus being able to leave the earth. To leave Earth and overcome gravity, the body has to travel at 11 km per second. However, Cabral continued his physics lesson, "any force acting on an object can only exist if there is an opposite force." Translated, the force was the Portuguese colonialists, the foreigners and occupiers who had exerted a force on the African people, who had "operated so that they should halt our history for us to remain tied to the history of Portugal." They did so after overcoming a force—Africans—through wars of occupation. Finding a deserted island and offering no opposing force, the Portuguese turned Cape Verde into "a storehouse for slaves," and as the slave trade became less profitable, they turned these dehumanized commodities into machines on agricultural plantations. They also exerted colonial force on Guiné, but "there was always our force which acted in the opposite direction," in the form of "passive resistance, lies, doffing one's cap, 'Yes, Sir,' to use all possible and imaginable stratagems to fool the Portuguese . . . as we could not challenge them face to face, we had to fool them." But the energy of this African force was undercut by backwardness, disease, and famine, something which the new force under PAIGC was addressing through struggling as motion, as the force ("Unity and Struggle" 1969, 33).

Hence, Cabral said, "the liberation struggle of the colonial peoples is the essential characteristic, . . . the prime motive force, of the advance of history in our times." In a 1965 speech, Cabral stressed that theory for the sake of erudition (or the "Marxist" label) did not excite him because people's lives were at stake: "*Always bear in mind that the people are not fighting for ideas, for the things in anyone's head. They are fighting to win material benefits, to live better and in peace, to see their lives go forward, to guarantee the future of their children*" ("Tell No Lies, Claim No easy Victories" 1965, 70; my own emphasis). Six years later he retorted to his London audience: "*The labels are your affair. People here are preoccupied with the question: are you Marxist or not Marxist? Just ask me, please, what we are doing in the field*" ("Our People" 1972, 20; my own emphasis).

The force of theory—in fact the theory making—was in the act of struggling, its author being the African, and the writing being the unmaking of colonial subjectivity and the making of ourselves as Africans into the new person, rehumanizing and returning ourselves back to the humanity and history from which the European colonizer had ejected us. Marx had asserted that "all history is a history of class struggle," perhaps following Hegel's view of Africans as people without history until Europeans arrived. Cabral rejected that:

> We are African peoples, we have not invented many things, we do not possess today the special weapons which others possess, we have no big factories . . . but we do have our own hearts, our own heads, our own history. It is this history which the colonialists have taken from us. *The colonialists usually say that it was they who brought us into history; today we show that this is not so. They made us leave history, our history, to follow them, right at the back, to follow the progress of their history. Today in taking up arms to liberate ourselves . . . we want to return to our history, on our own feet, by our own means and through our own sacrifices.* (Cabral 1974, 62–69; my own emphasis)

The question was how to create a force to oppose the state that was commanding history, a new force for self-rehumanization, and what returning ourselves to history entailed. And that was a "struggle of the people, by the people, for the people." Since history could not be given, Cabral said in 1969, the people must be the force that must wage that struggle, that must conquer freedom (as we shall again hear Sankara saying later), to write their own history in blood. That is what freedom meant:

> For somebody on the left, and for Marxists in particular, history obviously means the class struggle. Our opinion is exactly the contrary. We consider that when imperialism arrived in Guiné it made us leave history—our history. We agree that history in our country is the result of class struggle, but we have our own class struggles in our own country; the moment imperialism arrived and colonialism arrived, it made us leave our history and enter another history. Obviously we agree that the class struggle has continued, but it has continued in a very different way: our whole people is struggling against the ruling class of the imperialist countries, and this gives a completely different aspect to the historical evolution of our country. Somebody has asked which class is the "agent" of history; here a distinction must be drawn between colonial history and our history as human societies; as a dominated people we only present an ensemble vis-à-vis the oppressor. . . . When there is a developed national consciousness one may ask which social stratum is the agent of history, of colonial history; which is the stratum which will be able to take power into its hands when it emerges from colonial history? Our answer is that it is all the social strata . . . since unity of all the social strata is a prerequisite for the success of the national liberation struggle. . . . This brings us to what should be a void—but in fact it is not. What commands history in colonial conditions is not the class struggle. . . . It is the colonial state which commands history. ("Brief Analysis" 1964, 56)

A people's struggle was theirs provided the reason for struggling was "based on the aspirations, the dreams, the desire for justice and progress of the people themselves and not on the aspirations, dreams or ambitions of half a dozen persons" ("Struggle of the People" 1969, 75). The people were not fighting against all Portuguese; not all of them were oppressors. The ruling class in Portugal exerted the force; it was the (op)pressor (to press, squeeze) of the Portuguese people at home as much as Africans in Guiné. It did not act alone; it was part of the world system of class domination to extract raw materials for their industries and markets for their manufactures, which made liberation struggle in Guiné a fight against both external and internal enemies, a specific value system. Nobody could "come and struggle for us. . . . We are the ones who must buckle down to all the means of struggling." The people were waging the struggle by providing the water for the fish to swim in, the militants, leaders, combatants, and militia who were their sons and daughters. Lose the people's support and there was no struggle, no force. The people saw the weapons in the hands of their sons and daughters as confirmation of power being literally in their own hands. African nurses, doctors, engineers, and technicians were only white people and the odd black was the figure of knowledge. And now, by 1969, PAIGC was arming the people to act as a militia defending themselves to heal the sick and construct infrastructure ("Struggle of the People" 1969, 77).

The people, as distinct from *the population*, was a very specific historical definition, depending on a specific moment that the land was experiencing. "Population means everyone, but the people have to be seen in light of their own history," Cabral said. That is, the people born in Guiné and Cape Verde, "who want what corresponds to the fundamental necessity of the history of our land." Namely, ending foreign domination in the land of their birth. They were also defined as "the people" "in terms of the main stream of the history of that society, in terms of the highest interests of the majority of that society." That was the original meaning of democracy in ancient Greek society (*demo*/people's *cracy*/rule), but even then, the term was coined by those at the top, who were the legible people, while everybody else was a slave and illegible. That had not necessarily changed, especially in colonial societies where those with power (political, guns, the law) still made a democracy for themselves and anointed themselves the people or their spokespersons. Cabral wanted power to be in the hands of the majority, not the few ("Not Everyone Is of the Party" 1969, 90), what he called revolutionary democracy, wherein leaders and responsible workers "live among the people, before the people, behind the people" ("Revolutionary Democracy" 1969, 97).

Therefore, while every responsible worker was to bear their responsibility bravely, demand respect for their own, and show respect for others' activities, they were never to hide anything from or deceive the people. If the people came to purchase goods from the people's stores that PAIGC set up in the liberated areas, they should not be burdened with war material on their way back. That abuse of authority and faith was colonialistic; rather be honest and tell them they would be fetching war material. "This is better than lying, cheating and looking small in the people's eyes, for they, however wretched and suffering, are like any people, and they know the difference between the truth and a lie, justice and injustice, good and evil, and they are wise enough to lose respect for anyone who has lied to them," Cabral warned ("Revolutionary Democracy" 1969, 94). If they feel their trust has been betrayed, "they will chase us out, throw us out. . . . We must not deceive the people with fine words, with false promises. We must tell them frankly the difficulties. . . . You must remember that you are not the only ones with needs" ("Revolutionary Democracy" 1969, 97).

In a people's struggle, "we are all needed but nobody is indispensable," Cabral says, in a passage that proposes an ethos of leadership driven by communality of purpose and teamwork. The struggle continues even if one or several leave or die; if it stops or slows, then it was an individual project, not a struggle, and "it is because the struggle was worthless." Achievement is an achievement of many, who can continue even if a pair of hands is removed. Equally, nobody should ever be good at one job, or feel insecure or completely useless if transferred or is required to vacate it for somebody else. This fear of losing power, Cabral said, is what leads to rulers clinging on to power past their energy ("Revolutionary Democracy" 1969, 97).

Struggling itself was a winnowing process, separating good leaders from bad ones, revolutionaries from opportunists, out of which doing and making emerged a theory of leadership forged in the furnace of struggling. The stronger the struggle grew, the more the responsibilities on the leadership, and the more the power, which by 1969 began corrupting some leaders, who began "enjoying themselves," growing complacent, idle, soft. A liability to the struggle that must be ejected. "We cannot allow it, while the struggle advances, our people sacrifice themselves in the cause of our struggle, some comrades die and others are wounded or disabled, while we grow old in the struggle, giving our whole life to the struggle with so many folk putting their hope in us, inside and outside our land" ("Our Party and the Struggle" 1969, 68). The liberation struggle was "like the basket which separates clean rice from the husk, like the sieve which sieves pounded flour," uniting but also sorting out persons, a vigilant and selective process to show "who we are." Some would pass through the sieve, others not, and only those who "give everything and demand nothing," who "serve our people correctly," would pass through. The rest would be chaffed out. Only "aware men and women, whatever their origin, and wherever they come from," would be left. As Cabral warned, "Our struggle demands enlightened leadership and we have said that *the best sons and daughters of our land must lead*" ("Our Party and the Struggle" 1969, 69, 70; my own emphasis).

An enlightened leader is one who knows "where he may give way and where he must not give way, . . . what is essential and what is secondary . . . an essential problem or a secondary problem . . . a problem just for today or for always, . . . how to make concessions, to give way, to . . . give way a little . . ." ("Fidelity to Party Principles" 1969, 99). Such a leader was a necessity where the people were "underdeveloped—practically non-developed," their history "held back by colonialist and imperialist domination." They had started with nothing, began the struggle "almost naked," with a 99 percent illiteracy rate, no reading and writing skills, and with only fourteen university-trained men, a population underfed, victims of many diseases, its mind held back as much by European underdevelopment as it was by spiritual beliefs that mapped all kinds of superstitions on natural phenomena like rain, high tide, lightning, storms, typhoons, and tornadoes ("Practical Problems and Tactics" 1968, 115–116). An enlightened leader was one who carefully threaded the needle between what people have always known and what more they could know *en route* to making revolutionary choices.

Starting, organizing, and leading a revolution is a creative process of conscious designers mixing many, often unrelated, elements to forge a force for change. A struggle has no interior or exterior; hence rice is cooked in the pot, but that does not make firewood redundant. By joining strengths from inside and outside, a reality is transformed and another is born ("To Start Out from the Reality" 1969, 64). On the one hand were those petty bourgeoisie with a tendency to "struggle on the radio" just to be noticed for political gain, especially "the opportunists in Dakar" too coward to go inside Guiné to fight and who would, after the colonizer departed, "go straight to Bissau and sit down on a departmental director's chair." They sat outside, in the comfort of their office chairs, never understanding those inside Guiné "who are opening fire, or preparing the political ground and suchlike" ("Our Party and the Struggle" 1969, 65). Equally, those outside who thought that they "cannot dirty their feet in the mud, that they cannot be bitten by mosquitoes," and suffer through what combatants suffered, were mistaken. Thankfully, PAIGC could declare in 1969 that all its leaders had undertaken missions both inside and outside Guiné. Some had spent less time inside pleading to be redeployed back in, but the question was what compelling purpose their presence there would serve, "because tourism we can leave for later." Some inside pleaded to be sent on missions to Europe and looked the type who would stash their pockets with cash after independence to take leisure trips there. A leader's movements were struggle movements—always "respond[ing] personally to the needs of our struggle"—representing at conferences, meetings with heads of state and parties, and so on ("Our Party and the Struggle" 1969, 66). Credentials were not simply about learnedness; experience was itself an educative space far more profound. "The struggle is not a debate nor verbiage, whether written or spoken," Cabral declared the party position in 1969. "Struggle is daily action against ourselves and against the enemy, action which changes and grows each day so as to take all the necessary forms to chase the Portuguese colonialists out of our land" ("Our Party and the Struggle" 1969, 65).

As leader of PAIGC, Cabral was a very hands-on person—which the time demanded—who did not delegate something he could otherwise do himself. He was not an armchair leader sitting in the comfort of neighboring countries while sending other parents' children to die:

> But it is a great strength to me to have the certainty that there is no important operation in our war, no important political project of which I am not personally aware, which I do not study. There is no change or real development on the political plane or in the armed struggle which does not go through my hands. The trouble is that we have human limitations. Unhappily I cannot be everywhere at the same time but I have spent as much as possible at the side of our combatants and militants. ("Our Party and the Struggle" 1969, 66)

Cabral could only be a leader through the party, and the forging of a new unit of belonging (independent Guiné), requiring unity of all colonized and oppressed Africans, could only be done through one unifying mechanism—the party. All forces of division or choice of belonging in between PAIGC (as the representative of the oppressed) and the Portuguese (as the oppressor) were no longer permissible. Everybody had to choose one side or the other, to be a patriot or sellout. We can also now understand how, as the struggle unfolded, all areas that PAIGC controlled were being turned into a *de facto* one-party state ("At the UN" 1962). PAIGC "reeducated" the masses to understand that political participation in national affairs was only possible through the party. National identity was promoted, ethnic and class identity vilified. It was the duty of the party to train cadres "to obtain the necessary means for successfully pursuing the struggle," militarily, politically, and technically, at all levels, study and plan the groundwork for rapid economic progress as liberated areas opened up.

To be clear, leadership was not an armchair leadership of thinker-talkers, but a thinker-doer leadership. Thoughtful leadership ensured, first: that the objectives of the struggle were clear: immediate independence, democratization and emancipation, rapid economic progress and social-cultural advancement. Second, that an effective mechanism for achieving the objectives, namely, the party, with its structures root to branch, was in place. The struggle had only one political and military leadership, and it was the political leadership. PAIGC thus "avoided the creation of anything military," and as Cabral declared in 1968, "We are political people. . . . Our fighters are defined as armed activists." The political bureau of the party directed the armed struggle and life in liberated and unliberated regions through its activists. The political bureau had a war council as its instrument to direct the armed struggle, each war front its own command, each sector a sector commander, each unit its own commander. These guerrillas were based at bases, each base chief with a political commissar ("Practical Problems" 1968, 118).

Third, that such leadership was collective and not a monopoly, whether it was in the political, military, security, education, or health departments. Collective leadership does not mean "everyone must command" or complete absence of authority. The chain of command exists, but there is no need to show everybody who is boss; a good leader mingles with everyone. Leadership in armed struggle was about being humble to one's position. Cabral warned against colleagues who assumed that "by virtue of going into the bush for the struggle, they were kings and could walk over anyone around." Anybody who made a lord unto himself in the bush and forgot the party's watchwords was to "watch out, for we are going to forget the Portuguese a while to go and deal with him first" ("Our Party and the Struggle" 1969, 66).

Collective leadership derives its strength from the people, from their strength not weakness, from unity between leader and led, nobody turning their back on their fellow strugglers. A leader is a worker not wrecker—namely, commissar and commander disagreeing between themselves or disregarding the local leadership and people ("Our Party and the Struggle" 1969, 74). A leader is one who sets aside personal problems and focuses only on the "common problem, the problem of our people, . . . in the service of the liberty and progress of our people" ("Our Party and the Struggle" 1969, 75).

That common problem was the struggle that Cabral called "political work," and every leader, therefore, was a "political leader." As he put it, "every shot fired is . . . a political

FIGURE 5.2

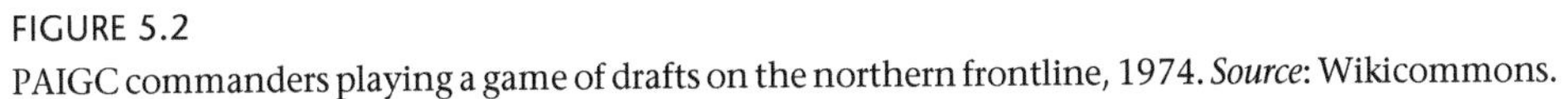

PAIGC commanders playing a game of drafts on the northern frontline, 1974. *Source*: Wikicommons.

act," so too every medical injection, all education, clothing, even goods exchanged. All military leaders were also political leaders, with double position in the army and party, and every unit or subunit had a commander and commissar acting as one, every military decision or act also a political one. Always, politics led the gun, never the opposite. "We can defeat the Portuguese at Buba or at Bula, we can enter and take Bissau, but if our population has not been well trained politically, is not fully wedded to the struggle, we shall lose the war, not win it" ("For the Improvement of our Political Work" 1969, 100).

The work of leading and the work of struggling was simultaneously a political and military act, involving winning the hearts of the people as a precondition for executing the objective of struggle. PAIGC established its guerrilla bases before it launched the armed struggle. Most of its bases were in the south, initially just a couple in the north. First, PAIGC created autonomous guerrilla groups, each linked to the party leadership, until 1963, when things unfolded quite rapidly. With the integration between the guerrillas and the people now complete, some guerrilla leaders became too autonomous in relation to chiefs in their areas of operation, "certain tendencies towards isolation developed, tendencies to disregard other groups and not to co-ordinate action." The 1964 congress took disciplinary measures, including the detention, trial, and sentencing of some guerrilla leaders; PAIGC now moved to "collective leadership of the guerrilla" under the party committee ("The Development of the Struggle" 1968, 91–102).

Zones and regions were now created, each under party committees, the party leaders became the guerrilla leaders. Sections of the guerrilla forces were turned into regular forces to extend the armed struggle into new areas. PAIGC did not see it necessary to mobilize everyone for the armed struggle, but only "a reasonable proportion of the population." Thereafter, regular forces were created that opened new fronts in the east (Gabu) and west (San Domingos and Boé). It became possible later to create fronts (not just regions and zones corresponding to party structures), northern and southern fronts first, then the eastern front ("The Development of the Struggle" 1968, 91–102).

This gradual evolution led Cabral to remark: "We consider ours as having developed like a living being, in successive stages. Often a stage was completed rapidly, sometimes slowly. We never rushed any stage: when one stage was completed, we moved on to the next. This gave an overall harmony to our struggle." By 1967, PAIGC forces constituted an army and all its guerrilla forces became regular forces. Until then, its guerrilla bases were actually villages; by then it had reduced the number of its bases, which was "extremely fortunate, because the Portuguese had pinpointed all of them on their maps and they intended to bomb them" ("The Development of the Struggle" 1968, 91–102).

For PAIGC, denying the enemy freedom of movement on roads was key; this was war by infrastructural engineering. PAIGC decapitated road transport as a device to lure the enemy into foot mobility, the guerrilla's forte. Here's Cabral, 1968: "We cannot fight riding in jeeps or trucks; we are the first to know that our country does not have good roads, since we ourselves have cut down the few existing bridges, we have destroyed many sections of highways, and our people have felled trees to block the highways. In fact, the enemy today can travel on almost no road in our country. Therefore, we do not have trucks, jeeps, etc., to travel along the roads that we ourselves mine. We have to move on foot within our territory" ("Practical Problems and Tactics" 1968, 115). Legwork was transport (conveyance of self, weapon, and supplies), combat (strategic-tactical mobility), and communication (information transmission) all rolled into one. It was struggling, what I have called elsewhere a transient workspace of self-liberation (Mavhunga 2014).

We have already seen a spectacular example of the strategic deployment of radio as weapon of self-liberation in Algeria. It was not until late 1968 that PAIGC set up a radio station to "play a role at least as important as, or more important than many guns." However, it hoped that friendly countries like Guiné Conakry, Senegal, Cuba, and others with broadcasts that reached into Guiné-Bissau could provide such a facility. "Once in a while we communicate the results of our armed struggle," Cabral said, while acknowledging

PAIGC could not put out communiqués with any degree of frequency "because communications are difficult between the different fronts of struggle and the centre that coordinates these communications." The communiqués were delayed and reflected neither the state of struggle nor the frequency of combat ("Practical Problems and Tactics" 1968, 1115).

Information management was key to an effective administration of the liberated areas. Section committees (*tabanca* committees), zonal committees, and regional committees held frequent meetings to explain to the people what was happening in the struggle, the party's intention, and the enemy's ("Tell No Lies" 1965, 70–72). In enemy-occupied areas, information management entailed movement of messages and instructions through clandestine work, mobilization and organization of the people, and preparing militants for action and to support the armed militants. Mass political education was not a one-way street; the party educated itself as it educated the people, "fighting fear and ignorance, to eliminate little by little the subjection to nature and natural forces which our economy has not yet mastered. Convince little by little, in particular the militants of the Party, that we shall end by conquering the fear of nature, and that man is the strongest force in nature" ("Tell No Lies" 1965, 70–72).

Hence the weapon of theory was also tangible products of writing and art in service of struggle, written material deployed as media of propaganda against the enemy. Cabral implored the cadres:

> Write posters, pamphlets, letters. Draw slogans on the roads. Establish cautious links with enemy personnel who want to contact us. Act audaciously and with great initiative in this way. . . . Do everything possible to help enemy soldiers to desert. Assure them of security so as to encourage their desertion. Carry out political work among Africans who are still in enemy service, whether civilian or military. Persuade these brothers to change direction so as to serve the Party within enemy ranks or desert with arms and ammunition to our units. . . . *Hide nothing from the masses of our people. Tell no lies. Expose lies whenever they are told. Mask no difficulties, mistakes, failures. Claim no easy victories.* ("Tell No Lies" 1965, 72; my own emphasis)

But theory, writing, or art could only be effective if grounded in thorough understanding of "the historical reality which these movements claim to transform," what Cabral called "the struggle against our own weakness." Lacking such historical understanding, many national liberation movements suffered an "ideological deficiency"—for Cabral "the greatest weakness of all" ("The Weapon of Theory" 1966, 75).

History in service of problem-solving enabled PAIGC to avoid copy-and-paste approaches and to tailor theory to specific local context. The movement launched the war in 1963 from the center moving to the periphery of Guiné, contrary to elsewhere in Africa. It was a "centrifugal" strategy that caught by surprise the Portuguese, who had amassed troops along the Guiné-Senegal borders ready to thwart any invasion ("Practical Problems and Tactics" 1968, 110). Tactically, no reinventing the wheel—PAIGC used guns old and new, combining methods the Chinese, Vietnamese, and Cubans had developed and deployed. Cabral only learned of Mao Tse-Tung's works on revolutionary warfare in 1961. Che Guevara's *Guerrilla Warfare* offered a basis, but Cabral is quick to caution: "Nobody commits the error, in general, of blindly applying the experience of others to his own country" ("Practical Problems and Tactics" 1968, 111). While "greatly admiring" Che's scheme derived from the Cuban people's struggle experiences, and a "certain application" of that scheme to Guiné's struggle, PAIGC doubted that it was "absolutely adaptable to our conditions." As a rule, the organization had reservations about "the systematisation of phenomena" ("Practical Problems and Tactics" 1968, 114).

Therefore, as we saw, the organization studied like a book "the geographical, historical, economic, and social conditions" of Guiné, and "basing ourselves on concrete knowledge of the real situation in our country . . . established the tactical and strategic principles of our guerrilla struggle" ("Practical Problems and Tactics" 1968, 111). Here was a small country 14,000 square miles (Guiné-Bissau) and 1,500 square miles (Cabo Verde) in size, the latter on the coast of Africa. PAIGC established its bases prior to launching the struggle, most of them in the south adjacent to its key host and ally Guiné Conakry (in Cobucare, Indjassan, Quinera, Gambara, Quitafene, and Sususa), and just a couple in the north adjacent to Senegal. Transporting material was a challenge, and once inside, it was placed under custody of local guerrilla groups acting autonomously but linked to the party leadership (until 1963). Coordinated action was threatened as some commissars and commanders disregarded other groups. The major problem: deploying guerrillas to basically turn their own villages into guerrilla bases and operate in their home areas where, soon after expelling the Portuguese, they abandoned the struggle and settled to village life.

Also as noted, PAIGC convened the 1964 congress to self-correct and chart a new course for the struggle. Certain guerrilla commanders were detained, tried, and executed. That is when the move was made to "collective leadership" of newly created zones and regions under party committees such that guerrilla leaders were simultaneously the party leaders. PAIGC also created a regular armed force to open up new "regions" and "zones" in Gabu (east) and Boé and Sao Domingos (west) corresponding to the regions and zones of the party. It was deemed strategic to mobilize "a reasonable proportion of the population" first and later mobilize the rest. The demarcation into northern, southern, and eastern fronts came later, armed forces units being (re)deployed to any of them as needs must ("Towards Final Victory" 1969, 130). By 1969, PAIGC's armed force was composed of regular forces charged with opening up new areas and other offensive operations against the enemy, and the people's armed militia responsible for security and defense of liberated areas. Two to three village bases were initially joined up before the idea was abandoned altogether. "There are our people's villages," Cabral remarked in 1969 in a statement acknowledging the reckless endangerment of the people, "and there are support points for our armed forces." And just in time, as the Portuguese had plotted nicely on maps, and by the time they started raining bombs on them, these villages had already been abandoned ("Towards Final Victory" 1969, 131).

As noted, Guiné was a flat country with no mountains. So PAIGC "had to *convert our people themselves into the mountain needed for the fight in our country*, and we had to take full advantage of the jungles and swamps of our country to create difficult conditions for the enemy in his confrontation with the victorious advance of our armed struggle" ("Practical Problems and Tactics" 1968, 111). To defend themselves, the Portuguese troops had to concentrate, so PAIGC scattered, drawing out and dispersing the enemy. PAIGC attacked. When the Portuguese concentrated, that allowed PAIGC "to occupy the areas that are left empty and work on them politically to prevent the enemy from returning" ("Practical Problems and Tactics" 1968, 112; my own emphasis).

We must return to Cabral using the theory of gravity—a force and another opposing force—and strategic deployments required to attenuate weakness against an almost impossible enemy force. By 1968, the Portuguese *modus operandi* involved aerial bombing and strafing of villages in liberated areas and suspected PAIGC bases; paradropping troops and encamping with massive air support; burning villages and destroying crops and cattle; airborne resupply of bases and operational units; and a large-scale, combined ground-and-air operation. Cabral remarked the lack of adequate opposing force from PAIGC: "This is understandable if one bears in mind the weakness of our means of anti-aircraft defence and our forces' lack of experience in this field" ("The Development of the Struggle" 1968, 92). But the enemy had its many weak points too. Their fortified camps totally cut off, the enemy bombed and strafed the riverbanks to cover its resupply boats, with mixed results ("The Development of the Struggle" 1968, 93). Sensing that the fish were getting out of hand, the Portuguese resorted to emptying the water. The strategic hamlet strategy was created in the Gabu area to deny PAIGC guerrillas the essential logistic and political support of the people, thereby breaking the force and spine of a people's war. These tactics were "a more or less exact copy" of the United States' in Vietnam, except not with the same equipment ("Towards Final Victory" 1969, 132).

To oppose this Portuguese force, PAIGC pampered the enemy barracks and fortified camps, especially those in the liberated areas, with a constantly heavy dose of mortars, artillery, and RPG-7 (bazooka) shells. The smaller or weaker camps were left to small arms and direct-fire weapons, in ambushes, firefights, and assaulting strategic

FIGURE 5.3
PAIGC troops inspect a Portuguese aircraft they downed in the forests of Guiné Bissau, 1974. *Source*: Wikicommons.

hamlets. Heavy weaponry was deployed against enemy river transport and to isolate and/or soften enemy positions for infantry assault through. Anti-aircraft base plate positions were set up at guerrilla base HQs to defend against aerial assault. Finally, PAIGC mounted ambushes and surprise attacks on the Portuguese in the partially liberated or contested areas ("The Development of the Struggle" 1968, 94).

The Portuguese use of winged and helicopter aircraft was not just a sign of force or capabilities but confirmation of, and response to, the amount of force that PAIGC was applying upon it. PAIGC had made all roads and rivers impassable, thus restricting the Portuguese exclusively to aerial resupply and assault. The Portuguese were forced to provide aerial combat escort to their river transport. In turn, the increased emphasis on airpower placed new demands for anti-air defense systems, while the fortified camps required more powerful artillery systems ("The Development of the Struggle" 1968, 94). PAIGC was inflicting the heaviest human and equipment casualties during ambushes and surprise attacks, given the Portuguese conventional warfare training and limited experience in guerrilla warfare ("The Development of the Struggle" 1968, 94). "If we have not had to invent a great deal in the course of our struggle, the Portuguese have invented even less," Cabral could say in 1968. Despite having state-of-the-art equipment from the United States, Germany, Belgium, Italy, France, and so on—napalm bombs, armored cars, B-26s, T-6s, P-2Vs, Fiat 82s, Fiat 91s, Sabres aircraft, and helicopters—the Portuguese kept losing ground ("Practical Problems and Tactics" 1968, 113). Cabral's theory of an opposing force defying gravity and going into space was being proven correct.

For Cabral, Guiné was a classic case of a people without technology confronting the technological might of colonial oppression. Guiné would prove that "peoples such as ours, economically backward, living sometimes almost naked in the bush, having no clothes on their back, living in poverty, with a diet lacking vitamins, fresh foods, and even meat and other protein foods, not knowing how to read or write, not having even the most elementary knowledge of modern technology, are capable, by means of their sacrifices and efforts, of beating an enemy who is not only more advanced from a technological point of view but also supported by the powerful forces of world imperialism" ("The Nationalist Movements" 1965, 64). The people of Guiné had equivocally answered the question: "Can we in fact wage a war like this?" ("Practical Problems and Tactics" 1968, 117).

And that creative resilience, Cabral maintained, was proof enough of Africans as a civilized people: "Were the Portuguese right when they claimed that we were uncivilized peoples, peoples without culture? *We ask: what is the most striking manifestation of civilization and culture if not that shown by a people which takes up arms to defend its right to life, to progress, to work and to happiness?*" ("The Nationalist Movements" 1965, 62–69; my own emphasis). Struggling "enabled the 'marginal' human beings who are the product of colonialism to recover their personality as Africans," a personality that was not just individual but "the personality of Guiné as an African nation," reawakened with "a feeling of confidence in the future" ("At the UN" 1962, 37).

The African struggle surprised the Portuguese into realizing that "we were not what they had supposed, . . . they discovered an African they had never imagined" ("The Development of the Struggle" 1968, 91–102). The desire for self-rehumanization had never left the Guinéan people, who never accepted Portuguese rule despite being "wounded in their human dignity" and "deprived of any legal personality." They had refused paying taxes and emigrated *en masse* to neighboring countries "to defend their dignity and give proof of their love of freedom and hatred of foreign rule" ("At the UN" 1962, 28) and "to present ourselves before the world as a worthy people with a personality of our own" ("Practical Problems and Tactics" 1968, 117).

For Cabral, culture was "the very foundation of the liberation movement." "We must preserve for our children the best of what we have learned," he had said. "They are the flower of our struggle" ("The Role of Culture in the Battle for Independence" 1973). The colonizer had succeeded at everything else but destroying the African's culture:

> "With a few exceptions, the era of colonization was too short, in Africa at least, to destroy or significantly depreciate the essential elements in the culture and traditions of the colonized peoples. Experience in Africa shows that (leaving aside genocide, racial segregation, and apartheid) the one so-called "positive" way the colonial power has found for opposing cultural resistance is "assimilation." But the total failure of the policy of "gradual assimilation" of colonized populations is obvious proof of the fallacy of the theory and of the people's "capacity for resistance. . . . Liberation is not the job of one man but rather the result of the combined efforts of many men, men who have a goal and are willing to sacrifice for it. ("African Independence Leader Honored by Lincoln" 1972)

PAIGC strategically deployed culture as a weapon and building block toward the creation of a new Guinéan personality, but the struggle also transformed and strengthened it by eliminating its weaknesses. Liberation struggle entailed innovating with culture, for example, using the tribal structure, the language, the songs, the dances, and customs as mobilizing devices. Cabral implored everyone to "defend these cultural differences with all our strength," while also fighting "with all our strength all divisions on a political level" ("The Development of the Struggle" 1968, 129). Cabral has a rather beautiful passage in this 1969 lecture:

> We must enjoy our African culture, we must cherish it, our dances, our songs, our style of making statues, canoes, our cloths. . . . There are many folks who think that being African is being able to sit on the ground and eat with one's hand. Yes, this is certainly African, but all the peoples of the world have gone through the stage of sitting on the ground and eating with one's hand. There are many folks who think that it is only Africans who eat with their hands. . . . We must be aware of our things, we must respect those things of value, which are useful for the future of our land, for the advancement of our people.
>
> No one should think that he is more African than another, even than some white man who defends the interests of Africa, merely because he is today more adept at eating with his hand, rolling rice into a ball and putting it into his mouth. The Portuguese, when they were still Visigoths, or the Swedish, who give us aid today, when they were still Vikings, could also eat with their hands. ("To Start Out from the Reality" 1969, 57)

Culture was a foundation but it was not a place of residence; and Cabral approached it through the eye of an engineer and saw engineering through the self-reflexive lens of a son of the soil, charged with leading the armed struggle. The Vikings of olden days wore great horns on their heads, amulets on their arms, heading to battle, as did the Portuguese, Franks (who became French), Normans, and Angles and Saxon (later the English), crossing the seas in canoes just as the Bijago ancestors of Guiné had done. Just as these Europeans made culture their strength, so too Africans. All peoples take strengths from culture but they move on: "We should not persuade ourselves that to be African is believing that lightning is the fury of the deity (God is feeling angry). We cannot believe that to be African is to think that man has no mastery over the flooding of rivers." Every leader in PAIGC had to grasp "what concrete reality is," to understand that while "our struggle is based on our culture, [and while] culture is the fruit of history and it is a strength," culture was also "filled with weakness in face of nature. . . . We can preserve the memory of all these things, to develop our art and our culture which we display to others. But as we have already gone beyond this, we know that it is we who rule in the forest, in the bush, we, human beings, and not any animal or spirit lurking there" ("To Start Out from the Reality" 1969, 58).

Culture must not be an obstacle to the struggle, Cabral emphasized. It must be understood as a reality of the land and even within PAIGC itself. Cadres who wore amulets for protection against or to elude enemy bullets still died. In a struggle, PAIGC recognized this was the reality it had to start from, rather than "order the comrades to tear off the amulet." *Even the Germans years before could not go into war without them, and they still carried an ikon of Our Lady of Fatima inside a small book, or a bible. The Portuguese carried a cross hung onto their chest, kissed it before and in battle. All these were amulets. PAIGC's leadership accepted the amulet as a "specific reality of war" but did not believe it to stop death. The only way to avoid loss of life was to prepare well, remain disciplined, and be tactical.* Said Cabral: "You can tell me a whole series of tales you have in your head: 'Cabral doesn't know. We have seen occasions when it was the amulet which spared the comrades from death, the bullets were coming and turned back in ricochet.'" But he implored the comrades to see it as a psychological armament over and above party doctrine, military training, and weaponry and a weakness because it may encourage careless moves in the belief that the amulet will protect flesh against bullet ("To Start Out from the Reality" 1969, 58–59). Many PAIGC leaders believed that the army should not camp or fight in certain bush areas belonging to the *iram* spirit. "But today, thanks to many *iram* spirits of our land, our folk understood, and even the *iram* understood, that the bush belongs to man and no one is afraid of the bush any more. We are even well established in the Cobiana bush, the more so because that *iram* spirit is a nationalist." His close comrades had told him when he gave them orders to deploy into the interior that "Secuna Baio or some other soothsayer told them: 'Do not go, I have cast your fortune and see great harm for you if you go into the interior.'" They would not go. Some did not spring ambushes because some soothsayer warned of death to one of them if they did ("To Start Out from the Reality" 1969, 59). In a separate speech Cabral says:

> We are proud of not having forbidden our people to use fetishes, amulets and things of this sort, which we call *mezinhas*. It would have been absurd, and completely wrong, to have forbidden these. We let our people find out for themselves, through the struggle, that their fetishes

> are of no use. Happily, we can say today that the majority have come to realize this. If in the beginning a combatant needed the assistance of a *mezinha*, now he might have one near but he understands—and tells the people—that the best *mezinha* is the trench. We can state that on this level the struggle has contributed to the rapid evolution of our people, and this is very important. ("The Development of the Struggle" 1968, 129)

Elders complained of their "sons" who did not consult them going into battle anymore, and PAIGC painstakingly explained why tactically that element of surprise and tactical advance should not be compromised. And Cabral credits the elders, not PAIGC, for the resolution of tensions between culture and tactics: "When all is said and done, the elders were the intellectuals of our society, of our genuine, real society. They were the ones who saw things clearly, who understood everything (our strengths and our weaknesses) and they soon shifted their ground, adapted themselves to the new situation" ("To Start Out from the Reality" 1969, 60). Here, then, the synthesis of two knowledge systems, two cosmologies, in the crucible of doing.

"DOING" was not just physically fighting the oppressor with guns but what Cabral called "re-Africanizing" the people's minds. By definition, the "liberated area" was an area denied to the colonial government's forces. *As a space, the liberated area was a critical laboratory where the process of transformation from revolutionary struggle to nation building began even while the war to free every particle of the country raged on. The hard life of revolution demanded creative resilience, living rough, with not enough clothing, little to no fresh foods or meat.* A transformative space, where women were now in charge of *tabancas* and interregional committees, conscious of their worth and revolutionary role.

Not a frictionless process. Some among the leadership expressed a "silent resistance." Instances of comrades erecting every barrier to women taking leadership even where their superiority of ability was obvious were not uncommon. The slightest weakness was seized upon: "We told you so." True, some failed to succumb to certain temptations, to exercise certain responsibilities. But so did the male comrades. The reality of struggle, that people's liberation was women's liberation, that people's sovereignty demanded women's participation as combatants not recipients, that women's liberation must be conquered by the women themselves, not given to them, was not well understood in PAIGC's leadership. Cabral was accused of being obsessed with putting women in positions of leadership. "Let him do it, but we shall sabotage it afterwards," they said ("Our Party and the Struggle" 1969, 70–71).

In a statement almost prophetic concerning women's liberation that anticipated the rise of the increasingly confident, unshackled, and self-driven African woman of today, Cabral answered his critical comrades in 1969: "They can sabotage today, sabotage tomorrow, but one day it will catch up with them" ("Our Party and the Struggle"

FIGURE 5.4
One of them, Titina Ernestina Silá, was head of the People's Militia Committee that organized and ensured passage of people and goods across the Cacheu River, northern front, before her death crossing the Farim River en route to mourn Cabral in Guiné-Conakry, 1973. *Source*: Museu Militar da Luta de Libertação Nacional.

1969, 71). The strength of the organization, he warned, "is only effective if we, leaders, are able to open the way for the youth to progress, . . . to take over and to bring the best forward to lead. My greatest joy is to see a comrade, man or woman, carrying out duties conscientiously and willingly without being pushed. . . . Everyone in the Party knows what friendship, what regard, what respect, what warmth we have for those who can

FIGURE 5.5
PAIGC female freedom fighters posing with Amílcar Cabral. *Source*: Museu Militar da Luta de Libertação Nacional.

carry out their duty. Everyone we see working with complete enthusiasm is like a part of ourselves" ("Our Party and the Struggle" 1969, 71). That could not be said of every leader in PAIGC: some political commissars turned promising juniors into messengers instead of mentoring them in struggle tradecraft. Others turned "a bright and fairly attractive girl" into a mistress instead of mentoring her into a teacher, nurse, a good militia fighter, thus spoiling the future of cadre and party alike. "We do not want servants, girlfriends, or children," Cabral warned. "Anyone who wants a girlfriend, today or tomorrow, can find her, woo her, but he must not use the authority of the Party to have all the women he wants. . . . This is also why we must be vigilant against opportunists . . . among us, who knowing that our leadership requires the best sons and daughters of our land to lead may pretend to be the best" ("Our Party and the Struggle" 1969, 72–73).

As practice, liberated areas sought to involve the *povo* in their own struggle; hence PAIGC's slogan, *Povo na manda na su cabeça* ("Let people do it for themselves"). The emphasis: people's production based on the village. The everyday was the catalyst for innovating self-liberation; the reality of everyday life was a teacher, a force that pushed the people into action, to be creative in search of freedom ("Practical Problems and Tactics" 1968, 118).

The village was the central, most important unit of revolution, as Cabral remarks in 1964: "We shall work together, we shall find new riches under the soil, we shall cultivate better. . . . That means more than cultivation. That means realizing what people can do, can actually do. That's a question of village democracy, of village schools, of village clinics, of village cooperatives. Living better isn't only eating better" (Davidson 1964/1974). Cabral found it problematic that the majority lived at the mercy of the rains, practicing a monocultivation of groundnuts ("Guiné and Cabo Verde against Portuguese Colonialism" 1961, 10–19). Hence in the liberated areas PAIGC prioritized a self-reliance and export-oriented economic development of crop production (rice and other crops), rubber, skins (crocodile, animal), and coconut ("The Development of the Struggle" 1968, 98). Its nascent industry was mostly artisanal and local, while the reopening of sawmills abandoned by European settlers and a proposed "small rudimentary factory" to make soap out of palm oil were being planned in 1968 ("The Development of the Struggle" 1968, 98). Also by 1968, two people's stores had been set up in the north and in Boé regions to supply the people with basic needs, with much supply difficulties despite help from friendly countries. The Portuguese countered them with greatly reduced prices in adjacent areas still under their control ("The Development of the Struggle" 1968, 98).

Strides were also being made in healthcare, and here PAIGC started from nothing. The Portuguese had built just three hospitals and a few dispensaries in all of the country; by 1974, PAIGC had four hospitals, two in the south, one in the north, and another in Boé, about two hundred beds total, and permanently stationed doctors with sufficient nurses and necessary surgical equipment and capacity, dozens of dispensaries in all sectors serving combatants and the people. Boé hospital was upgraded to include general medicine, surgery, orthopedics, radiology, anesthesia, and analysis. By 1967, eighty nurses had been trained, thirty within Guiné and fifty in Europe; thirty more were under training in 1968; and the rural orthopedics hospital was on the way ("The Development of the Struggle" 1968, 99).

Also impressive was PAIGC's development of education facilities in the liberated areas. The Portuguese had only 11 government and 45 mission schools (56 in all) with 2,000 black pupils when PAIGC launched armed struggle, 127 primary schools with 13,500 pupils in 1965–1966, ages 7 to 15 ("The Development of the Struggle" 1968, 100). Most elementary schools—which as we saw were Catholic run—had closed a couple of years into the war as teachers slipped into the bush to join the guerrillas ("The Development of the Struggle" 1968, 99).

Cabral was already looking forward to an economy in which all of Guiné would be free. PAIGC had since 1964 also been training a small cadre of technicians in various branches of the "practical sciences." With the foresight that there might be oil offshore, the guerrilla movement had also sent its personnel for oil-related engineering training (Davidson 1974). Going into independence, Cabral identified Guiné's problem as one of "achieving the organizational conditions under which this 'poverty' may be overcome by a balanced social development" (Davidson 1974). Here was a man already looking ahead to an independent Guiné.

"NO TEARS COMRADES": MOURNING A REVOLUTIONARY THEORIST AND A MAN OF ACTION

Amílcar Cabral was assassinated by the Portuguese on January 20, 1973, in his exile home in Conakry, Guiné. As the Pan-Africanist historian Basil Davidson put it, Cabral's death served to galvanize the determination of freedom fighters in Portuguese Africa as a whole. The Portuguese had killed Mondlane and still the liberation struggle continued, even intensified, in Mozambique. Just as Samora Machel took over from Mondlane, PAIGC would have its own Machels—the likes of Luís Cabral, Aristides Pereira, Vitorio Monteiro, Nino Viera, and Osvaldo Viera, among others, would pick up the baton stick and continue the charge to freedom (Davidson 1973). Remarkably, Cabral also becomes part of a long story in which dynamic black leaders whose ideas and works had started transforming or would transform the lot of their people died young—Malcolm X, Martin Luther King Jr, Frantz Fanon, Cabral, Thomas Sankara, Samora Machel, Peter Tosh, Bob Marley, Josiah Magama Tongogara, Chris Hani. These are just a few who had attempts made on their lives, most of whom died from assassinations. Sometimes that is what makes African youths fear to dare, and slither away from the limelight, and expend their talents on personal enrichment, quietly living away the good life. Cabral had fought Portuguese colonialism with the AK-47, the pen, Pan-Africanist rhetoric, and theory; these instruments defined the revolutionary community that assembled to honor his passing. They did not mourn him; they celebrated a life well lived. And spent. Guinéan president Sekou Touré set the tone for the revolutionaries: "No tears comrades; the only action that must count now is the revolutionary one" (Johnson 1973).

Cabral's body was draped in revolution: his coffin wrapped in PAIGC colors mounted on a gun carriage was led into Conakry Stadium where twenty-five thousand "revolutionary" people were packed. A "carry party" of Guinéan sailors armed with AK-47 rifles accompanied the hearse, the drums and horns pacing their movement—a somber "slow march." When Cabral's coffin was placed on a saluting dais, thousands of Guinéan army, police, and workers paraded past in their detachments carrying a panoply of Soviet-made automatic rifles and submachine guns. Overhead, six MiG fighters roared now and then (Johnson 1973).

On January 31, Conakry radio announced that among the crowd of twenty-five thousand in Conakry stadium were delegates from eighty "sovereign nations, guerrilla movements and supporting organizations from the United States, the USSR, China, Yugoslavia, Romania, Czechoslovakia, Bulgaria, India, North Korea, Libya, the Portuguese Communist Party and from the Canary Islands." At a symposium in his honor attended by Cabral's revolutionary, intellectual, and communist friends, American poet and playwright and Pan-Africanist Amiri Baraka called for "an effective Pan-Africanism reaching outside Africa to the United States, the Caribbean and South America" as the only fitting tribute to the deceased (Johnson 1973).

Meanwhile in independent Africa, flags flew at half-mast. In Lusaka, the five liberation movements based in Zambia—the Movimento Popular de Libertação de Angola, the Zimbabwe African People's Union, Frente de Libertação de Moçambique, South West Africa People's Organization of Namibia, and the African National Congress (South Africa)—issued a joint communiqué condemning the assassination and marking a week of pan-African mourning. The only way such a pan-African tribute to Cabral could be paid was through revolutionary means: the five Soviet-backed guerrilla movements "vowed to intensify their struggle to "wipe out colonialism, imperialism, and racism" ("Freedom Fighters Declare Week of Mourning" 1973). To wage revolutions required more trained men and more guns: Libyan leader Colonel Muammar Qaddafi in a cable to his Guiné Conakry counterpart Sekou Touré: "The condolences offered by the Libyan Arab Republic on the death of this militant is additional arms and even men to wrest the freedom of part of the African continent in loyalty to martyred hero Cabral and in pursuit of his struggle." All anticolonial/white minority wars on the continent became "Africa's wars" ("Gaddafi Sends Arms and Men to Aid PAIGC" 1973).

The lasting impression of Cabral as a critical thinker-doer is best summarized in Pulitzer prize-winning journalist Jim Hoagland's piece in the International Herald Tribune on January 25, 1973:

> Cabral was a rare combination—*a revolutionary theorist and a man of action*, marching with his troops through the steaming swamps and river-shredded terrain of his West African country. Inevitably, he became labeled as Africa's Che Guevara—and, in some ways, he was. . . . Like Guevara, Cabral possessed an independent, restless spirit that roamed across standard ideological boundaries. He was a dedicated leftist, but he rejected key elements of Marxism because it failed to take into account human will and individuality, especially in unindustrialized regions like Africa. (Hoagland 1973; my own emphasis)

6 GUERRILLA HEALTHCARE INNOVATION

From 1975 to the end of 1979, Zimbabweans were engaged in *chimurenga*, the war of national self-liberation against the colonial settler state of Rhodesia. An estimated twenty-five thousand people died in this conflict, which resulted in independence on April 18, 1980.

Struggles against colonial rule like chimurenga are potentially inspirational. In Zimbabwe as in several other African countries, the struggle for independence has remained a political tool and monopoly of the parties that led the anti-colonial struggle instead of a rallying moment of inspirational creative resilience for all Zimbabweans. This is despite that everybody (other than the Rhodesian regime and its supporters) fought to liberate themselves and the country. It was a communal effort.

The crisis of chimurenga constitutes a failure of narrative to capture and account for the amazing bravery and innovations of all those who took part in it and forced a recalcitrant colonial regime that had sworn there would be no majority rule "in a thousand years" to sue for peace. The monopoly of chimurenga by a few has robbed Zimbabwe of a nostalgic rallying point that could be mobilized to inspire the nation to deal with its postwar challenges. Chimurenga was a time when people eked out survival in spite of the government. They also had to survive the government that had turned against them and become the enemy.

The chimurenga spirit could be summoned in times of crisis, when the nation desperately needs a spirit of creative resilience in the face of insurmountable hardship. At the moment it is summoned to label everyone who disagrees with the ruling political elites as puppets of the West and traitors. The burden of chimurenga lies in its association with the worst characteristics of the postindependence nationalists who monopolize it.

How then does one rescue chimurenga from elite discourse and a discredited brand of African nationalism? The first task is to restore to the narrative the specific creativities and sacrifices of two foot soldiers of chimurenga who fought side by side: the guerrillas and the ordinary people. Today they are peripheral to the history and material rewards of that struggle, while the politicians who were living in the comparative safety of Maputo (Mozambique) and Lusaka (Zambia) monopolize its benefits. That is not to trivialize their political leadership; rather, it is to remind them that chimurenga was a collective, communal struggle and its primary theater was in the bush, not the street.

Told from the guerrilla and ordinary people's experience, *chimurenga becomes a story of innovation under conditions of extreme hardship. It shows that, even where death is the outcome, and where the situation is insurmountable, people do not just surrender to fate, but die fighting. It allows us to go beyond the dead bodies, mourning the victims, and feeling outraged about the killers, toward the innovation of survival that sometimes ends in defeat, most often in triumph over adversity. This spirit of creative resilience, the refusal to surrender and the creative work that goes into it, is perhaps Africa's most important resource, but remains subdued because analysis ends with the victim.* Creative resilience pays tribute to the intellection and sacrifice that went into struggling for independence against seemingly insurmountable odds. It involved not only the political leadership or the guerrillas with guns, but also the ordinary people often caught between and victimized by both sides. *Chimurenga is the ultimate site of creative labor, a vast laboratory.*

Very few accounts of the war focus on innovation; where they do, the focus is on the Rhodesians, reinforcing a reduction of innovation to white, to technology, and to commercializable products (Godin 2008). A significant literature deals with Rhodesia's use of chemical and biological weapons, with Africans as victims (Nass 1992, 1; Nass, 1992/1993; Sterne 1967; Lawrence, Foggin, and Norval 1980). In 1975 the Rhodesian Army's G Branch produced the *Soldier's Handbook on Shona Customs*, which it distributed to the units. It basically packaged local African culture into an operational strategy against the guerrillas. A decade after the war, the US Army commissioned the RAND Corporation to undertake a study into the valuable lessons to be learned from the Rhodesian counterinsurgency experience, many of whose aspects were based on the handbook (Hoffman, Taw, and Arnold 1991). Meanwhile, ex-Rhodesian military personnel and their biographers began publishing about their wartime technological innovations, including the retrofitting of light-skinned vehicles into landmine-proof troop carrier vehicles like the creepy "Pookie." In 1994 Australia's Department of Defence began plans to refurbish its fleet of armored vehicles. It subsequently commissioned a study of Rhodesian innovations in the protection of light-skinned vehicles against landmines (Lester 1996). It is hard to imagine that the US counterinsurgency strategy since the 1990s and Australia's turn to the Bushmaster Protected Mobility Vehicle post-2000 did not benefit from these lessons (ACIL Tasman 2009; Wood 2005).

The Zimbabwe war theater of 1975–1979 was not simply a destination where incoming technologies were used just as the manufacturer's manual said, but also a source of technological innovation for countries traditionally considered its sources. The war suggests that insurgencies are not just a security challenge. They are generative of counterinsurgency and counterasymmetric warfare innovations that the hegemonic countries (in NATO, Russia, and China) cannot generate at home. I entirely agree with Helen Tilley's portrait of Africa as a living laboratory to the degree that it constitutes a space fecund with experimentation. For her, the experimenter is the European or imperial scientist and the subject is "how modern science is being applied to African problems" (Tilley 2011). For me, the experimenter is the African guerrilla and ordinary person concerned with not only applying incoming resources to local problems, but also innovating independence from oppression by integrating targeted resources from outside into endogenously generated ones.

Thanks to a cross-section of Zimbabweans alarmed at attempts by a small minority of politicians to monopolize and distort chimurenga, who are writing about their struggle experiences, this story of creative resilience can now be told from the rural countryside and the guerrilla bases. We can no longer believe the fable that certain ruling party (ZANU-PF) politicians and senior commanders fought while everybody else just "sat there"! (Chung 2006; Tekere 2007; Bhebe 1999; Mhanda 2011; Mutambara 2014; Sadomba 2011). The evidence permits us to tell the stories of guerrilla foot soldiers and villagers creatively engaged in struggle, thus challenging the Rhodesian portrait of them as "terrorists" and victims of terror, respectively (Smith 1997, 2008). Which is not to say guerrilla-villager collaborations were always peaceful (Kriger 1992; Alexander, McGregor, and Ranger 2000; Bhebe and Ranger 1995); whatever they were a result of, the objective here is to analyze such collaborations as processes of innovation. Among other things, chimurenga shows how local actors weave disparate locals across the world into sources of resources that, upon arrival, they strategically and industriously deploy to operationalize their own dreams and yearnings into reality.

"Guerrilla healthcare innovation" is thus an example of the war theater (chimurenga) as a laboratory. Originally published in *History and Technology* (Mavhunga 2015), this chapter locates practices conventionally done on the bench inside rooms in universities and cities in the middle of the forest, on the move, in caves, and other unlikely places. So what becomes of practice when it is unmoored from its conventionalized, formalized spaces into the forests, in the midst of a raging, highly mobile guerrilla war? Among other things, it means that the operating philosophy, procedures, and even instruments change, as discussed in the next section. Whatever Africans learn about medicine abroad becomes in the field mere ingredients in the construction of an integrated healthcare system contingent to the war being fought. This conversation continues in the next two sections. The penultimate section solidifies around a theme already strongly hinted in the preceding ones: the role of ordinary people, not only as victims caught between two sides, but also active combatants without guns and fellow innovators alongside guerrillas, engaged in their own self-liberation. The essay ends with reflections on the implications of telling stories of innovation from such open laboratories.

CHIMURENGA AS MEDICINE: THE BIGGER PICTURE

Chimurenga must be viewed as bitter medicine that Africans resorted to after the failure of decolonization (peaceful handover of power from white settlers). It was an acknowledgment of the failure of top-down solutions to an African problem and the recourse to self-liberation or self-cure.

The dilemma was whether the colonial settler could simply give up the advantage assured through racist oppression that had enabled his prosperity since 1890. That was the year when the British South Africa Company, armed with a royal charter, occupied the lands between the Zambezi and the Limpopo Rivers. Africans rose in rebellion in 1896–1897 to reclaim their lands, without success (Mufuka, Mandizvidza, and Nemerai 1983; Pikirayi 2001).[1] Their leaders were captured, many beheaded, and some had their heads sent to Britain as evidence of capitulation. Africans were violently removed to overcrowded, infertile, and disease-infested lands called "native reserves." The settler administration now imposed a slew of taxes that forced Africans to have to work for a pittance on the white settler farms, mines, factories, and emerging suburbia. If they did not, they were arrested and turned into convict labor anyway. Or the settlers simply raided entire villages and force-marched them to their properties to work for nothing (Ranger 1967; Beach 1989; van Onselen 1976). The entire road, rail, urban, mine, and agricultural infrastructure of the colonial period was built using poorly paid, conscript, or convict labor. This is how Africans subsidized colonial settler prosperity, became the tools of empire, and built Europe.

In deciding to grant independence to its colonies in 1960, the British government overestimated its power over these settlers, who had forged a racist Rhodesian citizenship and nationality. They had no intention to leave or to free Africans because their entire enterprise could not sustain itself without Africans subsidizing it through miserly paid and forced labor. In contrast to other parts of Africa where Europe had direct control over its colonies, technically Rhodesian settlers had been self-governing since 1923. Through vigorous promotion of white immigration, they had steadily built a significant white population, relatively wealthy, and in love with the climate. The infrastructure—cities, roads, rail, electricity—were all designed with permanent residency and citizenship in mind, not temporary or occasional stay.

Things deteriorated rather rapidly from 1960. All peaceful means of achieving independence had been dashed. One African political party after another was banned as soon as it was formed. Then in 1965, the settlers declared unilateral independence from Britain as a white-ruled state. Africans had had enough. From 1961 onward, first the Zimbabwe African People's Union (ZAPU) and then the Zimbabwe African National Union (ZANU) started sending their young cadres for military training overseas and in newly independent Ghana, Algeria, Egypt, and Tanzania. They returned to set up training bases in Tanzania and later Angola, Zambia, and Mozambique. Reconnaissance and skirmishes followed. By 1971 two guerrilla armies had emerged to lead the self-liberation war. The Zimbabwe People's Revolutionary Army (ZPRA), the armed wing of ZAPU, was based in Zambia, had training bases in Angola, and mobilized military training and logistics from the Soviet Union, East Germany, Yugoslavia, Cuba, and Iraq. The Zimbabwe African National Liberation Army (ZANLA), the military arm of ZANU, operated initially from Zambia, then Mozambique, with training bases in Tanzania and military support from China, North Korea, Romania, and Yugoslavia.

Chimurenga was a self-curing process with emphasis on the restoration of pride in cultural and spiritual identity. This connection between the struggle for freedom and African cultural values can be traced back to the 1950s and early 1960s when the nationalists adopted Zimbabwe as the name for Rhodesia. All that was left was to fight for its liberation. Having found that their top-down, learned jargon had failed, the mission-educated, urban elites went back vigorously to their roots. They reclaimed their leopard-skin hats, shrines, drums, and ancestral spirits, their own Mwari (god), and their rituals, which the missionaries cast as symbols of the devil and the colonial settlers saw as dying remnants of the primitive tribesmen. Their spiritually anchored culture now became a weapon or medicine against the disease of *chirungu* or *svexilungwini* (the white man's corrupting ways). Without this self-cure, which was also an epistemic and ontological return, the gun and the syringes would never accomplish their assigned role of arming the comrade (as the guerrilla was called) and repairing his sick and injured body. Thus while various scholars have covered the intersection between nationalism, guerrilla war, and faith (Ranger 1999; Lan 1985; Daneel 1995), I am concerned with that convergence as an example of building on local, grassroots-invented idioms and integrating incoming resources to plug deficiencies.

At the level of personnel, guerrilla healthcare was an integrated system of indigenous and allopathic human resources—the spirits, spirit mediums, *n'anga* (healers), physicians, medics, and ordinary people. It reflects the tendency of Africans, locating themselves strategically and deliberately at the intersection of endogenous and incoming therapeutic systems, to exercise pragmatic choice, not limiting themselves to one or other, or just seeking the best in both, but integrating them into one. Whether one goes to a *n'anga* or a hospital or both, the balance of spiritual forces governing the body and soul must be in equilibrium before seeking therapy. Hence people consult the ancestors (and with Christian encroachment, the prophet, pastor, or simply prayer) before taking the patient to seek treatment (Chavunduka 1978).

ZAPU HEALTHCARE IN ZAMBIA

For clarity, the two principal geographic arenas, the "rear" (operational and refugee bases in neighboring countries) and the "front" (the war theater inside Zimbabwe), will be explored separately for the remainder of the essay. In this section, the discussion focuses on ZAPU/ ZPRA and ZANU/ZANLA healthcare infrastructure in Zambia and Mozambique, respectively, before turning to the front. The discussion covers 1975–1980 only. All sites referred to in the rest of the essay are indicated in the map below (figure 6.1).

ZAPU's politicians were headquartered in Lusaka, while ZPRA's command element and troops were based at the main army camps at Mulungushi (CGT 1–4), Solwezi, and Maheva. The forward bases into Zimbabwe were located in the Zambezi valley, with one entry into Zimbabwe via Kariba and Chirundu, the other between Kazungula and Batoka. Nampundwe was a transit camp for recruits coming from Zimbabwe and South Africa via Botswana. Everybody underwent a medical exam at Nampundwe to determine their suitability for military training or their needs if going to refugee camps and school. The tests covered visual, heart, respiratory, and other conditions. Sometimes thoroughness was deemed secondary to clearing the place and moving people out quickly to avoid bunching up and becoming easy targets for Rhodesian airstrikes (Ndlovu 2011b). After being processed (interrogated and medically checked), they were transported to Freedom Camp, Victory Camp, and Mkushi Camps (refugee camps for boys, girls, and women, respectively), or for training initially at Morogoro in Tanzania (prior to 1976), then later in Angola (1976 and after). Other children were accommodated at JZ, Works, and Makeni camps.

"We started small," recalls Benjamin Dube, the deputy head of the movement's health department. Initially ZAPU had only a "medical assistant," Jabulani Ncube, who ran ZAPU's first two camps, Nkomo and Luthuli, from the party's headquarters at Zimbabwe House, Lusaka. During training, the recruits were also taught basic first aid, hygiene, and other essentials. Meanwhile, ZAPU had farmed out some recruits to different countries (including Zambia) to train as medical assistants, nurses, laboratory technicians, and physicians.

FIGURE 6.1
The "rear" and "front" of Chimurenga, including Mozambican and Zambian guerrilla and refugee camps, 1976–1979. *Source*: Author.

Also from 1977 as the war intensified, more secondary school students were coming into Nampundwe via Botswana, and as the bombings intensified, recruits increased in tandem with injuries, hence the need for more medical personnel (Ndlovu 2011c).

The combination of increased health personnel and necessity for healthcare had significant infrastructural implications. Prior to 1979, ZAPU had no field hospitals and relied entirely on Zambian facilities. Dube explains:

> We had a very good working relationship with the University Teaching Hospital, Zambia, Lusaka. We also had a very good cooperation with the hospitals [in] Kabwe where our cadres were nursing; we used to get them admitted there. We had a good working relationship with the hospitals at Solwezi. When we were bombed [in 1978], I think over 100 of our chaps at one time they filled that hospital. And also in the Copperbelt, that is Kitwe and Ndola. (Ndlovu 2011c)

By 1979 ZAPU had two field hospitals, one at Victory Camp, the other at Solwezi, with a combined capacity of 250 beds. Dube says of the latter field hospital:

> We had a big field hospital there [VC] which had everything, as you can even see, with microscopes and all those things. Things were expanding. We were even in the process of establishing an x-ray sort of unit. In the big camps in Solwezi, we had a big field hospital which was donated by the Swedes. It had almost everything you can think of. It was a field hospital . . . with operating theatres . . . the laboratory was at Victory Camp. It was through the assistance of a German couple. The husband was a doctor; the wife was a laboratory technician. And she's the one who assisted us to train . . . to do in-house training for the girls who were there. (Ndlovu 2011c)

The laboratory was used to test mostly malaria, pregnancies, stools, blood, and so on. The medicines were donated "from all over the world," as were the equipment, testing materials, and supplies like cotton wool (Ndlovu 2011c).

The Swedish Air Force had delivered the mobile field hospital to ZAPU to alleviate the dire medical situation in the wake of the devastating Rhodesian aerial attack in 1978. John Landa Nkomo, a high-ranking ZAPU official and later vice-president of Zimbabwe, "received that hospital at the Lusaka International Airport, with the jeeps and so on." Later on this fully equipped mobile hospital was moved to Solwezi "because life had become very difficult there with malaria." ZAPU took it with them to Zimbabwe after independence, while some of its equipment was donated to the Zambian government ("Nkomo" 1995).

ZAPU/ZPRA's physicians were based in Zambia, moving between the party headquarters in Lusaka, the guerrilla bases, and the refugee camps. Gordon Bango (figure 6.2) headed the organization's Department of Health, with Benjamin Dube, trained in the German Democratic Republic (GDR), deputizing him from 1978. Bango trained as a medical doctor in Hungary before becoming a resident doctor in Scotland (Herald Reporter 2011). He was part of a scholarship scheme that Joshua Nkomo, leader of ZAPU and ZANU's precursor, the National Democratic Party, had secured during his visit to the Soviet Union in 1959. The students would be sent for university studies in strategic areas in Yugoslavia, the GDR, the Soviet Union, Czechoslovakia, and Hungary. Besides Bango, beneficiaries included Sydney Sekeramayi (later ZANU) and Stanley Urayayi Sakupwanya, who both specialized in medicine (Gwaunza 2014).

The most detailed account of ZAPU/ZPRA healthcare comes from Dube himself. Initially he was assigned to look after the refugees at Victory Camp, before being seconded to all of ZPRA's military bases. Diseases like scabies, tuberculosis, diarrhea, and malaria were a problem. Water shortage, poor hygiene, and malnutrition made them worse. At Freedom Camp, the malnutrition problem was one of "some mischief from the logistics people," Dube explains. "You would find that more food would go to female colleagues than to

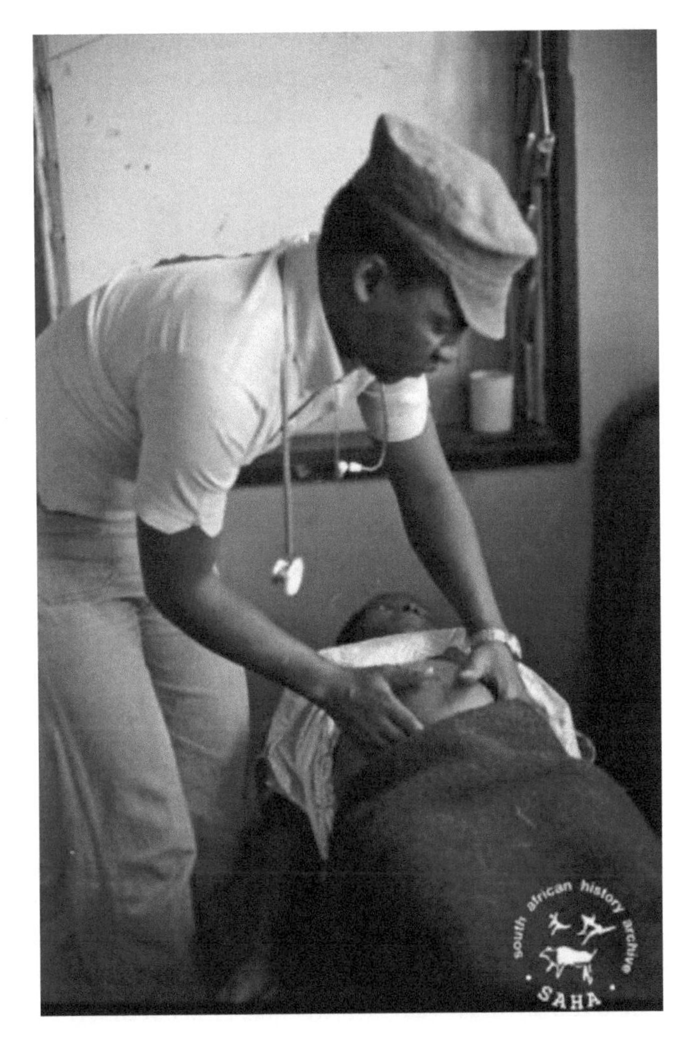

FIGURE 6.2
Dr. Bango busy at work at Victory Camp. *Source*: South African History Archive.

male colleagues yet the male colleagues outnumbered female colleagues" (Ndlovu 2011c). Tuberculosis was also a serious problem in the military camps, brought by comrades returning from military training abroad, especially in Libya. There was no water at Victory Camp; it was carted in a bowser. Only later, as classrooms and hostels were increased, was a borehole sunk. Whether it was water, food, or medical checks, each group of girls, boys, and soldiers was organized in sections, each with its own commander. Instructions and standards were easy to issue and maintain because it was a hierarchical system (Ndlovu 2011c).

Then there were the injuries from airstrikes: "You would get people . . . hit with fragments everywhere and some with mutilated limbs" (Ndlovu 2011c). The worst nightmare, Dube says, was napalm. The regime's engineers had developed a home-brewed version of napalm called "Frantan"—"gelatinized gasoline carried in large drum-like containers [dropped from light aircraft] that detonated the moment they hit the ground, showering everything and everybody within a fifty-yard radius with adhesive globules of burning petrol" (Thomson 2001, 280). Dube's own medical assistant was one of the casualties of napalm bombing. As a doctor, he draws a sharp distinction between burns from napalm and the injuries comrades brought from the front—gunshot wounds, bullets, sometimes fractures, and soft-tissue injuries. The medics in each unit could patch these up and get serious cases medevaced (from the term medical evacuation) to the rear (Ndlovu 2011c).

In the wake of the Rhodesian attacks, each resident of Victory Camp was mandated to dig her own pit to shelter so that "when [the Rhodesians] throw napalm on us, we could be safe, because the pit was such that you dig downwards and then to sideways, so that [the opening] is just an entrance, we would be behind [the cover when the napalm was discharged]" (Ndlovu and Nkomo2011b). This was obviously an outcome of research and careful engineering planning, indicating the factor of war as teacher, and guerrilla organizations as innovative and adaptive. Like in ZANLA, the inmates of camps went through training to sensitize them to the napalm threat and anti-napalm procedures.

For girls, being in Victory Camp presented two problems. One was the monthly period. Regina Ndlovu explains the predicament of a teenage girl having a monthly period, who did not have enough cotton wool, and was expected to go for *toyi-toyi* (military exercises) at 4:00 a.m. In those instances one had to get panties from friends, and wear several of them. It was not a valid excuse to say "I have a period therefore I can't go for exercises." "What would you do if you [were] in the bush and the enemy was following you behind?" Ndlovu asks. "Would you say, today I am menstruating, I will not shoot?" (Ndlovu 2011a). The second was teenage pregnancies, "a big problem" according to Dube. Being a traditionalist, Nkomo was strongly opposed to contraception. It says a lot that logistics personnel was sending more food to the girls camp and far less to the boys camp, despite the former being far fewer. This food-for-sex problem pushed the physician to write "a nasty report" to ZPRA intelligence chief Dumiso Dabengwa (Ndlovu 2011c). Dube says the issue was quickly solved, but the problem of "chefs" (superiors) abusing their positions is not unique to ZPRA; ZANLA too had its problems (Nhongo-Simbanegavi 2000).

The scale of the Rhodesian bombardment on October 19, 1978, left not only dead, but also injured and ultimately disabled bodies. Up until then, ZAPU had no facilities of its own dedicated to care of its disabled. For straightforward amputations and other war injuries, the party relied on the University Teaching Hospital at the University of Zambia. The evidence comes from Zakhele Ndebele's gripping account of injuries sustained during Rhodesia's attack on Freedom Camp, code-named "Green Leader." Already hit by shrapnel on the arm, he ran for cover, only for his foot to be shattered by a bomb. "So when I tried to stand up my foot was already shot. I stepped on my foot and I felt as if I'm falling into a pit, but it wasn't really a pit; in fact my leg was shattered so the bones came out a bit." He blacked out momentarily. The pain jerked him awake. Ndebele started crawling, toward safety, as the Rhodesian planes pounded above. When they were gone, somebody pulled him out of the trench he had holed into.

After the fog of bombs and gunfire had lifted, Ndebele and four others were taken to the University Teaching Hospital. Four or five wards were full of ZPRA. Not one Zambian. That was in October. He was hospitalized until December; in between, his leg was amputated above the knee. December passed. January. Although he says he had recovered mentally, Ndebele "no longer wanted to dress. . . . My arm could not move upward well, so when they were trying to dress me I was feeling pain, I was saying 'No, leave me like this.'" He was discharged that January, but ZAPU was determined to ensure his full recovery. A group of them were flown to Czechoslovakia for treatment. The Czechs fitted him with an artificial leg, but unbeknownst to him, Ndebele was also carrying shrapnel in his ear. He got it removed, only for the doctor to find another lodged in his throat; he carries it to this day (Ndlovu and Nkomo 2011b).

When the group returned from Czechoslovakia in 1979, they were taken first to Makeni and then to a camp for the disabled. The camp, says Dube, was located in Solwezi. Naison Ndlovu was in charge of this rehab center, working under Stephen Nkomo, who headed the social welfare department along with Thenjiwe Lesabe. As a teacher and counselor, Ndebele received comrades coming from the front with serious injuries. Some recovered and went back to the battlefield. Others did not and were accommodated until the end of the war. Those cadres who suffered from mental illness—Dube concedes "depression" was a serious problem—were admitted in Zambian government institutions for treatment (Ndlovu 2011c).

ZANLA HEALTHCARE IN MOZAMBIQUE

Situated some fifteen miles north of Chimoio town was a complex of twelve ZANLA camps called Chimoio, in the Mozambican province of Manica. Two of them are important for this essay: Parirenyatwa, ZANLA's hospital, and Percy Ntini, its rehabilitation center (Mutambara 2014, 73). The third site discussed here, Chibabava, was the scene of devastation by the jigger flea.

Named after Zimbabwe's first African physician and nationalist Simon Tichafa Parirenyatwa, Parirenyatwa Hospital was a one hundred–bed hospital at the time of the Chimoio bombing in November 1977. It was ZANLA's major health facility for handling injuries and conditions except critical cases, which were referred and transported to the Mozambican government hospitals at Chimoio, Gondola, and Beira (Mutambara 2014, 150). The admission wards were a barrack structure made of wood and grass. The living quarters for hospital staff were round huts made of similar material—the traditional style for constructing housing structures in Zimbabwe. Parirenyatwa had at least one mobile operating theater on wheels and three ambulances ("The Grim Reality of War" 1978). It also had three theaters stationed outside the camp complex in tents in an area commonly called *kukavha* (at the cover). As a security precaution, injured guerrillas were taken to *kukavha* during the day and brought back into the ward at night until they recovered because the wards were in the open and susceptible to aerial bombing (Huni 2013d).

Herbert Ushewokunze, Sydney Sekeramayi, Felix Muchemwa, and James Chideme Muvhuti were among the physicians to join ZANU in the Mozambican rear bases in 1977. Sekeramayi trained first in genetics in Czechoslovakia and then medicine at the University of Lund, Sweden ("Sydney Sekeramayi" 1995). Ushewokunze had attended the University of Natal, South Africa. Muvhuti had a medical degree from Sofia University in Bulgaria, while Muchemwa was coming from Birmingham Medical School in the United Kingdom. Sekeramayi was charged with monitoring all bases for standards and compliance, Ushewokunze was assigned the task of sourcing medical supplies from Europe and North America, and Muchemwa was appointed the chief medical officer of Parirenyatwa. Little is known of Muvhuti thus far from the available archives. The main duty of all four doctors was to serve the guerrillas and refugees and to provide direction to ZANLA's medical corps, dynamic and responsive to the rapidly unfolding situation at the rear and the front. All four were civilians when they joined the struggle; they underwent basic military training afterward (figure 6.3) (Huni 2013c, 2013g).

As head of ZANLA's health department, Herbert Ushewokunze did not rely on allopathic medicines alone. On the contrary, one of his assistants at Parirenyatwa was a herbalist named George Muchemeyi, who was killed during the Chimoio bombing on November 23, 1977 (Mutanda 2015d). To be clear, these herbalists were usually spirit mediums and healers whose task went far beyond treating the body to preventing the causes of injury or affliction. Said Tendai Zvichapera, who survived the Rhodesian raid: "A number of *zvapungu* (bateleur eagles) prevented enemy warplanes ('birds') from bombing Chimoio armoury in 1977. . . . On the battle fronts, combatants got war ethos and chaplaincy from mhondoro" (Ruvando 2015). Adds former ZANLA political

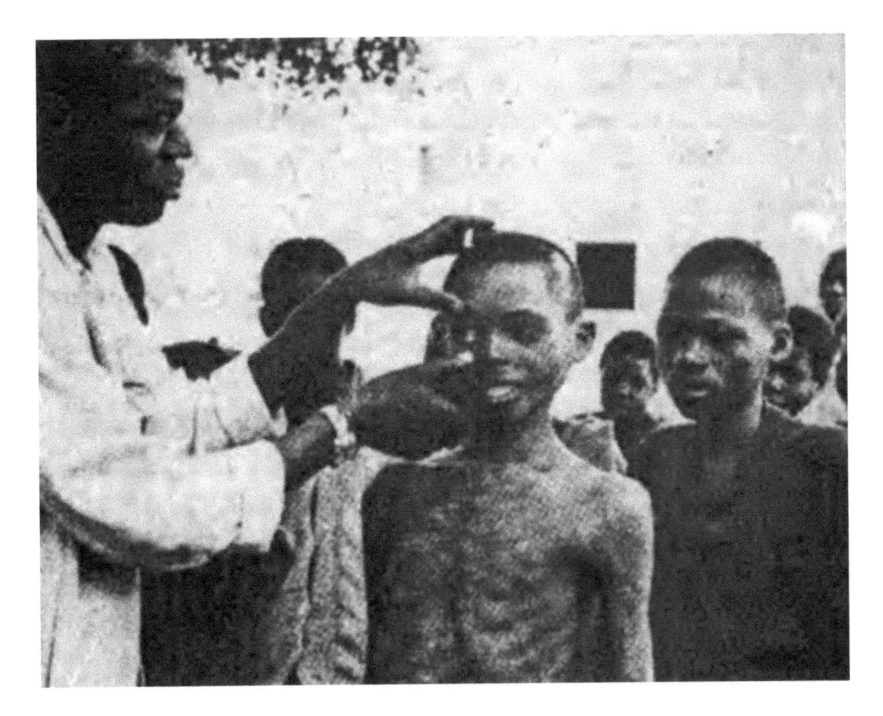

FIGURE 6.3
One of the medical doctors (who looks like Dr. Ushewokunze) busy at work at Parirenyatwa. *Source: Zimbabwe Not Rhodesia.*

commissar Wilfred Mhanda (chimurenga name Dzinashe Machingura): "I had personally worked with very senior spirit mediums from 1975 . . . in Zambia . . . and enjoyed their full respect. We had fighters in our camps . . . who were also spirit mediums" (Karsholm 2006, 7–25) ZANLA's operational headquarters at Chimoio was named after Nehanda, the great spirit of the 1896–1897 chimurenga, to whom the anthem "*Nehanda Dzika Mudzimu*" (Nehanda descend your spirit) was dedicated (Ruvando 2015).

Like ZPRA, ZANLA trained its own medical corps. At the end of the six-month basic military training of recruits, candidates that had excelled in first aid exercises were selected for medic training. The syllabus covered the treatment or containment of gunshot and shrapnel-inflicted wounds and diseases associated with crowded conditions and common to villagers in the Zimbabwean countryside. It was standard practice to recall personnel from the front (at all ranks) for advanced medical training in Eastern Europe. This was particularly so in the last two years of the war as it became essential to acquire capabilities to roll out a public healthcare system from Zambian and Mozambican rear bases into the newly liberated zones on the borderlands (Huni 2013d, 2013a, 2013b).

The medical corps was represented at all levels of command: the section, platoon, detachment, sector, the general staff, high command, and in the revolutionary council and ZANU. At every level there was a commander (in overall charge of operations), a commissar (political reorientation of the rural population), a quartermaster (logistics), and a medic (all health matters). A detachment medical officer was in charge of about five hundred medical details distributed to various sectors, battalions, and platoons. In a platoon, each medic moved with his AK-47 rifle, a *bhandreya* (bandolier) or magazines, and a first aid kit for his troop. He answered to the platoon commander (Huni 2013c, 2013d).

The medics at the front were usually recalled to the rear for refresher courses after serving one or two deployments. Their stints in the field enabled them to contribute their practical experience to the organization's training. They brought back important empirical evidence on what was new, what was working, what was going wrong, and what needed to be changed and how. Thus ZANLA met the Rhodesian use of weapons of mass destruction with a three-month training conducted in Maputo aimed at managing the effects of napalm, toxic chemicals, anthrax, and cholera.

Versatility was required of medics because situations changed suddenly in war. Look no further than Texen Chidhakwa, who one day found himself undertaking a very delicate surgery after his boss Sekeramayi was summoned to Nehanda by ZANLA commander Josiah Magama Tongogara. All along he had been assisting the physician as he operated on a guerrilla who had a piece of shrapnel lodged in one of his testicles. "You get hold of the testicle," the quietly spoken doctor told his medic. "You put pressure over the foreign body, don't tamper with the inside tissues, cut a good opening, get hold of the foreign body, manoeuvre it out slowly, don't disturb the surrounding tissues, scoop the debris surrounding the foreign body, clean it up with anti-septic and then finish the operation as we always do" (Huni 2013d). "We learnt on the job," Chidhakwa would say three decades after independence. "I was confident that I would succeed and indeed it was successful. That guerrilla is still alive today. That's when I got the nickname, the Bush Doctor" (Huni 2013d).

* * *

"To this day we live with men, women, children with all four limbs gone, two gone, one gone. Some people are just vegetating," said Herbert Ushewokunze in his appeal for medical aid in Durham, North Carolina in 1977 ("The Grim Reality of War" 1978). He told his US audience of people who could not be operated on because shrapnel, bullets, and other foreign bodies were lodged in very delicate areas of their bodies. If they were operated on, they could die. All such comrades with serious physical and mental disabilities were accommodated at Percy Ntini Rehabilitation Centre (named after the ZANU nationalist) and at Mupata Wegwenya near Katandika (Huni 2013d; Kadungure 2014).

Recalls Texen Chidhakwa: "Percy Ntini was a special area where physically fit combatants were not allowed to visit because you could see comrades without limbs and . . . with severe injuries from the war" (Huni 2013d). The leadership feared that seeing their debilitated colleagues in such a state might demoralize the guerrillas to the detriment of the struggle. Depression during the war was so serious that some guerrillas deliberately requested to go to the war front as a way of committing suicide (Huni 2013c). Especially as the war escalated, the demands of Percy Ntini for medical supplies like prosthetics and wheelchairs also increased to crisis level.

While many comrades were injured inside Zimbabwe in combat, some at Percy Ntini were casualties of the massive bombing raids at Nyadzonia (August 9, 1976) and Chimoio (November 23, 1977). Increasingly, Rhodesia was using Frantan-packed cluster bombs. Again Ushewokunze at Durham describing the devastating aftermath of the Rhodesian attack on Parirenyatwa Hospital:

> One of our biggest medical centers was at Chimoio, a hundred-bed hospital. In it 25 were incinerated, including 15 nurses who were trying to evacuate patients. Our whole health transport system was disrupted, the self-reliance trucks and cars sent to us by support committees, the only mobile clinic we had. Our operating theater-on-wheels, which was on its way to a nearby town and carrying big red crosses on it, with eight patients inside, was bombed by the enemy. The eight patients died plus three nurses who were accompanying them. All our libraries, both medical and educational, were completely destroyed. ("The Grim Reality of War" 1978)

Anybody who inhaled the napalm fumes bled heavily from the nose and mouth. Few survived (Huni 2013e). The Rhodesians usually started the bombing upwind so that the wind drift carried the napalm fumes toward the target to induce sleep and asphyxiation. The guerrillas thus became too weak to return fire or flee. Some were found with the whole body charred, the breathing nose the only sign of life. A guerrilla base became "a burning furnace" within minutes, napalm cooking life out of human bodies, slowly and painfully sucking the life out of them (Huni 2013c, 2013g, 2013d). Ushewokunze showed his Durham audience another picture from Chimoio: "This is wood covered by napalm porridge! Imagine somebody dipping his arm in a jar of sulfuric acid—the pain, the agony, the disintegration of tissue, the death. If that porridge lands on you and you try to wipe it away, you're in trouble—your hand falls off, gets eaten away" ("The Grim Reality of War" 1978).

Anthrax and cholera were also introduced just upstream of and directly on rivers than supplying camps. At the onset of the first rains (October–November), the chemicals

FIGURE 6.4
Chibabava Camp in 1979. *Source*: Zimbabwe Solidarity Committee 1979.

dissolved into and contaminated runoff and ground water, the soil, plants, and the entire food chain, with deadly consequences. No wonder why strange diseases began to occur. At Chibabava, ZANLA's refugee or holding camp in the Gaza province of southern Mozambique, there was "the hurricane." A person with this illness suffered "weakness of the knee joints and moved like a chameleon—take a step forward, hesitate midway before completing the step." It had no known cure and left no scars (figure 6.4) (Dhliwayo 2012, 58, 73, 76; Mutambara 2014, 51).

Chibabava was notorious for another reason: jigger flea. The insect had arrived from Brazil on Portuguese ships in the 1870s. The inmates called them *zvitekenya* or *zvimatekenye* (painful ticklers). ZANLA military instructor Agrippah Mutambara (aka Dragon Patiripakashata) lived at Chibabava in 1975:

> The females of these tiny wingless insects burrow into the skin, causing painful sores. The usual entry point . . . is the area between the toenails or fingernails and the soft skin beneath. . . . To take them out soon after entry using a pin causes a lot of pain. Some comrades advised that it was better to leave them in one's skin and wait for them to come of their own volition. Heeding such advice was the most terrible blunder one could ever make. Inside one's foot or finger the jigger flea would create a sacksful of eggs that would hatch into tiny fleas. When hatched, twenty or more young ones would seek freedom by leaving their birthplace and coming out of the skin, but only to burrow their way back into the foot or fingers independently and in many different places. Many comrades lost their toes and were permanently maimed because of these jigger fleas. These tiny creatures thrive and multiply in sandy and dirty conditions and the effective way to curb their multiplication was to keep the floors of the barracks and the surrounding ground wet. For many comrades, that lesson came too late. (Mutambara 2014, 51; also Dhliwayo 2012, 76)

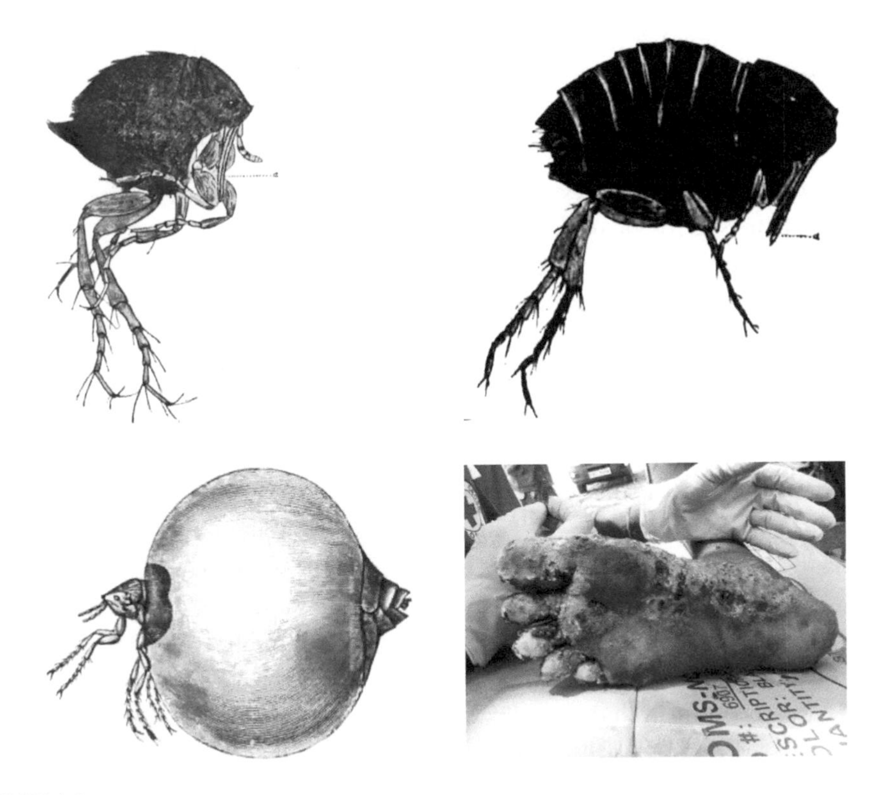

FIGURE 6.5
Top: Illustrations from microphotographs of male and female jigger flea specimens. *Source*: Cooper 1900–1901, 659. Bottom left: A newly hatched jigger leaving the sac, soon to feed on flesh around it. *Source*: British Museum of Natural History 1909. Bottom right: The damage caused by jigger flea on human feet. *Source*: R. Schuster, Wikimedia Commons, 2013.

An adult jigger flea was about a millimeter long. The pregnant flea would enter the skin under the toenails, lay hundreds of eggs in a small sack, which in turn developed into a white, pea-sized structure, up to 10 mm in diameter. Other problems like malaria, diarrhea, cholera, and so on were virulent, but nothing beats napalm and *zvitekenya* in the war memories of refugees and guerrillas who survived.

MEDICAL SUPPLIES TO ZPRA AND ZANLA

Guerrilla healthcare defies the conventional association of the African guerrilla movement with the Sino-Soviet blocs. To set up healthcare infrastructure in Zambia, Mozambique, and liberated areas required medical supplies. The communist countries supplied guns and military training but limited medical assistance materially. Many groups throughout Western Europe and North America filled this critical void. The assistance included building hospitals and clinics in Tanzania, Zambia, and Mozambique, medical equipment (ambulance vehicles, surgical equipment, microscopes, stethoscopes, crutches, first aid kits, stretchers, and syringes), and medical supplies (drugs principally), tents, water pumps and filters, kerosene-powered refrigerators, and so on (ZIMA 1979, 2–3).

At least in the context of the United States, Canada, and much of Western Europe, where government and the media characterized the liberation movements as "communist terrorists," blacks and progressive organizations saw "freedom fighters" waging a just cause against fascist and racist oppression. Sweden in particular openly and actively assisted ZANU and ZAPU with "humanitarian" resources. In large part, the capacity of NGOs to raise medical and other nonlethal resources for the same guerrillas their governments criminalized or did not support owes to the work of ZANU and ZAPU's diplomatic efforts. In ZAPU it was George T. G. Silundika, Kotso Dube, and Calistus Ndlovu. ZANU had the likes of Reward Kangai, Tapson Mawere, Sydney Sekeramayi, and later Herbert Ushewokunze. Often these Zimbabwean representatives started out as students in North American and European universities, organizing campus protests, fundraising events, and chapters of "liberation support movements." It was to such forums that heads of specific departments within the two liberation movements were invited to articulate the aims, needs, and progress of the struggle.

By the end of the war, some of the European support groups raising medical supplies and shipping them to ZAPU and ZANU included three British-based organizations, International Defence and Aid, War on Want, and the Anti-Apartheid Movement's Health Committee. The organizations from the Netherlands included Dutch Anti-Apartheid, Holland Committee for Southern Africa, and Medical Committee of Angola. The Swedish government offered direct assistance through its agency Swedish International Development Agency. Other organizations in Norway, Belgium, Germany, France, and Canada also gave medical assistance on a smaller scale (ZIMA 1979, 2–3).

ZAPU and ZANU submitted their requirements to these organizations, emphasizing the numbers of refugees in each camp, initially just those outside Zimbabwe, then, as war intensified, Africans inside "liberated zones" as well. Says one memo commenting on a ZANU (PF) request presented at a meeting of medical NGOs in Leiden in June 1979:

> Medicines list remains large, [only] trade names [are] used, and [the list is] really a pharmacopeia. It looks as if parts of drug company lists in MIMS and some of the British National Formulary have been combined randomly. Contains dangerous and out of date drugs (e.g.

> Mersalyl—a mercury diuretic!—discontinued in Britain 15 years ago as it poisons the kidney). Dangerous reflection of drug company imperialism in Africa generally. (ZIMA 1979, 4)

Also attending the same medical indaba was Benjamin Dube of ZAPU. The list he presented is captured in the minutes:

> ZAPU list is less serious than [ZANU's] in that it is shorter but suffers the same regretable (sic) pharmacopoeia (sic) problem. No doubt of ZAPU's major needs, especially due to ZAPU's very own limited supplies. Both PF wings emphasize rehabilitation needs etc. Clear list of bedding, dried foodstuffs and clothing plus a summary of projects. (ZIMA 1979, 4)

* * *

In the United States, several organizations were composed mostly of African Americans, churches, students, and other "progressive" forces, fundraising and collecting medical supplies, principally—if not exclusively—for ZANU (ZMD 1978–1979). As often as they could, ZAPU and ZANU heads of health departments traveled overseas to make this argument and solicit materials. Money and supplies were raised through all manner of activities, from fundraising dinners, garage sales, monthly pledges, and blanket and clothing drives to benefit concerts. Among the artists involved in the concerts were African musicians in exile (like the Zimbabwean mbira maestro Dumi Maraire, South African jazz gurus Hugh Masekela and Abdullah Ibrahim, aka Dollar Brand) and Jamaican outfits, including Bob Marley and the Wailers. Some of these shows were endorsed by prominent antiapartheid supporters, like the boxer Muhammed Ali, singer Harry Belafonte, and Rev. Wyatt T. Walker (Martin Luther King Jr.'s top aide) (ZMD 1978; 1978–1979; 1979). The Champaign-Urbana Coalition Against Apartheid was one of many groups that collected recreational materials, clothing, blankets, tents, and school supplies for ZANU. As the war intensified, they also prioritized "antimalarials, antibiotics, cardiovascular medicines, and painkillers, and such medical equipment as microscopes, water purifiers, prosthetic limbs, mosquito nets, detergents, [and] disinfectants" (C-U Coalition 1977).

In their appeals for donations the organizations skillfully deployed the stories of individual contributors to inspire those who wished to make a difference:

> Medical students have donated their stethoscopes to ZANU. An owner of a drug supply house donated several thousand dollars worth [of] drugs. A hospital that was closing donated hospital beds to ZANU rather than have them sold for scrap. Retired doctors and dentists can donate equipment. Many examples of this type show that a little hunting can locate a good deal of medical material. . . . Cake sales, rummage sales, entertainment events, cultural shows are all well-used techniques for raising money. (ZSC 1979, 16)

Every appeal was quantified to assure a statistics- and accountability-sensitive US public that the donation would be used "solely to buy medicines and medical equipment."

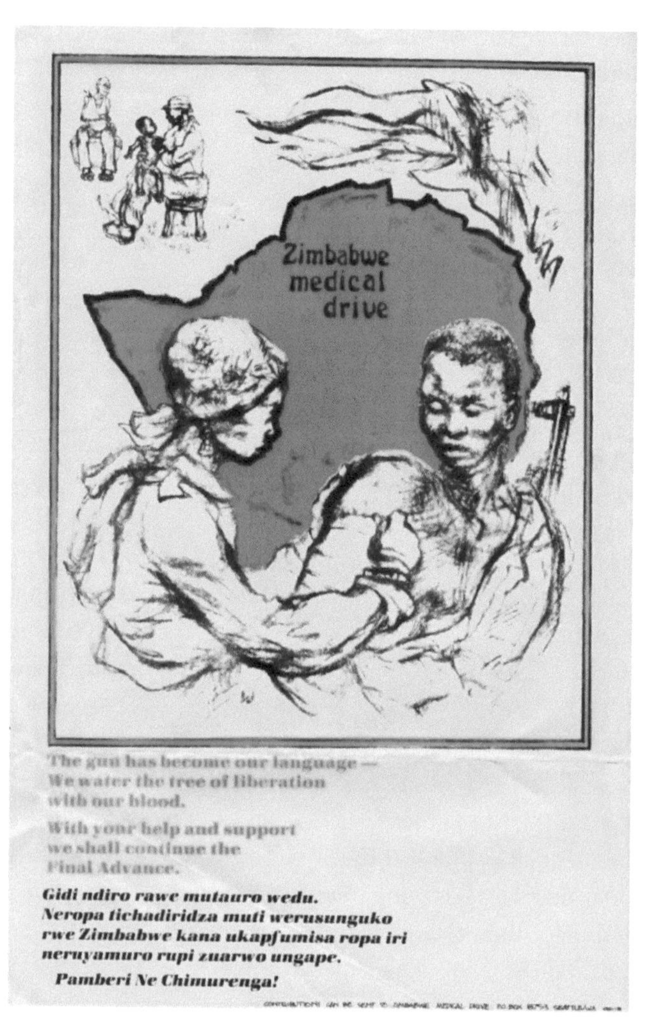

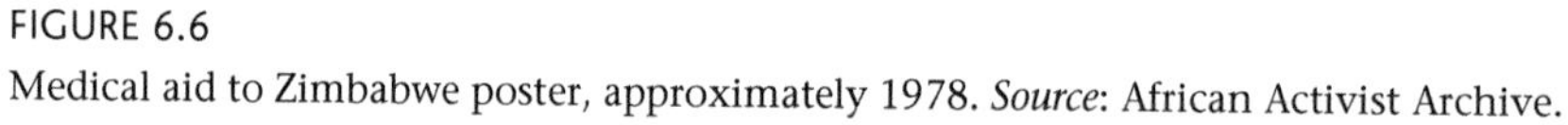

FIGURE 6.6
Medical aid to Zimbabwe poster, approximately 1978. *Source*: African Activist Archive.

Medical Aid to Zimbabwe, a Bronx-based organization of health workers and community activists raising and supplying material aid to the liberation struggle, used this strategy to perfection. Here is their poster in 1979 (figure 6.6).

GUERRILLA HEALTHCARE ON THE FRONT

Since the advent of colonial rule, Africans had grossly inadequate or no access to allopathic healthcare facilities. For example, in 1976, the state provided one hospital bed for every 255 whites compared to one for every 1,261 Africans (Kriger 1992, 62). Whites enjoyed "state of the art care at a ratio of one doctor for every 1800 people" compared to "one doctor [for] no less than 100,000 Africans" (ZSC 1976). By 1979, the official policy was to shut down rural district hospitals that served Africans, leaving only mission hospitals or nothing at all. Already, by 1976, the missions provided over 75 percent of healthcare to Africans in the rural areas, compared to government's allocation of just 9 percent of the total health budget to rural Africans (Kriger 1992, 62).

Then there was Rhodesia's chemical and biological warfare, launched in 1974. Chemicals that had been intended for problem insects, birds, and animals were extended to kill a new problem animal: "the terrorists" and villagers "running with terrorists." One was thallium sulfate, a tasteless pesticide with delayed toxicity, so that a guerrilla might eat a meal laced with it in one place and travel far, before the effect kicked in two or three days later. Organophosphates like fenthion and parathion, and the organochloride isobenzan or telodrin, killed through skin contact; many guerrillas died after wearing denims received from trusted businessmen (Mavhunga 2011, 166–168; Mutambara 2014, 202). If the guerrilla died in the bush, the vultures, eagles, hyenas, jackals, and lions that ate his remains also died. The poisons affected not just the "terrorists"; children too were dying after eating candy, wild fruits, and tinned foods left as bait to kill "terrorists" (Daneel 1995, 9).

The Rhodesians also extensively used biological materials as poisons. Guerrillas suffered from Rhodesia's use of cholera bacterium, which induced acute diarrhea and death. The Selous Scouts planted the cholera in water reservoirs. The guerrilla leaving for the front or to conduct other errands then spread the bacterium over a wide arc (Parker 2006, 170–171; "The Grim Reality of War" 1978). No cholera outbreaks had occurred in the country from 1890 to 1974, but since then it had become a constant problem, when the Rhodesians began to systematically "pollute the wells and water supplies with cholera and typhoid . . . then rush back to vaccinate the whites in case the thing boomerangs" ("The Grim Reality of War" 1978; ZSC 1979). Thus by late 1979, anthrax had affected one-third of Rhodesia, with about eleven thousand cases in 1978 alone. Pools and pastures were contaminated with the deadly spores to kill African villagers and their livestock to deny guerrillas a protein diet, and on filter caps of cigarettes to kill the "terrorists" individually (Nass 1992, 1992–1993; Sterne 1967; Lawrence, Foggin, and Norval 1980; Davies 1982, 1983; Parker 2006, 176).

In a PBS *Frontline* interview in 1998, Zimbabwe's former health minister Timothy Stamps admitted that the Rhodesians, colluding with South Africa and other Western allies he did not name, transformed the guerrilla-infested countryside into a testing ground for weaponized Ebola and Marburg viruses, which are not endemic to Zimbabwe. Just like cholera, the Marburg virus outbreaks were focused, not generalized, in an area of intense fighting between the South African army and Mkonto weSizwe guerrillas. The

Ebola outbreak was limited to the Zambezi River, leading Stamps to conclude that it was an experiment to test whether the newly developed virus was capable of directly infecting black people ("Interview: Dr. Timothy Stamps" 1998; Siamonga 2014).

* * *

On the front, the *mhondoro* and *n'anga/izinyanga*, mission hospitals, African physicians, and the *mujibha/chimbwido* were the four major factors in guerrilla healthcare. The rural majority believed in ancestral spirits, and most guerrillas were rural people despite their secular military and ideological training. Many had been born and raised in these very lands, worshipping Mwari (God) via the ancestral spirits (Lan 1985, 5).

Shrines, caves, pools, trees, forests—all these were therefore available to the guerrillas as sacred associates and signs of the ever-presence of the ancestors. Caves had served traditionally as burial places of chiefs, bunkers where entire communities retreated and fought from during the first chimurenga of 1896–1897. The injured would be carried there and the healers would work on them until recovery. Or death. Now the comrades did the same. Certain pools and springs were sacred, believed home to *njuzu* (mermaids) and *majukwa* (water spirits), and sources of potent underworld medicines. There were sacred forest groves where long-departed ancestors were buried, where nobody was allowed to venture, cut trees, or burn grass. The trees grew into thick forests, accessible only to spirit mediums, elders, and healers. Injured or sick comrades were now treated and nursed there. And some of the guerrillas were spirit mediums.

The cave became the guerrilla hospital; here, allopathic and indigenous medicine met. It was too dangerous to keep an injured comrade in one's house; sooner or later word would go out. Traditionally, shrines like the *muchakata* or *muhacha* tree are sites where people make beer offerings to *makombwe* (rainmaking spirits). Now guerrillas went there to ask *makombwe* to create the right atmosphere to enable the therapeutic process of chimurenga, which would heal the land of the scourge of oppression, to succeed.

Guerrilla healthcare infrastructure was thus not always an outcome of human modification or fabrication, but people's strategic deployment vis-á-vis the available mountains, caves, and so on as weaponizable material in the service of chimurenga. *By strategic deployment in this very specific Zimbabwean context I mean the way people use their intellect to locate themselves in the environment in such a way that it works advantageously for them without significant or any modification. One example is turning a mountain into a weapon against the enemy by deploying and fighting from inside, under, in front, or behind it. Under—because in chimurenga, infrastructure ceases to be just terrestrial, human made, or material and physical.* Through *njuzu*, the medicines to treat wounded comrades could be obtained from underneath the pool's surface, in the pale blue underworld whose "subterranean villages and cities" stretched into the horizon, its pastures teeming with cattle (Daneel 1995, 157). I seek the practitioners' reasons, not ours; that's how one could read this view of medicine.

* * *

The ZANLA infiltration of Dande in 1971–1972 set the tone for the role of ancestral spirits and spirit mediums throughout the war. The area was home to numerous powerful mediums—Chiwawa, Musuma, Mutota, Madzomba, Chipfene, Chiodzamamera, Chidyamauyu, and Nehanda. The guerrillas sought them out to gain legitimacy among the population, which firmly believed in ancestral spirits; to access the safest, most secret routes and sites for weapons movement and caching, usually located in sacred places, which were sources of herbal medicines known only to spirit mediums and healers; and to obtain war medicines not just for wounds and ailments but, as per *dzimbahwe* traditions, sacred and spiritual medicines for protection against misfortune and evil spirits as well. For allopathic substances and herbs to work—to be medicine—the spiritual and psychological conditions had to be well aligned, all misfortunes that shut out the medicative elements from working cast away. Then, and only then, could the academy-trained physicians' medication work.

As physicians or *n'anga/izinyanga* coming to cure the land of the disease of white minority oppression, guerrillas' primary medical tool, the gun, could only be efficacious if armed by the ancestral spirits. Thus the *mhondoro* became privy to the minutest of guerrilla operational details, including their state of wellness, routes, and modus operandi. That is why in 1972, the medium of Nehanda, now very old, had to be stretchered out of the northern district to a ZANLA base in Mozambique. The guerrilla commanders—at least most of them—also believed that the departed ancestors communicated battle directives to the fighters through spirit mediums, birds, and animals (Lan 1985, 5; Daneel 1995).

The African population in the operational zones generally deferred to their indigenous healthcare system composed of ancestral spirits, *n'anga/izinyanga*, and apostolic prophets, who not only healed the medical conditions (a wound, a disease), but also gave spiritual healing and protection to the community and kin within it. Hence leaders like Joshua Nkomo, a Methodist immersed in the faith of his Kalanga ancestors, implored guerrillas to obey *imithetho yabadala* (isindebele) or *mirau yavakuru* (ways of the elders). To obey the ancestors was itself a self-medicating move: the spirits used animals and dreams as vehicles to warn and guide the fighter in times of danger (Damasane 2014).

ZPRA guerrillas regularly consulted the ancestral spirit mediums and *izinyanga* to heal them of afflictions, injuries, and misfortunes. Common conditions treated included persistent headaches and stomachaches (especially from poisoning), injuries, dizziness, fits, bomb shock, and possession by evil spirits. The guerrillas took part in cleansing and propitiation rituals at shrines whenever required by the spirit mediums in their tactical areas of operation. During these rituals, especially when just arriving into an area, the guerrillas would be required to place their weapons in the hands of the healers or spirit mediums as part of the process of informing the ancestral spirits of their arrival. The reason: so that these tools and those who carried them could receive spiritual armament and lethality during the firefight against the enemy. Thus informed, the ancestors would throw a protective shield around the community and guerrillas, that they be insulated from enemy discovery, bombardment, and death (Alexander, McGregor, and Ranger 2000; Brickhill 1995).

That was the medicine needed to cure ailing operations. Jairos Muumbe was a *mujibha* in Muumbe village in the Chief Musikavanhu area of Chipinge:

> When the boys came into the Musikavanhu area, the spirit mediums went with them into a *ngòmé* (the sacred hut) for the ancestral spirits to help the comrades access the area. For some time, every time there was a contact, a comrade would lose his/her life. Thus, the Chief and the spirit mediums led in the propitiation rites alongside the comrades. From that day on, nothing of that sort happened. (Mutanda 2015a; Lan 1985, 4)

The ZANLA guerrilla Solomon Ndunduma (chimurenga name Jeffrey Muridzo), operating in Buhera, says that before the propitiation procedures "mine detectors easily swept away the mines put up by the guerrillas. . . . Afterwards landmines did the business of sabotaging the economy and killing the enemy." He confirms that things turned after his section heeded Chief Makumbe's instruction to "avail a black goat, black and white clothes," and traditional beer offerings to the ancestors. The landmines they planted began to blast the enemy's vehicles in earnest (Mutanda 2015c).

As the guerrillas testify, the spirits worked in different ways. Can See Mapuranga's statement echoes in hundreds of guerrilla accounts: "At times, we got communication from eagles. If they fought each other, a fight was imminent. Definitely, our spirits and indeed our ancestors helped us get our independence" (Mutanda 2015b). Sydney Mukwenje emphasizes the critical role of spiritualized materials as medicines and weapons in chimurenga: "We also had snuff [*buté*] and *chifumuro* (a herbal tuber) in our pockets which we continuously rubbed onto our [fore]heads which according to tradition could ward off whatever evil intentions the soldiers had" (Mukwenje 2013). He negates to say *chifumuro* means "the exposer"; its purpose is to expose the adversary's hidden plans. Marijuana was also used for a similar purpose, in addition to galvanizing its smoker into doing fearless and daring things. The male guerrillas and refugees in particular were frequently punished for venturing out to source it from surrounding Mozambican villages

(Dhliwayo 2012, 79). *Buté*, by contrast, is the ancestors' all-purpose medicine and weapon against afflictions of any kind, including bad luck.

The Rhodesian Security Forces' *Soldier's Handbook on Shona Customs* (1975)—the field manual every soldier had to master as part of the counterinsurgency operations—also acknowledged the widespread use of charms among Africans as follows:

> Almost every African wears a charm, either round the neck, round the wrist, or [keeps it] in the pocket. These are usually very small, wrapped in cloth, worn on a string. Tiny beads are also worn, again round [the] neck, waist or arm. One such charm or "medicine" is *mangoromera* . . . the skin or armband which has python skin or crocodile gall, etc., stitched into it. It is a charm vesting super human strength, courage or audacity in the wearer. It was, and may be even still, the prized magic of the boxer, criminal and tough guy of the urban areas and would be ideal for the terrorists. (Rhodesian Army 1975, 40–1)

As can be seen, the Rhodesians recognized that "the terrorists," regardless of their secular training in communist countries, remained steeped in their own idioms and extended them to their personal health and well-being and the conduct of insurgency.

This pragmatic healthcare philosophy is apparent in *Guerrilla Snuff*, wherein Martinus Daneel tells the story of the *svikiro* Lydia Chabata who, possessed by *shavé renjuzu* (mermaid spirit), went underneath the pools to search for special, powerful medicines to cure the guerrilla commander, Weeds Chakarakata (aka Cosmas Gonese), of poisoning. Chabata nursed the commander for over two months in a cave working with Chakarakata's medic to treat him, carrying him from cave to cave in the Bikita area, where proguerrilla physicians would sneak in to examine and treat the patient. The commander was suffering from poisoning—the festering sores all over his head and body having caused his hair to fall out, muscles to shrink, and eyes to sink into their sockets. He was a moving skeleton (Daneel 1995). Cosmas Gonese recovered and saw independence, and thanks the ancestral spirits, the *njuzu* medicines, and allopathic treatment in equal measure.

* * *

So far we have seen *mhondoro* and *n'anga*/*izinyanga* as vital cogs in ZANLA and ZPRA's war fronts. The third element is the mission hospital. By and large, ZPRA's relationship with the church was strained to say the least because of the issues that fall outside this book. The literature on the church's role in chimurenga is rich, but it does not deal with healthcare in any detailed way (Bhebe 1999; McLaughlin 1996). Thus at present, our evidence for the guerrilla initiative to creatively subvert the church into a critical pharmacy or dispensary for drugs and other medical supplies comes from the ZANLA front. The story, however, should not begin there.

By remaining open, the mission hospital thus became available to guerrilla initiative as an important healthcare infrastructure. Guerrilla "MOs" (medical officers) replenished medical supplies at mission hospitals and clinics and collected "donations" from teachers toward the struggle. In Gutu, for example, the priests of the Catholic Order of the Bethlehem Fathers at Mukaro and Mutero were not just replenishing the guerrillas' medical supplies, but also treating them when injured or ill (Daneel 1995, 38). A summary of this relationship between guerrillas and the Catholic priests in particular is contained in Bishop Donal Lamont's *Speech from the Dock* (1977, 68–70):

> I drove from Regina Coeli Mission to Avila in Inyanga [sic, Nyanga] North on 21st April [1976], and shortly after arrival was informed that a letter had been handed into the Mission requesting medicines. I was shown the note [which] asked for anti-malarial tablets and medicines for diarrhea. . . . The letter had been delivered by a villager, a man, on behalf of [the guerrillas]. I was asked what ought to be done about it, and I replied that we ought to give medical aid to anyone who asked, and that the nurses should not argue about the matter. . . . As far as medical help was concerned, no missionary should inquire about the religion or politics of those who asked for help. . . . If the Security Forces came looking for medicines, they too were to be given whatever they needed and whatever the Mission could afford to give . . . The second decisive motive was the safety of the personnel involved. What would happen, then, were they, few in number, without any means of defence whatever, without even a telephone closer than a two-hour journey—what would these helpless people do were they to refuse things demanded of them? . . . If our missionaries deny the medicines, there is nothing to prevent armed men from invading the Mission at any hour of the day or night and forcibly taking what they want. (Lamont 1977, 68–70)

Hence Lamont's act of "charity" also derived from fear of the consequences of refusing to help the guerrillas, a desire to preserve the lives of church staff and property because of missions' isolation in the war-torn countryside. The mission in question, Avila, was one of several isolated Catholic stations in the diocese of Mashonaland.

There were also Africans trained as medical doctors who were inside Zimbabwe who lent critical services to the struggle. When guerrillas were injured, these physicians would go into the bush to attend to them, and then direct that they be medevaced to caves where they would come and treat them. I do not yet have evidence for the ZAPU/ZPRA-aligned physicians, but there were three prominent ZANLA-aligned medical doctors rendering assistance to guerrillas inside Zimbabwe in 1976–1978. Drs. Simon Mazorodze and Oliver Munyaradzi had both trained in medicine at University of Natal Medical School. Stationed in Masvingo, they would sneak out to treat injured or sick comrades in the caves of Shurugwi, Zaka, Mwenezi, and Masvingo. Edward Munatsireyi Pswarayi was also helping out guerrillas in Zaka and Bikita. Pswarayi had graduated with a degree in medicine from the University of the Witwatersrand in 1956, just after Samuel Tichafa

Parirenyatwa. Ex-ZANLA guerrilla field commander Henry Muchena, later an air marshal in postindependence Zimbabwe's army, remembers all three physicians well: "I first met [them] in 1976 in Machingambi-Harava in Zaka [and again a year later in Bikita] when Pswarayi came to the battle front to deliver medical supplies. A number of our comrades had been injured during combat and the late Simon Mazorodze brought Pswarayi and the late Oliver Munyaradzi and would treat the wounded comrades" (Herald Reporter 2014).

SUICIDE STAGE AND GUERRILLA MEDEVAC

"Suicide stage" is the circumstance at which an injured or sick guerrilla could be left behind, when he "volunteer[ed] to be left behind to die." As ex-guerrilla Texen Chidhakwa explains, it was a stage "where you see that a fellow guerrilla has lost both his legs and an arm but they are bleeding heavily." There was nowhere to get an ambulance or stretcher bed to medevac the guerrilla to hospital—in fact, no hospital even existed anywhere near! "We would make the painful decision that 'comrade *vava kutosara pano*' [comrade is now going to be left here]." The medic first certified that the guerrilla was dying, then took away his gun, and left him with a grenade to blow himself up instead of being captured and tortured. At least suicide was a faster way to die than torture (Huni 2013d).

Usually the guerrilla would grab his grenade, remove the firing pin (but not throw it away), and wait for the enemy to come. Once the guerrilla removed the safety pin, a grenade became a self-firing explosive, and when it exploded, it disintegrated, the shrapnel tearing the guerrilla into pieces. A grenade, once fired, was no longer usable or there. By contrast an AK-47, an RPG-7, or a mortar could be taken by the enemy and lost to living comrades. Then, as Rhodesian soldiers (which the people called *masoja* or *masotsha*) closed in, the guerrilla let go. The rule was "*unofa nemurungu*" (you die with a white man) rather than be captured, tortured—then sell your comrades and chimurenga out—and still get hanged afterward anyway (Huni 2013d). The grenade was a way for a comrade to die fighting and with honor; it left no trace useful to the enemy.

Sometimes the medic pronounced "suicide stage" too quickly, leaving a guerrilla for dead when he was alive. Two cases illustrate this. The first is that of ZPRA regional commander (southern Matabeleland) Comrade Thadeus Parks Ndlovu, narrating the events of August 1978:

> We went back to Botswana where we used to get our [ammunition] supplies. Coming back . . . we made a contact . . . around half past four on the 8th of August. . . . After some time I thought, now I'm failing to handle the gun, why? I said "Hey, comrades I've been hit." I tried to skirmish but I saw something which was unusual to me . . . those were intestines. . . . Then my intestines came out. I had to push them back in again. I said to my comrades, "Comrades, . . .

> I don't think I will proceed. What you can do now, take out the shoes." Ah luckily . . . Boers [Rhodesians] . . . assaulted that area. . . . I don't know how they by-passed me. It was . . . very lucky I survived that. I went to sleep somewhere. In the morning . . . a villager took me to another village. . . . There was this home-brewed beer. There were a lot of people there. . . . They took me to a house. They tried to make me some porridge. . . . Then the headman took his new bicycle, gave it to another young boy: "Take this man." And then I went . . . back to . . . a hospital [at Phikwe]. (Ndlovu and Nkomo 2010)

For ZANLA guerrilla Onias Garikai Bhosha (chimurenga name George Gabarinocheka), "suicide stage" was when his fingers were blown off by a bomb *masoja* had put in a transistor radio, which detonated as soon as one member of the guerrilla section turned it on. This radio had come from a local businessman in the Kaitano area of Rushinga named Nyamupfukudza, who said it was a gift. Two guerrillas, Tafirenyika and Pedzisai Madyiwa, died on the spot; Gabarinocheka survived only because he was a few meters away, but the shrapnel perforated his face and head and tore into his fingers. The surviving guerrillas concluded he was dead, buried him under tree branches, and proceeded with the struggle for self-liberation.

Gabarinocheka regained consciousness three days later, looked around, and started crawling, his sight partially blinded by the blast, hitting against trees and rocks. Afterward he managed to get to Kaitano School and asked the schoolchildren to go and tell their elders to come and fetch him. They then informed the guerrillas, who took him to one of the local ZANLA bases called Mukoma, and subsequently medevaced him to Chifombo in Zambia. In this case, medevac was also self-evacuation, even in extreme pain, with shattered leg or bullets lodged inside one's body. This was usually the only option when one was left for dead by his comrades (Huni 2012b, 2013d; Sunday News Reporter 2014).

Whenever possible, injured comrades were stretchered out of the combat zone to mountains or rear bases. That did not always happen, for example, when faced with "the practical problem of having two stretcher cases, one with a fractured leg and the other with a swollen ankle and suspected fracture of the collar bone—in addition to the punishing weight of the munitions" (Mutambara 2014, 105). The only option was to take up positions, administer first aid, wait until it got dark, and then medevac the comrades to the rear. "This make-shift stretcher was also our ambulance during the liberation struggle," Chidhakwa remembers (Huni 2013d).

It was made from two straight poles driven into, along, and out of the longer sides of a disused grain sack. A bicycle, where it could be found, acted as a very good "gun carriage." This gun carriage could also be converted into a platform for carrying the stretcher bearing an injured guerrilla. Carrying the injured guerrilla on a makeshift stretcher and under guerrilla escort, the *mujibha* would walk on foot, taking cover in the caves and under trees by day, resuming by night. After crossing the border, they either walked some more or, on the rare occasion, took a ride on the train or bus inside Mozambique or Botswana to the hospital. Sometimes they used donkeys, very rarely motor vehicles, scotch-carts, or wheelbarrows because they left a signature very easy to track. Whenever they used a bicycle, the guerrillas and *mujibhas* would put a tree branch behind the rear wheel to wipe out their footprints and confuse *masoja* as to the direction they had taken (Mwale 2014).

It was sometimes in the course of acting as a field ambulance that *mujibha* and *chimbwido* began their journey to training in Zambia and Mozambique. This was certainly the case of ZANU chairman Herbert Chitepo's bodyguard Sadat Kufamazuva (birth name Bensen Kadzinga). It was in the Rushinga area, in the wake of a gun battle in 1973. The section commander, David Zvinotapira, had been shot in the leg. The surviving guerrillas assigned the then untrained Kadzinga to organize a bicycle from the local villagers, medevac the injured guerrilla to safety, and look after him while they provided armed escort. The pursuing *masoja* caught up with the group, and after another contact, the other two remaining guerrillas gapped, leaving Kadzinga alone with the injured guerrilla. Zvinotapira handed the young Kadzinga his submachine gun and remained with his pistol, and began training the *mujibha* on the move how to fire it. Next he took off his camouflage and remained with civilian clothes, then started walking painstakingly but inauspiciously through a maize field pretending to be a farmer. Meanwhile, his young protégé and evacuator was crawling on the ground. When the two at last got to cover, they stayed at a village near Karanda for three months, before escaping to Zambia when their trail had become cold. Sadat was not the only one (Huni 2012c; 2013e; Chipamaunga 1983).

By the end of 1978, ZANLA in particular had made significant inroads into Zimbabwe and declared large swathes of areas near the Mozambican border "liberated zones." Pswarayi was behind bars. The fates or whereabouts of Mazorodze and Munyaradzi are not yet apparent from available archives. We know that ZANLA was busy setting up mobile health clinics and field hospitals, as described in a pamphlet published around 1979:

> The ZANU guerrillas operating inside Zimbabwe are fighters, teachers and nurses all at the same time. The villagers feed and shelter the fighters while the guerrillas bring medicines, political and practical education to the people.
>
> The fighters travel with portable schools—a blackboard and chalk, pencils and paper. Their classroom is under trees, close to the village. Fighting disease is a common political and health lesson taught in these outdoor classes. . . . The guerrillas instruct the people in ways they can protect themselves from [malaria and bilharzia]. People should build their houses away from the rivers so that their sewage does not seep into the river; they should boil any water they use for cooking. . . . The guerrillas help the villagers build wells that will reach clean water. (ZSC 1979, 13)

That is why, in its request to European donors in June 1979, ZANU specifically requested "relevant texts that could be translated into Shona for distribution in camps." Besides offering material medical supplies, the donors also suggested to ZANU that it should translate parts of David Werner's *Where There Is No Doctor* as a manual for self-sufficiency in times of crisis (ZIMA 1979, 5).

TOWARD A NARRATIVE OF INNOVATION AGAINST THE ODDS: THE END

I have sought to show that the ordinary person is not merely a victim, beneficiary, or bystander in innovation. I have also stressed that in the context of chimurenga, it is impossible to make sense of creativity absent the spiritual.

The Africans we see here are not merely victims of tools of empire or beneficiaries of benevolent nationalist leaders. They are creative beings who die fighting, innovating, and making and triumph in their struggles because they are creative and resilient. We have to take their philosophy of innovation seriously: they persevere because they trust that they are not walking alone, but under the protective wing of their ancestors. And because, in the case of chimurenga and other African struggles, agency does not end in the secular realm but is always spiritually anchored, the secular-based framework that animates analyses of technology does not adequately account for the forces at play. Perhaps technology in such narratives is too secular, too neat, too tekkie. It is impossible to understand what is technological to Africans like these without first understanding the spiritual forces that give context to their mortal or physical/material activities. The most important aspect of science, technology, and innovation in this case is not the science, technology, and innovation itself, but the African spirit of creative resilience born of the specificities of struggles people are enduring. The exploration above centralizes people as the critical agent of history, with intellection and spirituality as critical elements. Technology comes downstream of worldview and the forces people collectively believe to populate everyday life and the way the world works.

III TO RENDER FREEDOM FRUITFUL

7 HEAD, HAND, AND HEART

This chapter will perform two tasks that advance the theme of daring to invent the future through the creation of spaces for and modes of knowledge in the service of and through problem-solving. First, it revisits the Tuskegee model of industrial education through selected writings of its founder and first principal, Booker Taliaferro Washington, with the privilege of hindsight, in search of ideas for the crafting of genuinely community- and people-centered problem-solving knowledge. Second, it explores the impact of Washington on Africa, not so much as evangelical work on his part, but the creativity and sense of daring on the part of Africans who sought out Washington and Tuskegee as an ingredient for creating similar spaces of education as Washington had done.

In both examples I see hope, possibility, and an invitation to dare to invent the future of education beyond bookish knowledge. He inspires those who have an idea and shows them clearly how to set about turning it into a reality with little or almost nothing, in a century where we think that big ideas need big money to become successful start-ups. Tuskegee is also a story about an education model, one where a poor boy could pay fees with labor in place of money, that draws to it talented and determined but poor children.

This chapter does not focus on all of Booker T. Washington, or embrace all of his views, or all of Tuskegee, whether interpreted from the view of the present reading back into history, or as interpreted contemporaneously while he lived. I declare no interest in his comments and the high value he placed on teaching black people how to sit at a table and eat properly, the uses of a napkin, tablecloth, knife and fork, and toothbrush. I do not agree with him when he says "I have no faith in any race, black or white, which has not learned the use of the tooth brush" (Washington 1902b, 73). I certainly do not believe that people in the community are "wholly ignorant" and must be taught "right

FIGURE 7.1
Booker T. Washington. *Source*: Washington 1901b.

living" defined in hubris by myself; on the contrary, I see them as intellectual agents engaged in everyday innovation, and hence partners in knowledge who see from a different experiential and cosmological location than the still-Western-and-white-centric academy (contrary to Washington), which I engage from a view of knowledge informed by relevance to community (like Washington). I have read his writings well, and understand that, for his time, and given his life experiences, Washington made compromises and held certain public positions that I disagree with. I understand them as the price he determined to be strategically necessary to pay in order to provide black people with an education that transformed their lives.

Booker T. Washington is big enough for us to share, even to take what inspires us about him or what upsets us, while leaving enough of him to pass around. I am interested in the critical thinker-doer who built an institute based on a philosophy, pedagogy, and practice of "head, hand, and heart," whose reasons I find inspirational as one who is heir to the black critical thinker-doer tradition. His reasons for founding Tuskegee constitute the only Booker T. Washington that matters to this book; they are summarized in an important article in 1906 titled "Tuskegee: A Retrospect and Prospect":

> The idea uppermost in my mind, when I began the work of establishing the school at Tuskegee, was to do something that would reach and improve the condition of the masses of the Negro people in the South. Up to that time, and even to-day to a large extent, education had not touched, in any real and tangible way, the great majority of the people in what is known as "The Black Belt". . . . Any institution which was to be of real service to the millions of the lower South must not confine itself to methods that were suited to some distant community where conditions of life were vastly different. . . . I found that the teaching of books alone was not sufficient for the students who came to Tuskegee to be trained as teachers and leaders. . . .
>
> I wanted to give Negro young men and women an education that would fit them to take up and carry to greater perfection the work that their fathers and mothers had been doing. I saw clearly that an education . . . without ability to change conditions, would leave the students, and the masses . . . worse off than they were. . . . *It was my aim to teach the students who came to Tuskegee to live a life and to make a living, to the end that they might return to their homes after graduation, and find profit and satisfaction in building up the communities from which they had come, and in developing the latent possibilities of the soil and the people.* (Washington 1906, 514–515; my own emphasis)

School was more than studying books; it existed not simply to study or learn about things, but to "do" things. The teaching of farming derived not from books but through actual and tangible work. To meet the need for food to eat, to come out of the hand and mind. To start from the reality of the land, sweat it into a farm. Not as I say, but professor in front, "my coat off, ready to begin digging up stumps and clearing the land" (Washington 1904/1915, 40–41).

As with other chapters, this Booker T. Washington is similarly not an object of intellectual curiosity or a resource for theoretical bombast, but, rather, a précis for producing a knowledge in the service of and through problem-solving, and to dare to invent a space to produce and apply it. *The goal is that, where I find people in tears, I leave smiles on their faces. Where I find the maligned and self-doubting, I leave the exalted and self-confident. Where I find problems, I leave opportunities.*

I merely seek to read his writings and understand his work for a purpose, my chosen life mission, and legacy to my own people: an institute to serve as a catalytic space and knowledge production and application center dedicated to lifting up their lives, our lives as a people. The goal is not simply an academic one. This is Booker T. for the critical thinker-doer, writing from the realm of what it feels to be—the experience of being—a slave. And that felt or experiential location has its analytical advantage, not least for those in search of tools for self-rehumanization, subjects of impoverishment by others who seek to empower themselves with tools for making wealth, especially those whose goal, after success, is to lift others out of poverty. What could be worse than being born a slave? Or more inspirational than self-making oneself into the founder of the jewel institution of a race enslaved?

In that story I find myself, and I may not be alone to confirm, through him, that present station does not define one's destination in life, but that, instead, hardship can inspire creative resilience. While some were born with a silver spoon to their mouth, Washington was born into rags and fought against impossible adversity to be a pioneer in mechanical arts education. The experience of passage from property to person, as a journey from bondage to freedom, from coercion to choice, looks at a girl and boy in the eye, clothing them with a confidence that the tattered clothes, bare feet, and nothing-but-corn meal that is their daily life are a season for the present that shall pass, replaced by greatness *if* they persevere.

So they lift up their eyes and chin up and walk on water and thorns, no obstacle too great to stand between their surmountable present and their future. Each of these children is a story of sweating to earn and save, with every help from parents, siblings, and kin for bus fare, that experience of leaving home for the unknown, the nervousness of a journey of daring to invent one's own future, and that of family, all too familiar to those of us who have traveled it.

There will always be those who give us a chance, who dare take risks with us, trusting us, not always guaranteed a success, but when we deliver, it is a good feeling. An even better sensation engulfs those who, after the individual success, do not consider it the end of the journey, but the beginning of another driven by remembering others.

After graduating from Hampton, Booker T. Washington could have chosen to forget, but remembered that his journey was not that of an individual but the race. And the result was Tuskegee Normal and Industrial Institute.

BOOKER T. WASHINGTON, BRIEFLY INTRODUCED

When we say that Booker T. Washington was born a slave, we need to pause and reflect on slavery as an experiential location for a black person. As a system and a moment, slavery prevented the development of family life, with devastating consequences for generations spanning 250 years. A family member could be sold anytime; domicile as a family or community was always uncertain. The concept of home as a stable institution did not exist; to the slaveowner it was simply a biological structure, a breeding mechanism, black personhood merely reproductive organs, fertile womb, and muscle for labor. The idea of a "home" insinuates that one has control and ownership over where one lives, including the house, which gives one confidence. But being black made one a tool of production; we heard Césaire calling it thingification earlier.

Upon attaining freedom, one started in absolute poverty, then embarked on building a home and a family life, "to collect the broken and scattered members of his family," sold as slaves to white owners far and wide. "I do not know who my own father was," Washington explains in 1900. "I have no idea who my grandmother was; I have or had uncles, aunts and cousins, but I have no knowledge as to where most of them now are" (Washington 1900, 223). The enslaved had no "honoured and distinguished ancestry" to rely on, be judged by. As Washington said, "When a white boy undertakes a task, it is taken for granted that he will succeed. On the other hand, people are usually surprised if the Negro boy does not fail." And he attributes this to the influence of ancestry, of memories, family reputation built over generations, and known family (uncles, aunts, cousins): "If a white boy fails in life, [he] disgrace[s] the whole family record, extending back through many generations" (Washington 1901, 36). With the black, there is nobody to disgrace, no temptations to resist, no history or connection to ancestry. The white boy has no limit on aspiration. No obstacles (Washington 1900, 223).

Booker T. Washington was born a slave on a Virginia plantation in 1857 or 1858 in a typical 14- by 16-ft square log cabin in the slave quarters, which he, her mother, his brother, and his sister shared. He knew nothing about his ancestry, except that, as told him by his mother, his maternal ancestors had endured the middle passage of the slave ship from Africa to America. Under slavery, black family history and records received little attention, so there was no way to trace roots. His mother was purchased—along with him just like cow or horse; his father was a white man living on a nearby plantation, who never took any interest in him. Growing up as a slave was not easy. The usual diet for the enslaved was corn bread and pork. His first shoes were wooden soled with leather on top, stiff on the foot. His first shirt was made of flax, which felt like "a dozen or more chestnut burrs, or a hundred small pin-points, in contact with his flesh" until it was "broken in" (Washington 1901b, 11, 12).

The emancipation occurred in 1865 when Washington was seven or eight years old. As a boy he worked in a salt furnace as a packer under horrid conditions; it was back-breaking work, but he had to look after his mother. After leaving the salt furnace, Washington worked in a coal mine, a wretched, dirty workplace underground, "in the blackest darkness." It was hard work, dangerous labor, punctuated by premature powder explosions and falling slates. Too much for a child. After some months, he left the coal mine to work for one and a half years in the house of General Lewis Ruffner, who owned the salt furnace and coal mine. His wife was a strict boss; she demanded cleanliness, promptness, honesty, and efficiency (Washington 1901b, 36–44).

It was while working as a house servant that Booker T. Washington's introduction to education came, via a Webster's "blue-back" spelling book his mother had bought him. When a school was opened locally, his stepfather refused to spare him from work to attend. So he took night lessons, an idea he would institutionalize at Tuskegee later. Washington was finally allowed to go to day school, rising early to work in the furnace, then attending school at 9:00 a.m., returning right after school to resume work (Washington 1901b, 40).

When he was still working in the coal mine, he had overheard two miners talking about Hampton, an institute where poor boys could work for their education. His mind was set; he started saving up, helped by his mother and brother. In 1872 he set off for Hampton on a journey full of difficulty, his money running out before arriving and forcing him to look for work on a dock offloading pig iron from a ship by day and sleeping under the sidewalk by night. After earning enough to get to Hampton, Washington was soon on his way, energized by "the chance for me to work for my education." And he was determined to pay it forward—to "remember," by returning to black communities to offer the same chance for self-help to the youth.

The biggest influence upon Booker T. Washington was his principal at Hampton, General Samuel Chapman Armstrong. To this "superhuman" influencer he paid the ultimate compliment: "One might have removed from Hampton all the buildings, class-rooms, teachers, and industries, and given the men and women there the opportunity of coming into daily contact with General Armstrong, and that alone would have been a liberal education" (Washington 1901b, 54). The principal was "worshipped" by his students. Besides General Armstrong, Washington was indebted to Mary F. Mackie, the head teacher, who offered him a job as a janitor and became his strongest, helpful friend (Washington 1901a).

BOOKER T. WASHINGTON'S PHILOSOPHY OF INDUSTRIAL EDUCATION

In the post–Civil War United States, the term "industrial education" constituted a fundamental tension and a contradiction. Slavery had made labor undignified; in the experientially grounded and felt perspective of Southern black people, to work was to slave for the white man, to be exploited. The role of black people in the South's—and the United States'—development had begun with their arrival in chains in 1619. The white man had tried enslaving indigenous people and fellow whites, but neither proved adequate.

In the South, no plantation functioned without black labor—we dug the ditches, cut the forests, built the railways, opened up the mines.

Washington identified the dignity of labor as the major transformation that education had brought on black life:

> The Negro learned in slavery to work but he did not learn to respect labor. On the contrary, the Negro was constantly taught, directly and indirectly during slavery times, that labor . . . was the curse of Canaan, he was told, that condemned the black man to be for all time the slave and servant of the white man. It was the curse of Canaan that made him for all time "a hewer of wood and drawer of water." The consequence of this teaching was that, when emancipation came, the Negro thought freedom must, in some way, mean freedom from labor. The Negro had also gained in slavery some general notions in regard to education. He observed that the people who had education for the most part belonged to the aristocracy, to the master class, while the people who had little or no education were usually of the class known as "poor whites." In this way education became associated, in his mind, with leisure, with luxury, and freedom from the drudgery of work with the hands. Another thing that the Negro learned in slavery about education was that it was something that was denied to the man who was a slave. Naturally, as soon as freedom came, he was in a great hurry to get education as soon as possible. He wanted education more than he wanted land or property or anything else, except, perhaps, public office. Although the Negro had no very definite notion in regard to education, he was pretty sure that, whatever else it might be, it had nothing to do with work, especially work with the hands. (Washington 1913, 226–227)

Therefore, to be a slave, "to be worked," was "a badge of degradation," of inferiority, while after 1865 black people were now emancipated "to work." "There is a vast difference between working and being worked," Washington said. "Being worked means degradation; working means civilisation" (Washington 1904/1915, 16–17). While Washington credited Armstrong with "mainly" inaugurating "the systematic, industrial training of the negro" in the 1860s, it was still hugely unpopular among black people because of the perception of labor as being worked, of education as relief from labor, as what educated people do not do.

During slavery, black people were banned from learning how to read or write during the days of slavery; after it, the education was too bookish and white-centric. The earliest black schools and colleges were intended to complete the work the Civil War had started (i.e., to "free the slaves"). After physical freedom, moral and intellectual freedom was the next goal. The defender of slavery, John C. Calhoun, had said he would only revisit his views when black people gained mastery of Greek—the marker of humanity and civilization for him. The post-1865 black schools therefore sought "to prove the capacity of the Negro to study and learn everything that the white man had studied and learned." As a justification for slavery, the white man had designated the black intellectually inferior, and proceeded under slavery (including the colonial one in Africa) to deintellectualize and dehumanize them. A myth was created that the black could not do what the white man did, so the post-Civil War black education project aspired to give blacks those elements that whites considered markers of high intellect, never mind their irrelevance to black everyday life.

Education meant freedom from manual labor and knowledge of Greek and Latin languages, which made one "a very superior human being, almost supernatural" (Washington 1901, 81). The black majority had hardly been "touched by education," with 1.4 million children of school-going age not enrolled, while those enrolled usually attended just five months a year. Five dollars was spent on educating one Northern white child, compared to fifty cents on one black child in the South. It would be impossible to educate the children of ten million black people (Washington 1908, 111).

In his "Talented Tenth" essay in 1903, W. E. B. Du Bois had argued for an education that started with "exceptional men," "the Best of this race," to "guide the Mass away from the contamination and death of the Worst, in their own and other races." Du Bois believed in an academic education that made "manhood the object of the work of the schools' intelligence," higher education as a foundation to "build bread winning, skill of hand and quickness of brain" (Du Bois 1903, 33). The Talented Tenth, not "the blind worshippers of the Average," an aristocracy of talent and character, not the fantasy of civilization bottomup—that was how black renaissance could be built. Teachers.—Clergymen.—Physicians.—Lawyers.—Bureaucrats in government.—Business people. All these were products of liberal arts education, and without them even the industrial colleges could not exist. Du Bois conceded being "an earnest advocate of manual training and trade teaching" for black and white boys, but insisted that the aim of true education was not "to make men carpenters," but "to make carpenters men" (Du Bois 1903, 62). That could be done in two ways: either give the community liberally trained teachers and leaders or offer him just enough intelligence and technical skill "to make him an efficient workman." The former required "a few of excellent quality, . . . enough to leaven the lump, to inspire the masses"; the latter good schools, "well-taught, conveniently located and properly equipped" (Du Bois 1903, 63).

The men who founded post–Civil War black colleges were expected to lead and teach their race in learning things of the book that had little or no connection with their everyday lives. These trainees had to receive "the very highest and best education" available, but there were far more relevant things to learn than "the little smattering of Greek and Latin that they obtained" (Washington 1909, 3). They needed to know less the languages and more the political history of Greece, of Rome, and of Europe, where slavery had existed and that had made transitions to free labor. Here they were, people enslaved just the other day, suddenly plunged into Greek! All they needed was to learn to work for wages as free men and save money (Washington 1909, 3–4). "I doubt that much reliance can safely be placed upon mere ability to read and write a little as a means of saving any race," Washington said in one of his disagreements with "Talent Tenth" thinking. "Education should go further" to teach the black, "in connection with thorough academic and religious branches, the dignity and beauty of labor, and to give him a working knowledge of the industries by which he must earn a subsistence" (Washington 1900, 226–227).

Booker T. Washington agreed that Greek and Latin opened the path to a career as a lawyer, teacher, or church minister. But what was the point of producing lawyers, academicians, doctors, and so on who managed, critiqued, or litigated what others built, what they could not build? Haiti's tragedy underscored the limits of mental and religious education alone. Its intellectuals had received a most thorough mental and Catholic education in Haiti and France in philosophy and the languages. And yet Haiti remained underdeveloped. A country that exported millions of pounds of coffee and precious woods and was France's seventh biggest trade partner. So many intellectuals distinguished in scholarship and science, writing beautiful poetry, conducting abstract science, and plotting revolution, in a country without engineers, architects, and agronomists badly needed to build much-needed roads, bridges, streets, railroads, and processing industries. Such a richly resourced island, yet importing much of its food and clothing. Graduates from top universities in France, yet "content themselves with wearing clothes imported from Europe." All thought, no doers. All hat, no cow—just a "worthless idle class" of gossipers, always drowned in politics and conspiracy to overthrow the government (Washington 1904/1915, 18–22).

For Du Bois the important question about college-educated blacks was whether they earned a living or not (Du Bois 1903, 51). But for Washington, the purpose of education was to teach them not just how to make a living, but to have a comfortable life, with "bathtubs, carpets, rugs, pictures, books, magazines, a daily paper, and a telephone" (Washington 1904/1915, 32). Rather than enter politics or become an oratory or street protest kind of activist, Washington chose to provide "service which would prove of more permanent value to my race." Namely, a solid educational, industrial, and economic type of foundation, whereas political life was "a rather selfish kind of success" (Washington 1901b, 93).

What is the meaning, purpose, and end of education? For Washington, it is not enough to simply memorize dates of key battles.—Recall events accurately.—Grammatical rules.—Arithmetic formulae.—Locating place names on maps. True—such education welded a strong, orderly mind. And "pitied . . . is the individual who is all heart and no head." But such should not be the end of education; rather, "to give us an idea of truth," change us for better.—To make us more thoughtful. Broader, to think and feel beyond self, to "help all people, regardless of race, regardless of colour, regardless of condition." That our educatedness reflects in our being the kindest.—Gentlest (Washington 1902c, 113, 114).

All told, "an education that educates.—Equips.—Galvanizes for action.—To solve problems—Not just theorize about doing, but actually doing.—Not speculating about, but building the world.—Real-life problems, not hypothetical.—Our problems, felt by us.—An education that engineers us into self-confident, self-reliant actors.—Doing with mind and hand.—That awakens the mind, our faculties.—Guides our hands.—An education that cajoles us to put our education into visible, tangible, usable, and transformative form. "We are trying to turn out men and women who are able to do something that the world wants done, that the world needs to have done," Washington says in 1902. "We are preparing you for a place in the world, . . . so that when you get to that place, if you fail in it, the failure will not be our fault" (Washington 1902c, 99).

Industry and education could be brought together, the one giving education tangible tools for living life, the other being deployed as a tool to rid the black mind of the idea of labor as biblical curse. The task of Tuskegee was to change the mindset and opinion of black people concerning labor and education, to connect education with "practical daily interests of daily life," to apply what is learned in school to "the common and ordinary things of life," to understand education as a means of dignifying labor, not an escape route from it, build the race from bottom to top, not top to bottom, "to lift the man furthest down, and thus raise the whole structure of society above him." Hence the necessity of teaching the masses that all labor was honorable and all idleness disgraceful, to prove to them through living example that investing in education rewards immensely (Washington 1913, 227, 228).

Critics said blacks did not need to be taught the dignity of labor, which they had performed for 250 years. But this was different: now the black would not work for nothing, but would enjoy the fruits of their labor. Tuskegee curriculum was based on "studying the needs of our people," starting with agriculture "since the majority . . . earn their living by farming." As Washington argued, "I don't believe it is right to teach people everything in heaven and earth and keep from them the knowledge of the means by which they earn their living" (Washington 1902b, 73). Rather, equip students to see both the bright and dark sides of life, and plan for both.—To prepare not just for the rainy or sunny days, but the dry and cold ones.—For pleasure and sadness (Washington 1902c, 19).—For wealth and poverty.—To accept discouragement as part of life, there for a purpose: as obstacles to be overcome, that make us stronger.—An education that does not make one claim to know more than one knows, is sure of what one knows, enough to deploy it as assured doing. Early at Tuskegee, "the larger the book and the bigger the words it contained, the more highly it was revered" (Washington 1902c, 24, 54; 1904//1915, 50).—Elitist jargon.—Persuasive argumentation. "Set a high example for every one in your community," Washington had implored his students. "You must remember that the people are watching you every day" (Washington 1902c, 32).

A knowledge to sculpt a responsible, reliable, and humble human being. That takes ownership not merely for the successes, but failures, and learns from them.—Builds upon them. That realizes that actions have consequences, not just to self. Others too. Especially others.—Who lives for others too, not just self.—Who touches the lives of others.—Who is touched by the lives and plights of others.—And recognizes happiness of self only through that of others first.—Who turns every tear into smile.—Whose yearning, when possessing something good, is for others to have it too. What Washington called "the habit of realizing one's responsibility to others . . . and [to] govern yourselves accordingly" (Washington 1902c, 204). To be relied upon, where the race is caricatured as undependable.—People who think, map out, take initiative, not those who "are not worth anything [and who] ought to pay rent for the air they breathe" (Washington 1902c, 13).—People who persevere. To be thorough, smart, educated, yet remain simple.—Humble.—Approachable. "I am not trying to discourage you about wearing nice collars," Washington said in 1902. "I like to see every collar shine. I like to see every collar as bright as possible. I like to see you wear good, attractive collars. I do not, however, want you to get the idea that collars make the man. You quite often see fine cuffs and collars, when there is no real man there. You want to be sure to get the man first" (Washington 1902c, 39). Let the actions speak for themselves, so that in the end when you ask yourself the question "Have you done your best?" your heart will answer affirmatively (Washington 1902c, 44).

Industrial education was learning through doing. An industrial education inclusive, not exclusive, of other forms of education, not "offered the Negro because he is a Negro," or that reduced the black person to being capable of only training to use only the hand, plow, or hoe (Washington 1905c, 8–10). An education that started at the bottom up, nourishing tree from roots, the black majority. An industrial mastery that built on academic skills, agricultural chemistry, mechanical engineering, the liberal education training the mind to use the hand. An education system whose ultimate test was its actual achievements (Washington 1905c, 12).

Industrial education was for Washington a knowledge anchored in and talking to everyday realities. The curriculum involved making work, community, and everyday life a natural part of learning. The student now found it as much a part of their education to spend the day on the farm, kitchen, machine shop, or laundry as they would studying algebra, chemistry, or literature. The school and the community, the curriculum and everyday life, fed into each other. Washington explained in 1909: "The boy who learns about rods and furlongs and acres in the class-room learns out on the farm to measure off actual furlongs and actual acres. The boy who learns something of botany and something of plant life and something of the chemistry of the soil in school puts all he has learned into practice when he goes out to work on the soil" (Washington 1909, 5; 1906, 521). The country school buildings must be similar in location and appearance to a farmer's cottage in the field, with fruit, flowers, and vegetables growing all around, a stable with horses, cows, chickens, a good well, hay and fodder, a repair shop attached to the barn, and, inside the school, an assembly room, kitchen, bedroom, dining room, and the requisite furniture (Washington 1910d, 243). In short, the farmhouse must be a place of good life too, attractive to young people with skill and ambition, and passion for developing their own community.

Hand and heart was not exclusive of head, or work from spiritual matters. The mind of the technically trained student must also be "thoroughly awakened and developed by a severe and systematic training in the academic branches," Washington insisted. Something more than industrial or academic training was needed if students were to "make themselves useful in changing the conditions of the masses of their people." Namely, an education that went beyond making them "sympathetic" to the masses, but which made them "actively and practically interested in constructive methods and work among their people" (Washington 1906, 516). The teacher led by example, beyond ordinary classroom duties, out there in the community, among their people, leading monthly farmers' institutes or local farmers' conferences, showing the farmers how to buy land, get out of debt, to "cease mortgaging their crops, and to take active interest in the economic development of their community" (Washington 1906, 516).

The "heart" part was about remembering. That education was not the individual's alone, but to equip one with "intelligence, skill of hand, and strength of mind and heart [so] that you may help somebody else" (Washington 1902c, 11). That was why the student paid only a small part of their expenses, and why "everybody is here on trust, . . . *every day here is a sacred day, that it is a day that belongs to the race*" (Washington 1902c, 12; my own emphasis). Service, not servitude, lest some get spooked by the ghost of slavery.

While emphasizing industrial training, the curriculum had a rich balance of "training of the mind with that of the hand, so that the one will complement the other. . . . Our industrial system," Washington explained in 1905, "is intended not merely to put students on their feet in the bread-winning occupations, but also, as it does to a surprising extent, to serve as the most effective medium for mind training" (Washington 1905a, 397). Industrial education did not mean a lesser "hand training" for hewers of hay (i.e., intended for blacks only because they were an inferior class), but "because the undeveloped material resources of the South make it peculiarly important for both races." The industrial institute did not render the "negro college" redundant; rather, to "co-operate in the common purpose of elevating the masses" (Washington 1905a, 397). Even preachers, taking classes in ministry, were required to "learn a trade, something of agriculture, so that they can give the members of their congregations an example of industrial thrift" (Washington 1905b, 23).

The purpose of being educated was not simply personal but to acquire a tool for elevating the race and serving the community. For Washington, industrial education served not only an individual or his family, but was about elevating the race and building and serving the community. Tuskegee insisted upon the importance of service, imbuing its graduates as leaders of their people, lifters of the race; if not religious work, certainly humanistic. That was Tuskegee's "most important work" and accomplishment (Washington 1902a, 292; 1905b, 23).

Washington laid out the principal elements of a curriculum and institute in the service of problem-solving as follows:

> *The school in a farming community should get its arithmetic problems from the farm.* The reading lessons, the grammar lessons, the lessons in history and science should be ordered, arranged, and taught from the point of view of the farmer, with a view to enlarging, enriching, and improving, not merely the farms, but the homes and country life generally. A model country school should be the centre, not merely of the intellectual life of the countryside, but of all the efforts that are now being made by the county, state, and national governments to improve farming conditions. *It should maintain, when possible, in connection with the school, a little experiment station and laboratory where new methods could be publicly demonstrated and tried out. It should maintain a library. It should provide lectures on subjects of special interest to the community; it should maintain a school bank and teach the art of saving and investing money, and constantly strive in every way to widen the circle of its light and its influence among the people.* (Washington 1910a, 149; my own emphasis)

TUSKEGEE

Having completed his training in industrial education, in 1881 Booker T. Washington arrived in Tuskegee, having been recommended by Armstrong as the ideal person to set up a normal school for black people there. Tuskegee was a town of two thousand residents located in the "Black Belt," a term referring both to the color of its people and rich soil, where enslaved Africans were taken to turn it into cotton, thus inevitably outnumbering white people three to one. His first task was to find a building in which to open the school; all he had to work with was a small church, a shanty, and a hen house, without property, bar a hoe and a blind mule (Washington 1899/1972).

The first thing Washington had to do was to know the Black Belt well enough to design curriculum that answered to its needs. That meant traveling through Alabama, examining the "actual life" of its people. On the country roads with mule and cart he traveled, giving no advance notice. *Eating what the people ate. Sleeping in those little cabins where they slept, and "seeing the real, everyday life of the people"* (Washington 1901, 111; my own emphasis).

He found the condition of the people to be rather dire. Entire families sleeping in one room, surviving on a diet of fat pork, corn bread, and black-eyed peas cooked in water, bought from stores in town when they could grow them themselves, yet they planted nothing but cotton "up to the very door of the cabin." The farmer sold the cotton seed instead of enriching the soil, impoverished every season, top soil eroded for lack of contours, and mules underfed for want of feed, which had to be bought. Black folk in the South were used to cotton as "the money crop," and moneylenders dissuaded from growing food crops; come January, they had to buy from the store food they might have grown. The storekeeper "advanced" them such provisions from money borrowed from the local bank, which borrowed from New York, with each adding interest and passing it on (15 to 30 percent in all) to the farmer. Failure to pay, the farmer lost the farm, animals, and tools, was forced to rent, or found somebody else to "run him"—that is, to sharecrop for (Washington 1904/1915, 33–35).

When the farmer sold their crop, the storekeeper and lender were waiting, and whatever crumbs were left were spent on consumables that did not generate new value. Money came from crop sales only in fall, in time for Christmas, when whiskey, cheap jewelry, cheap buggies, and other junkets consumed whatever was left. Washington was shown sewing machines bought each on instalments for $60, but they remained unused.—Showy clocks for $12 to $16, which kept incorrect time. An organ for $60 in monthly installments, which "nobody had skills to play" (Washington 1901b, 113). Most farmers in debt mortgaged their crops. No schoolhouses, so classes were held in churches or log cabins. The investigative journey in the "Black Belt" left Washington "with a very heavy heart" (Washington 1901b, 118). A curriculum was needed to address the realities he had observed and experienced, that held black people back. Where the farmer had raised just twenty bushels of cotton on one acre, Washington wanted to double the output with less labor (Washington 1904/1915, 37).

When he opened Tuskegee on July 4, 1881, Washington was the sole teacher for the initial thirty students, increasing to fifty by month's end. The classroom was the leaking church. When it rained, a student had to hold an umbrella over him lest his books became wet while he lectured. Fellow Hampton alum Olivia A. Davidson (later his wife) joined him in the second month, having just completed a two-year degree at Massachusetts State

FIGURE 7.2
Tuskegee in 1901. *Source*: Washington 1901b, 110.

Normal School at Framingham (now Framingham State University) (Washington 1901a, 110; 1901b, 124–125).

Everyone who has ever attempted or built a not-for-profit or for-profit institution will know how difficult it is to fund projects. It is not like big money is just waiting there to be accessed; usually there is not money. It is not like just after establishing your website or registering your organization the money will come pouring in. It is head scratching and sleepless nights all the way. Friends will get fewer, if you are lucky enough to be left with any. Spouses can leave. It can be a very lonely, depressing, and foolish thing to take the risk of daring to invent the future. But that is when you know who your really true friends, true loves, and humane beings really are. I know. I have been both in those situations and stuck by friends in them.

People marvel when the buildings begin to come up, criticize the font of the website, and give the sort of worthless advice long after the fact. *That is when you know that all the positive stories you're told about start-ups, bootstrapping, elevator pitches, taking the risks to fail and learning from failure, and—my favorite one—investors that will listen to you if you have tried and failed before, are all lies. Trust me, I have been there; it is one of the reasons why I had to read the history of Tuskegee*, to listen to a fellow black man tell me the unwashed story about trying to establish an institution while black.

Founding Tuskegee was a case of having to "make bricks without straw," which Washington turned into an opportunity for "pluck, self-help, and perseverance" (Washington 1905a, 35). Together Washington and Davidson started conniving the future institute, then just the little old shanty and the abandoned church loaned them by the black folk of the town for classes. An initial loan from his friend the Hampton treasurer General J. F. B. Marshall's personal savings set the institute on its way, but it needed to be repaid. Davidson hosted festivals and canvassed black and white families for cakes, chicken, bread, or pies, and so on to sell at the events. The first months of fundraising were to pay off the loan of $250 General Marshall had advanced to purchase the one-hundred-acre farm (Washington 1901b, 126–136).

"No one knew where the money was to come from," Washington remembered the first years of Tuskegee well. "I recall that night after night I would roll and toss on my bed, without sleep, because of the anxiety and uncertainty which we were in regarding money. I knew that, in a large degree, we were trying an experiment—that of testing whether or not it was possible for Negroes to build up and control the affairs of a large educational institution" (Washington 1901b, 145). Washington and Davidson ran up against a general assumption of white success as a given, whereas black success was greeted with surprise. "I knew that if we failed it would injure the whole race," Washington concluded.

Booker T. Washington had true friends, and General Armstrong was one such. In the hardest moments of indebtedness, it was the general who helped Washington by introducing him to his networks in the North. "Meetings were held in New York, Brooklyn, Boston, Philadelphia, and other large cities, and at all of these meetings General Armstrong pleaded, together with myself, for help, not for Hampton, but for Tuskegee," and these meetings were successful. Thereafter Washington's stays became longer, even more successful (Washington 1901b, 178–180).

I now know from experience that money to turn vision into reality is the single most important challenge in daring to invent an institute. It is not like you first collect all the money until it is all there and then you start the project; if you do that you will never start. On the contrary, visions are launched as projects with little to no money, as a leap of faith, which acquires the status of a reality in the eyes and ears of those you tell about it. That is what inspires, cajoles, and shames you into having to work yourself into the ground to raise the funding, not to turn it into a reality because it already is (in the eyes and ears of those you have told), but to extend such reality to new dimensions.

In the course of fundraising, whether by working or knocking doors, you learn a lot. And that was also Washington's experience on the East Coast of the United States. When dealing with wealthy and noted men, Washington's strategy was simple: to make the institute's work known to specific people and organizations, and not worry about the results—something hard to do when up to the neck in debts and bills without a cent.—Never get excited.—Never lose self-control.—Always be calm.—Self-possessed.—Patient.—Polite. Washington did not have patience with people that criticized rich people just because they were rich or that did not give more to charity. "How much suffering would result," he asked, "if wealthy people were to part all at once with any large proportion of their wealth in a way to disorganize and cripple great business enterprises?" (Washington 1901b, 182). Knocking doors and sitting in the offices of the rich was "a rare opportunity to study human nature," up close. Some were nice, thrilled at the "privilege to have a share in . . . [and] help a good cause," and indebted to Tuskegee for "doing *our* work" (Washington 1901b, 184–185).

Boston was especially dear and responsive to Washington; its donors felt being asked for funding and giving as an honor conferred upon them. He also did not like begging, and never "begged," for money; presenting facts of what the institute was doing was "all the begging that most rich people care for" (Washington 1901b, 183). In the early days he would go for days, on many streets, and many miles without getting a dime. The fact that he received no donation on the spot sometimes was not a sign of mean people; some would digest the request, then, like one Stamford, Connecticut, gentleman, send a check—$10,000 in this case—two years later (Washington 1901b, 186). The railroad mogul Collis P. Huntington had given Washington a miserly $2 on their first meeting, only to bequeath $50,000 toward an endowment in his will. Such donations were not luck but products of hard work: Huntington gave because he had seen tangible results.

Tuskegee had a permanent endowment fund, which in 1902 stood at $299,759.02, with an investment plan handled professionally by an investment committee in New York's financial district. A budget was also set aside for "permanent improvement" of the institute's infrastructure, such as new buildings, industrial equipment, the grounds, fields, and so on, and for "special purposes." The money was coming from both ex-masters and ex-slaves alike, in small quantities of fifty cents each and upward. Anonymous donors gave bigger amounts (Washington 1902a, 292).

THE TUSKEGEE MODEL

Booker T. Washington's 1881 tour of the "Black Belt" told him that the institute he was establishing would have to meet conditions under which people lived, not theory, not what he wanted the students to do, but what they were already doing. The early students were fond of memorizing grammar and mathematics, and "had little thought or knowledge of applying these rules to the everyday affairs of their life"; Washington wasted no time in informing them that this was not the right place to dabble in that (Washington 1901b, 122).

The buildings and facilities of Tuskegee emerged to meet the tangible needs of students, staff, and curriculum. Teaching was the medium for building such infrastructure, meeting such needs. Washington was convinced that education would eventually solve the race problem. That is, not narrowly as school teaching and learning, but relevant and practiced at home, industrial, thrifty, and disciplined mentally, morally, and spiritually, answering to the needs of the community in which we live (Washington 1900, 221; 1904/1915, 38).

* * *

The next step was to increase cultivation on that land, to make it pay back, the venue and material resource for training students in agriculture. Every industry at Tuskegee grew out of its needs, beginning with farming, because students and staff needed to eat.—With industry, so students could earn enough money to remain in school for nine months a year.—Livestock farming, because Tuskegee needed draught, pack, meat, and milk animals, numbering two hundred horses, colts, mules, cows, calves, oxen, sheep, and goats and seven hundred hogs and pigs by 1901 (Washington 1901b, 139). Washington explains the coevolution of the institute's infrastructure and needs and its problem-solving curriculum well in this 1906 passage:

> We needed food for our tables; farming, therefore, was our first industry, started to meet this need. With the need for shelter for our students, courses in house building and carpentry were added. Out of these, brick-making and brick-masonry naturally grew. The increasing demand for buildings

> made further specialization in the industries necessary. Soon we found ourselves teaching tinsmithing, plastering and painting. Classes in cooking were added, because we needed competent persons to prepare the food. Courses in laundering, sewing, dining-room work and nurse-training have been added to meet the actual needs of the school community. (Washington 1906, 515)

The economics of self-sustainability were built into all aspects of each program. When a donor gave money for a building, students made the materials and built it, saving money that might be lost to an outside contractor. Hand work was the furnace within which character was forged.—Civilization.—Self-help.—Self-reliance.—Teaching students how to construct their own buildings, put them in a position to make mistakes, to learn "valuable lessons for the future" (Washington 1901b, 149; 1904/1915, 43). Nineteen years after its opening, Tuskegee had forty buildings, all except four built almost completely by students under examinable and paid class conditions.—All the institute's wagons.—Carts—Buggies. The entire local market got its vehicles from Tuskegee. Communities did not eat Greek sentences, but they needed bricks, houses, and wagons (Washington 1901b, 149–155).

Initially the methods were primitive and crude, steadily improving with time. The initial land had poor soil, the best for learning how to make poor land rich. The first two hours of the school day were spent in the classroom, the rest outdoors stumping and felling trees, building fences, making ditches, and plowing. There were few, mostly borrowed, tools to work with, and it was hard to convince students that this was knowledge worth acquiring. Later, more farmland, better methods of fertilization, and labor-saving machinery were acquired and developed (Washington 1904/1915, 45–46). The classes stumped new acres, diversified crops, both increasing as more students were admitted.

Booker T. Washington was a beneficiary of the system of earning through learning at Hampton but the specific design of the Tuskegee iteration emerged organically. Initially all students lived with local families off campus because no accommodation existed at Tuskegee. The financial and time value of that mass of students was being surrendered to the town; a boarding house had to be built on campus. "The moment the idea of 'making it pay' is placed uppermost," Washington explains the concept, "the institution becomes a factory, and not a school for training head and hand and heart" (Washington 1904/1915, 63). The students would have classroom activity for four days and outdoor work two days a week, earning $2 to $3 a month that would go toward paying the monthly $8 charge for campus accommodation. Washington also started a night school for students unable to pay for day school. Thus the school was divided into two types of students: day scholars on two days industrial work and four in the classroom, and night scholars that worked all day and attended evening classes. The process of working with the hands ten hours a day to study two hours at night the ultimate test and proof of commitment, a "sifting-out process" (Washington 1904/1915, 52–63).

* * *

From the beginning, Washington ensured that students learned not just agricultural and domestic skills, but also the construction of their classes, dorms, and workshops, and to provide all their needs and, once that was achieved, to the produce to sell. This section samples brief examples of this curriculum. The courses were designed for students to turn the forces of nature—air, water, steam, electricity, and horsepower—into catalysts for forging knowledge into buildings, crops, and tools (Washington 1901b, 147). By 1902, Tuskegee had thirty-four such industries: carpentry, blacksmithing, printing, wheelwrighting, harness making, carriage trimming, painting, machinery, founding, shoemaking, brick masonry, plastering, brickmaking, sawmilling, tinning, tailoring, mechanical, architectural, and freehand drawing, electrical and steam engineering, canning, plain sewing, dressmaking, millinery, cooking, laundering, housekeeping, mattress making, basketry, nurse training, agriculture, dairying, horticulture, and stock raising (Washington 1902a, 292).

To be sure, industrial education had to do more than just make an efficient mechanic who earns a living and adds to society's productiveness. It had to arouse intellectual interest in art, literature, and spirituality. School was more than studying books; it existed not simply to study or learn about things, but to "do" things. The teaching of farming derived not from books, but was done through actual and tangible work.—To meet the need for food to eat, to come out of the hand and mind.—*To start from the reality of the land, sweat it into a farm.—Not as I say, but professor in front, "my coat off, ready to begin digging up stumps and clearing the land"* (Washington 1904/1915, 40; my own emphasis).

The tradesperson was in touch with real issues and things, their daily work involved building and equipping buildings. This was true even in times of slavery—there had existed on the plantation mechanics, blacksmiths, carpenters, crop experts, and other black tradespersons. For Washington, every large slave plantation of the South was "in a limited sense an industrial school," where young black women and men were trained as farmers, carpenters, brickmasons, blacksmiths, bridgebuilders, wheelwrights, dressmakers, plasterers, engineers, cooks, housekeepers, and so on. He is very clear that he does not wish to "apologize for the curse of slavery" but "simply stating facts," qualifying the training as "crude" and "given for selfish purposes," and devoid of literacy to equip the mind to independently direct the hand (Washington 1904b, 5).

These tradespersons, Washington recalled, had "their own secret processes for doing the work assigned to them," and they "carefully guarded their craft secrets, and handed them down to their children or whoever followed them in the trade" (Washington

1910b, 125). He recalled Lumsden Lane, enslaved in Newberne, North Carolina, in the early 1800s, who learned from his father (through doing) a secret tobacco-curing method; the "Hemp Brake," a machine that beats the hemp stalk to separate the fiber; and Benjamin Montgomery, inventor of a steamboat propeller that replaced paddle wheels and which the Confederate Navy adopted. In Baltimore and Charleston, numerous slave mechanics bought their time to work, save money, and, eventually, buy their freedom. Frederick Douglass, a ship caulker, was one of them (Washington 1910b, 127). Of course, being property, the slave had no right to intellectual property, and Confederate law made slave inventions property of their master.

The education of black people in literature, mathematics, and the sciences completely ignored this plantation education, and these experienced artisans started dying out, with their skills and knowledge, with nobody to replace them. Instead, Black America had many young men fluent in Greek and Latin, but few bridge builders, carpenters, mechanics, architects, engineers, machinists, or agronomists and no contractors or builders (Washington 1904b, 5). After emancipation, there was no need to now throw away the baby (the intertwined intellectual and industrial and the capacity of black people to invent, to be acknowledged for it, to benefit from it) with the dirty bathwater (slavery and theft of black inventions and profit from them).

In the Tuskegee curriculum, mechanical labor was indivisible from intellectual labors; the academic was the foundation and driver of the industrial. The humanities at Tuskegee were tools for physically involved problem-solving. The academic department was composed of twenty-eight black men and women faculty with degrees from Michigan, Nebraska, Oberlin, Amherst, Cornell, Columbia, and Harvard (Washington 1904/1915, 84).

The classroom work discussed the theoretical elements, and the field grounded them in practice. English was taught as an art of talking simply and naturally and communicating ideas effectively. Reading was specifically a means to furnish the mind with knowledge. The first course in history was biography of great (white) men—Washington, Jefferson, Adams—while the advanced course focused on the black person's peculiar position in US history, from the start of the slave trade to the turn of the century, among other topics. During senior year, the students took a course on the State History of Alabama. Having built character, the course now developed the power of observation and imagination. Also in the first three years, students were taught geography and nature study, and in that last year, geography was combined with history (Washington 1904/1915, 85–89). All this academic education was for purposes of turning theory into tangible products and services.

The mathematics curriculum covered arithmetic, algebra, geometry, trigonometry, and surveying, designed to equip the student with practical knowledge of things associated with figures instead of just figures in abstract, measures and multiples applied to trades. Students were taught how to and were required to deploy spelling, mathematics, grammar, and English composition skills into operation, showing that academic teaching was just as critical as "skill with the plane, the saw and the miter-box" (Washington 1904/1915, 71). The carpenter cutting boards to size when making furniture, the dressmaker yards of cloth to make dresses. Mathematics deployed as an instrument of economy. In carpeting.—Lathing.—Plastering.—Brickwork. The final term of senior year was devoted to elements of civil engineering, specifically surveying—a much in-demand service in the South. Nature study trained students in observation and knowledge of plants, animals, minerals, natural phenomena, and the human body (Washington 1904/1915, 92). The instruction was by lab work, textbooks, lectures, and reference readings. It covered farm implements, root systems of crops, seed germination tests, soil acidity and fertility tests, milk tests butter and cheese, studies, and so on.

As part of its agriculture programs, the institute maintained a well-equipped lab for studying germination of seeds, fertilizers, manures, soil bacteriology, cotton, and corn. It ran an experimental station with fields and orchards for class demos, with an assortment of animals, such as draft and coach horses, Jersey, Ayrshire, and Holstein cattle, Southdown sheep, Berkshire swine and a dairy division with a cream separator for butter and cheese, which I elaborate on shortly (Washington 1904/1915, 93).

The chemistry and physics courses "not merely taught how to do, but to do." Students took soap apart and put it back together. Made polishes, lacquers, and chemical cleansers, not just studied formulae. All this, to "develop their power of doing things" (Washington 1904/1915, 95–96).—To make flour, bran, and baking powder.—Test fertility of fertilizer.—Destroy the cabbage-devouring worm.—Physics applications to building construction and farm work. Agriculture students studied and observed pests infecting peach trees, formulated ideas about the scale or borer insect, and examined them in the orchard during spring, not just in written papers. The instructor convened class among the peach trees, asked students to hold the branches and comb them for, to identify, borers and treat the trees per the procedures and pesticides from the books and lectures (Washington 1904/1915, 96).

Tuskegee is synonymous with Booker T. Washington, but its most famous scientist is George Washington Carver, the institute's director of agriculture. Agricultural scientist. Inventor. Developer of hundreds of products out of peanuts, sweet potatoes, and soybeans. Carver was born an African American slave near Diamond, Missouri, in 1864, and went on to earn a masters in agricultural science from Iowa State University. As a boy, Carver developed a close passion for plants that blossomed during his bachelor of science degree at Iowa State, where he researched fungal infections in soybean plants. After graduation with a master of agriculture degree in 1896, Carver chose Tuskegee because it offered him an opportunity to establish an agricultural school to implement his vision. Carver's

FIGURE 7.3
The Jesup wagon. *Source*: *The Encyclopedia of Alabama*.

research revealed the average cotton yield per acre across the South was 190 pounds, but his experiment raised almost five hundred pounds of cotton on one acre of poor soil by crossing varieties of cotton. Based on that proof of concept, Carver was now teaching farmers how to improve the quality and yield of cotton without increasing acreage. He invented several products from sweet potatoes (flour and vinegar, stains, dyes, paints and writing ink) and peanuts (over three hundred food, industrial, and commercial products such as milk, Worcestershire sauce, punches, cooking oils and salad oil, paper, cosmetics, soaps, wood stains, antiseptics, laxatives, and goiter medications). The latter earned him the name "The Peanut Man." Carver taught black farmers the technique of crop rotation and soil chemistry.—That cotton monocropping depleted soil nutrients.—That soil could be improved naturally—the ancient Egyptian had "rested" the land to allow it to recover, improve the next season's crop quantity and quality (Washington 1904/1915, 165).

He set aside forty acres for the experiment, analyzed the soils well, identifying deficiencies of the important elements—nitrogen, phosphoric acid, and potash. He deep-plowed, rebuilt terraces across the fields, and then grew nitrogen-fixing plants (peanuts, soybeans and, sweet potatoes), restored nutrient balance, and boosted cotton yields. The

FIGURE 7.4
Margaret Murray Washington. *Source*: Wikicommons.

results were spectacular. Having followed the sweet potato experiment intently, Booker T. Washington paid Carver the ultimate compliment: "If we had a hundred such coloured men in each county in the South, who could make their education felt in meeting the world's needs, there would be no race problem" (Washington 1904/1915, 136, 168–169).

Carver also invented the Jesup wagon (Fig. 7.3) in 1906, a mobile (horse-drawn) classroom and laboratory that moved around communities teaching farmers and poor sharecroppers methods of growing sweet potatoes, peanuts, soybeans, pecans, and other crops. The wagon was named after Morris Jesup, the New York banker who financed the project. The US Department of Agriculture and the Alabama Cooperative Extension System then adopted the mobile outreach model for educational purposes. The latter extended the concept to the Water Wheels Mobile Conservation Laboratory, which educates Alabama residents on water conservation, and to the Nutrition Education on the Move Bus (Nutrition Bus) equipped with a mini-kitchen to teach basic nutrition classes.

One of the most pathbreaking projects at Tuskegee was the Industries for Girls program. As related by Margaret Murray Washington (Booker T.'s third wife from 1893 to 1915), Tuskegee's director of industries for girls in 1905, the idea was to teach the female students "other means of livelihood besides sewing, housework, and cooking" (Washington 1905c, 68). Learning to sew and being a full-fledged dressmaker was the height of ambition for girls brought up by their "horny-handed fathers and mothers" (Washington 1905c, 69). Nonindoor opportunities for women in the South were limited due to traditional prejudice and absence of training for women. Specialization was inevitably limited to cooking, sewing, laundering, art, music, and general literature. In the north, in New England and the Middle States, women had begun to make inroads in manufacturing, something impossible in the South because of lack of factories and the dominance of agricultural production on plantations. Booker T. Washington was therefore determined to take women out of the kitchen to claim these outdoor specializations too. Therefore, he placed Margaret Washington in charge of female students' outdoor work at Tuskegee (Washington 1904/1915, 107–109).

The Industries for Girls program covered a whole range of courses such as horticulture, training kitchen, housekeeping, dining room, hospital, kitchen gardening, poultry raising, tailoring, dairying, printing, broom making, mattress making, upholstering, laundering, plain sewing, millinery, and dressmaking. In the dairying course (also offered to men), for example, students learned all aspects of dairy cow care and dairying, including utensils and their deployments and care, gravity, creaming, churning, working and salting of butter, and its preparation for market. In poultry raising they learned about poultry economics, breeds and breed mixing, henhouse construction, nests, and runs. In the horticulture class they mastered orchards and fruit farming, especially peaches, pears, apples, plums, figs, grapes, and strawberries (Washington 1904/1915, 110–111).

Margaret Washington also restored the indoor as not simply a space of consumption, social gathering, or resting, but also an industrial one of manufacturing. For example, black women had always made brooms and were experts at sweeping houses and yards, so a broom-making industrial unit was created for female students. Carver had introduced broom corn as a crop on the institute farm and thus made broom making the value addition to the farming. He went further, by requiring the female students to write descriptive essays of their work in both, joining head and hands together (Washington 1904/1915, 67).

The wood components for furniture making came from sawmilling, which was a course on its own covering machines and their uses and care.—Timber trees and their defects.—How to fell trees, cut them into logs, and transport them.—Grade and plane lumber.—Belt

FIGURE 7.5
The industrial unit showing textiles (top) and the dairy (bottom). *Source*: Washington 1905c, 69, 256.

FIGURE 7.6
The sawmill (top) and wood-turning machinery (bottom). *Source*: Washington 1904/1915, 73, 90.

care.—Filing and caring for saws.—Designing and making cutters for moldings.—Calculating pulley speeds.—And so on. "Theory and practice in this department are dovetailed in the finished work, where the wood work is the product of the school saw-mill and planer, the carpenter shops and the paint-shop" (Washington 1904/1915, 74). In carpentry, for example, students learned to turn wood and make cabinets, to make the desks, chairs, and workbenches they used in class, and beds and wardrobes for their dorms, and to repair all broken furniture. From the onset, Washington had intended that students themselves erect the buildings—and the furniture too, especially in their dorms. Initially the bedsteads were rough and weak. The longstanding problem of providing mattresses was solved by sewing pieces of cheap cloth together into large bags, which were then filled with pine needles. From these humble beginnings, Tuskegee's mattress making grew steadily into store-quality products (Washington 1904/1915, 78; 1901b, 172–173). In the wheelwrighting division, wagons, drays, horse and hand carts, wheelbarrows, buggies, and road carts were built not just for institute needs, but to supply farms and merchants in Tuskegee and Montgomery.

All mechanical equipment repairs were done in the institute's machine shop, which had the latest machinery, powered from an Atlas engine generator. Steam pumps. Steam engines. Woodworking machines. Printing presses. Metal-working machines. Spare parts for repairs were cast in the foundry. The castings for the cotton press the Togo expedition took with them was forged in the school foundry. In the blacksmith shop, students learned the ironing of carriages, buggies, and wagons, the shoeing of horses, and the repair of farm implements (Washington 1904/1915, 78).

The institute taught "everything in connection with the construction of the buildings," to students whose pass grade depended on them building actual structures. The starting point was a three-year architecture course covering drawing, building construction, and design. In the third year, students were provided real-life office administration and management training in trade shops and attended heating, electrical lighting, and plumbing classes. It was in such classes that Tuskegee's best and biggest buildings were designed. The construction (brick masonry, plastering) and carpentry student learned not just how to build and equip a building, but also to design it. This work of technical drawing started with simple geometrical drawing.—Then projection or drawing objects to scale (Second Year).—Then making blue, solar, and black prints (Third Year).—Then field trips to shops, buildings under construction, brickyards, etc. They made the bricks, saved Tuskegee money, while learning the art of brickmaking. They laid the bricks.—Plastered the walls.—, The carpentry work.—The painting.—The tin-roofing.—All the electrical engineering (Washington 1902b, 73; 1904/1915, 57, 76–77).

The full extent of Tuskegee's learning through doing is captured in a powerful passage in *Working with the Hands* that summarizes this section:

FIGURE 7.7
Digging a foundation (top), and building a new dorm (bottom). *Source*: Washington 1904/1915, 57, 59.

> The visitor, who wishes to inspect the Tuskegee Institute, is met at the station by a carrage built by the students, pulled by horses raised on the school farms, whose harness was made in a school shop. The driver wears a trim, blue uniform made in the school tailor-shop, and shoes made by student class work. The visitor is assigned to a guest room in a dormitory designed, built, and furnished by the students. His bathroom plumbing, the steam heat in his room, and the electric lighting were installed by students. The oak furniture of his room came from the shops. The young woman who takes care of his room is a student working her way through the Institute. After supper, she will change her wearing apparel to a blue uniform dress and a neat straw hat, all made in the school. The steam laundry sends over to know if the visitor wishes some washing done, and girl students send it back, proud of the snowy polish of shirts and collars. The visitor is asked to be a guest, in the teachers' dining-hall. . . . The ham, roast beef, vegetables, corn-bread, syrup, butter, milk, and potatoes [in his meals] are products of the school farms, raised, cared for and produced by student labour. (Washington 1904/1915, 79–80)

THE TUSKEGEE SPIRIT: AN INSTITUTE THAT IS FELT IN ITS COMMUNITY

In the introduction to *Tuskegee and Its People*, Booker T. Washington defines faculty and in particular graduates as "a leavening influence upon civilization" (Washington 1905c, 15). He expands on an earlier address to his students, wherein he says the teacher must settle down in the community, to make it home, move around it to learn, not work for salary alone but sacrifice for the community good. To build a "convenient school-house," the teacher must mobilize the community, get each household to give something to buy lumber. The teacher must organize fundraising dinners, festivals, and church offerings, and each community member must volunteer labor to build the schoolhouse. To build a literate community, the teacher must organize the people and create an educational club that meets regularly. Each parent must pay ten to fifty cents a month to extend the school term. Those with no money could pay in kind—eggs, chickens, butter, sweet potatoes, corn, and so on—toward the teacher's food requirements. To that end, each farmer must set aside a portion of land, where their children could cultivate for such school fee contributions (Washington 1902c, 182–183).

That was not simply advice to students; it was the Tuskegee spirit, and its faculty led and lived by that code. For example, outside the institute, Washington's wife, Margaret, also organized the Women's Meeting, inspired by the first Negro Conference of 1892 that brought farmers to Tuskegee to encourage them to buy and own land, save money, enter into business, and get an education (Washington 1904/1915, 119).

The question that prompted Margaret Washington to start the Women's Meeting was what poor farmers could do if equipped with new techniques, hopes, and aspirations in the absence of women. How then to reach and assist such women, who lived in the countryside, with its monotony? If only there was a place they could meet The Women's Meeting dealt with all kinds of issues affecting women and girls.—Grooming moral girls.—The mother's example and duty.—Women and children's dressing.—Raising poultry.—Fruit canning.—Home management.—Housework, cooking and sewing.—Teaching men and boys to respect women "by teaching them to respect their mothers and sisters."—Diet and economy in the home.—Race pride. The tips on "How to become prosperous" is hilarious: "Keep no more than one dog. Stay away from court. Buy no snuff, whisky and tobacco. Raise your own pork. Raise your own vegetables. Put away thirty cents for every dollar you spend. Go to town on Thursday instead of Saturday. Buy no more than you need. Stay in town no longer than necessary" (Washington 1904/1915, 127–128).

Booker T. Washington himself lived by that code. Early at Tuskegee, Washington found that people earned a lot of money and worked hard, but needed financial education. Through mothers' meetings, conducted by Mrs. Washington; teacher and student visits to communities far from Tuskegee; local Negro conferences meeting monthly in various parts of the South; the annual Negro Conference at Tuskegee attended by 1,200 to 1,400 representatives discussing "the conditions and needs of the race"; the Workers' Conference, a meeting of officers and teachers of the leading schools for blacks one day after the annual Negro Conference; the County Farmers' Institute, the Farmers' Winter Short Course in Agriculture, and the County Fair in fall; the National Negro Business League, to "do for the race as a whole what the local business leagues are doing for the communities in which they exist." Added to that, alumni leaving to set up their own Tuskegees (Washington 1906, 517).

Graduates carried the Tuskegee spirit, just as Tuskegee itself was an outgrowth of Hampton Institute. By 1904, Tuskegee graduates, men and men alike, had established sixteen new schools, large enough for the state of Alabama to provide them charters. For example, in 1888, Cornelia Bowen founded the Mount Meigs Colored Institute in Waugh, Alabama, which was taken over by the state in 1911, and became the Alabama Reform School for Juvenile Negro Law-Breakers (a prison), then the Alabama Industrial School for Negro Children in 1947, and the Alabama Industrial School in 1970. In 1893, Tuskegee graduate William James Edwards founded Snow Hill Industrial Institute in Snow Hill, Alabama (now defunct). William Henry Holtzclaw left Snow Hill to found Utica Normal and Industrial Institute, Utica, Mississippi, in 1903. It is now the Hinds Community College-Utica Campus. Elizabeth Evelyn Wright, class of 1894, founded the Vorhees Industrial School at Denmark, South Carolina, in 1897; it is now Voorhees College. Charles Phillip Adams, established the Colored Industrial and Agricultural School at Ruston, Louisiana, in 1901, which is today's Grambling State University. John Wesley Oveltrea, class of 1893, established the East Tennessee Normal and Industrial Institute at Harriman, Tennessee, in 1898, but it shut down in 1912 due to indebtedness. Charles L.

Marshall, first-class honors in 1895, was invited back to Tuskegee as an assistant, declined, and headed for Wiley University in Marshall, Texas to take up a teaching position, before assuming the position of principal at Christiansburg Industrial Institute in 1896.

As many Tuskegee graduates fanned out into the plantations to work with and among the people, they found blacks in poverty, misery, and debt, drowning in punitive interest rates. These people worked hard, but had no knowledge of "how to use the results of their labor," and the two thousand students (male and female) that Tuskegee sent out to "uplift their race" taught not only skill, but also thrift and entrepreneurship (Washington 1902b, 73). Still, the black had made more progress since emancipation than had the serfs in Russia despite being freed at the same time, and the serfs having received large land grants. One-fifth of land in Virginia was in black hands as of 1902. In Georgia, taxes on black property racked in up to $5 million, with much more wealth not entered for taxation, as learned from whites. Washington predicted that in the new century blacks would buy more land and property and need more industrial schools. Italian unification had occurred in 1861, yet 80 percent remained illiterate. In Spain, for long an imperial power, 63 percent were illiterate, 78 percent in imperial Russia, and 80 percent in Latin America. Yet among the blacks of the United States, just 44 percent! (Washington 1902b, 73; 1904a, 72).

The entrepreneurial ethos that Washington embodied is evident among Tuskegee's graduates. For example, the blacksmithing graduate William M. Thomas, who worked to earn his tuition fees, then borrowed $25 to start his home business in Greensboro, Alabama. By 1902 he owned a decent blacksmithing shop, employed an assistant, and served a mostly white customer base. Also in Greensboro were located two Tuskegee graduates, who ran successful tailoring and tin-smithing shops. Another Tuskegee student, Dennis Upshaw, graduated into a life of farming with "practically nothing," but by 1902 owned 115 acres of land, a cute farmhouse, barn, and outbuildings. Upshaw grew cotton, corn, a variety of vegetables, fruit (he had a rather fine peach orchard), livestock, and fowls. Like his fellow graduate tailors, blacksmiths, and tinsmiths, Upshaw was engaged in beneficiation, boiling the farm-grown sugar in his small sugar house to make syrup. "His home and farm are models for other farmers," Washington stressed, emphasizing the advantage and function of a community having an industrial institute graduate in their midst (Washington 1902a, 292). And Thomas and Upshaw were two out of hundreds of such examples.

Tuskegee graduates were contributing to and tapping into the rising black spirit of entrepreneurship of the 1890s through 1910s that also saw the emergence of Black Wall Street in Tulsa, Oklahoma, and the Hayti community of Durham, North Carolina, home to farmers and mechanics. By 1910, blacks owned 19,057,377 acres, or 30,000 square miles of land in the South, or the equivalent of Massachusetts, Vermont, Connecticut, and Rhode Island put together. They were paying mortgages for 375,000 homes, where in 1866 they had owned none. They had come a long way from 1866 when they owned just $20,000,000 worth of property to 1910 when that inventory stood at $550 million. In 1870, almost no blacks owned a business; in 1910, no less than fifteen Southern blacks ran dry goods and grocery stores, drug stores, and other businesses. In 1900 when the first National Negro Business League conference was convened in Boston, just two black-owned banks existed that succeeded the "building and loan associations"—the True Reformers' Bank in Richmond, Virginia (est. 1881), and the Alabama Penny Savings and Loan Company (1889). By 1910 there were forty-seven black-owned banks and (since 1906) a National Negro Bankers' Association representing half of them. The banks were small (the oldest capitalized to the tune of $100,000 each) but growing rapidly with savings awareness and home building rising among black people. The availability of credit made it easier for black people to acquire property and settle down, to set up business, look for a job, marry, and raise a family (Washington 1910c, 129–130).

As a competitor for the black customer, the black business owner had complete advantage: blacks saw it as good economics and commonsense to give each other business rather than whites who had enslaved and made money out of their coerced, unpaid labor. Official segregation policies presented opportunities for black businesses. For instance, the black schools and churches were owned by and large by Southern blacks. Half a million black Baptists relied on the National Baptist Publishing Company of Nashville—annual business worth $200,000 by 1910—for their church and Sunday school literature. Reverend R. H. Boyd had founded the company in 1896 with "almost no capital and very little experience" and grew it into one of the biggest denominational publishing houses in the South. By 1910 Boyd had extended his portfolio to a thriving church furniture manufacturing business. Two other examples of black people giving business to each other were the Negro Calendar Company of Louisville, Kentucky, and the Negro Doll Company of Nashville, Tennessee. The former capitalized on an increasingly educated and self-reliant black middle class proud in their own skin color that fueled demand for calendars and advertising depicting black people and speaking their experiences. The Doll Company tapped into a black population seeking to buy their children dolls that looked like them and creating pride in black girlhood, womanhood, and motherhood (Washington 1910c, 131).

By 1910, black physicians, dentists, pharmacists, and drug stores had entered the scene. Whereas the white man was in it for the money and knew blacks as patients, the black doctor was more invested "in the welfare of his own people," a duty not just to his own race, but "to the community as a whole" (Washington 1910c, 132). Servicing as family, as a member of the community, the black physician pulls people to his practice and creates business for the black-owned drug store that dispenses the prescription. And because the community makes no distinction between his profession and his kinship, the physician, dentist, pharmacist, and drug store owner "meets his customers in their homes and in the churches" in ways that the white man cannot (Washington 1910c, 132).

These are the opportunities that Washington wanted his graduates to create, to tap into. That is what he inspired his students with, not sad stories of families that would rather buy a $300 piano while owning no land, living in a rented one-room cabin, in debt for food supplies, their crops mortgaged, and living without a bank account. In the North, only white people had such skills, and blacks found themselves without employment. Washington was determined to prevent that in the South with every ounce of his mind and hand. As he put it in 1904, "Mental development alone will not give us what we want, but mental development tied to hand and heart training, will be the salvation of the Negro. . . . The only way that we can prevent the industries slipping from the Negro . . . is to push forward, in a whole souled manner, the industrial or business development of the Negro, either in school or out of school, or both" (Washington 1904c, 6–7).

In his "Atlanta Exposition Address" in 1895, therefore, Booker T. Washington implored blacks to accept the reality that they had nothing and embrace the dignity of work and thrift. Washington gave the analogy of a ship at sea, its crew (black people) sending a desperate signal for help: "Water, water; We die of thirst!" Whereupon a friendly vessel, passing by, replies: "Cast down your bucket where you are." The distressed vessel's captain lowered his bucket, and fresh, sparkling water was duly delivered. Washington implored his fellow blacks to cast down their bucket where they were—in agriculture, mechanics, commerce, domestic service, and the professions, to "live by the productions of our hands," to begin "at the bottom of life, . . . not at the top" and never to "permit our grievances to overshadow our opportunities" (Washington 1895/2005, 19).

White people too were to cast down their bucket among the blacks, "who have, without strikes and labour wars, tilled your fields, cleared your forests, builded (sic) your railroads and cities, and brought forth treasures from the bowels of the earth." Washington implored whites of the South to invest—cast down their bucket of money—in "head, hand, and heart" education, offer their surplus land to black purchase, lease their untilled fields to them, and allow them to run white-owned factories. In return, he promised them good neighbors, "the most patient, faithful, law-abiding, and unresentful people that the world has seen." Humble, loyal, devoted, "ready to lay down our lives, if need be, in defence of yours, interlacing our industrial, commercial, civil, and religious life with yours in a way that shall make the interests of both races one" (Washington 1895/2005, 19).

Washington offered the white-dominated political and demographic order a stark choice: the almost sixteen million black hands would either be a positive force or a negative one depending on whether whites accepted to work with them. "We shall constitute one-third and more of the ignorance and crime of the South, or one third to the business and industrial prosperity of the South, or we shall prove a veritable body or death, stagnating, depressing, retarding every effort to advance the body politic" (Washington 1895/2005, 20). He was throwing down the gauntlet on white people: work with us or we are your problem! The blacks of the South had started since 1865 from sporadic ownership of "a few quilts and pumpkins and chickens" to "the inventions and production of agricultural implements, buggies, steam-engines, newspapers, books, statuary, carving, paintings, the management of drug-stores and banks" by 1895, "as a result of our independent efforts" (Washington 1895/2005, 20).

To his black kindred, Booker T. Washington was very clear: "Freedom can never be given. It must be purchased." Bottom up, not trickle down, from a very few to the rest of us. Freedom would only come "by our learning to exercise that patience, self-control, and courage which will make us begin at the bottom and lay the foundation of our growth in the ownership and skillful cultivation of the soil, the possession of a bank account, the exercise of thrift and skill, and the application of the highest culture of the hand, head, and heart to the thing which the times need have done" (Washington 1904a, 72). Washington elaborated this in *Industrial Training For The Negro*:

> Every white man will respect the Negro who owns a two-story brick business block in the center of town and has $5,000 in the bank. When a black man is the largest taxpayer and owns and cultivates the most successful farm in his county, his white neighbors will not object very long to his voting and to having his vote honestly counted. The black man, who is the largest contractor in his town and lives in a two-story brick house, it is not very liable to be lynched. The black man that holds a mortgage on a white man's house, which he can foreclose at will, is not likely to be driven away from the ballot-box by the white man (Washington 1904b, 10).

(Pause. That was before Jim crow and Tulsa Massacre of 1921, when Oklahoma's Black Wall Street was burned to the ground and its affluent black builders were massacred by racists. At the same time, I reflect on Washington's words from 100 years later, when I am faced with a generational mandate of my own, as a builder.) Washington was talking of industrial education and wealth not as ends in themselves, but means to the ends of the race. And having money and keeping it sure didn't hurt; without such means, other struggles would be more daunting. First though, black people could use some financial muscle. It started with economizing one's time.—Saving one's time.—Making the most of every minute and hour of one's existence.—Bucking the perception of black people as undependable to offer a faithful, regular, efficient service. Yeah, that hurt the race. Saving money was key.—Of course, the charge of blacks becoming too materialistic and industrialized followed (Washington 1902c, 268). Yet the race owned no single steam railroad!—No single street-car line!—No bank!—No housing block!

"In order to get hold of the spiritually best and highest things in life," Washington advised in *Character Building*, "there are certain material things that we are compelled to have first." The foundation is to "get hold of money and the saving of this money," so as to have a comfortable house, without which one cannot feel adequate or confident enough to do one's best work and to serve others diligently (Washington 1902c, 269). And he counseled on how to have one: save small amounts of money weekly, monthly. .—Buy a piece of land.—Mind the location—certainly not in a "crowded, filthy alley." (Washington 1902c, 57–60). After that, money could also enable one to build schools.—Churches.—Hospitals. To do one's part. But first, one must have self-control.—Discipline. House, money, and bank account imbued one with self-respect. "An individual who has a bank account walks through a street so much more erect; he looks people in the face," Washington went on. "The people in the community in which he lives have a confidence in him and a respect for him which they would not have if he did not possess the bank account" (Washington 1902c, 273). Flashy clothes were a sign of lack of self-discipline. The purpose of clothes was comfort and neatness, not showing off. Finally, the Tuskegee principal advised that while it was their right, "it is hardly the part of wisdom for [black people] to waste money in buying musical instruments before they get a home in which to put them, and before their children have had enough education to use them properly" (Washington 1906, 517).

This description, what Washington called the "exceptional" man or woman, sums up the kind of person that Tuskegee's philosophy of education was designed to create:

> The leader, the exceptional person, is never satisfied with the old way of doing things. No matter what it is, if it is washing dishes, sweeping a floor, cooking, ironing, working on the farm, in a garden, or teaching school, no matter what it is, the exceptional man or woman is never satisfied with present methods, but is looking out for new and better ways of doing his (or her) work. He or she is always anxious to meet and talk with a person who is supposed to know most about that kind of service, is always reading every book, every magazine, every newspaper, that he can put hands upon, anything that will give him a suggestion concerning new ideas, concerning a new method, that can be employed in furthering that work. That person becomes the exceptional individual by not being satisfied with old methods. If you will give heed here and hereafter constantly to all these little suggestions, you will not be the failure in life, but you will be the exceptional individual. (Washington 1917, 484)

Those who dare to invent the future not only ask questions of the past and the present, but also turn the present into raw materials for the invention of the future. They are not careless, contemptuous, or ignorant of the past and present. They do not make the past a place of residence, or keep their eyes down enamored by the present. They are contented only with not being content. They are builders, not simply dwellers. They do what they will talk about, not just talk about what they will do. They know in order to do, know through doing, and do in order to know. They see life and futures beyond their own lifetime, understand that life is finite, and therefore live not just in their own lifetime, but live for others. That way, long after they have breathed their last, they continue to live in others, as legacies of work done in their lifetime. Great black men and women who give not to receive, but seek and receive in order to give others. Seek to know so that others may know.

BOOKER T. WASHINGTON IN AFRICA

Booker T. Washington's influence on higher education went well beyond the borders of the United States. Whereas Hampton had already begun industrial education well before Washington arrived there to begin training, it was not the institute but Tuskegee instead that made it a famous model of learning. And in Africa and its diaspora, it was Washington, not General Samuel Armstrong. The reason is simple: Washington was a black man who led by the power of example, the passion for his people, a builder of institutions to turn his passion for his people into catalytic spaces to change their lives through thinking-doing and entrepreneurship. The students, like their mentor and inspiration, went on to establish institutions in Africa, taking that which was good about Tuskegee, as additions to their own knowledge systems, and putting it in the pot (their context), and designing institutions of their own. The thorough identification of what was useful overseas, mastering how it worked through experiencing it on-site, and then bringing it home, not to suggest it to somebody else or implement it uncritically, serves as a precedent to how the present, ubiquitous African diaspora can best make use of its global omnipresence for the good of the continent while living wherever they choose.

To be clear, the two builders of Ohlange and the University of Fort Hare (both in South Africa), John Langalibalele Dube and Davison Don Tengo Jabavu, respectively, were neither the first nor last Africans to build institutions of higher education. The earliest concept of a university in Africa is traceable to the University of Al Quaraouiyine (Morocco, 859 CE), Al-Azhar University (Egypt, 970 CE), and Timbuktu (Mali, 1100 CE), connected to Islamic influences that rendered most visible, but may not have invented, the practice of knowledge production in a secluded, built environment by an elite circle of privileged men. (Systems of knowing and knowledge systems is a subject for a sequel that demanded me to write a separate book on the subject.) This did not necessarily mean that Islam deliberately criminalized or delegitimated other modes of knowing, specifically within everyday life; Blyden's observations on this question are very clear.

The white colonial university was established as an outpost of Western-based institutions (churches and universities) without latitude to deviate from syllabi or doctrines laid out in Europe. The earliest ones were Fourah Bay College (Sierra Leone, 1827), University of Cape Town and Stellenbosch University (South Africa, 1829 and 1866), University of Khartoum (Sudan, 1902), Cairo University (Egypt, 1908), and University of Algiers

(Algeria, 1909). In most of Africa, no universities were built, and if university colleges (subsidiaries of European universities) were available, they were for whites only. To get a college or university education during the European colonial occupation, black people were forced to seek admission in South Africa, Sierra Leone, Liberia, or overseas. In addition, up until the 1990s, most schools in Africa were still having their examinations set by European universities.

Ohlange and Fort Hare, by contrast, were intended to offer education that met black people's needs; they were founded either by black people themselves (Ohlange) or black people commissioned by the white government to research, build, and operate them (Fort Hare). What now follows are aspects of the colleges that exemplify the meaning and inspiration of Washington and Tuskegee in Africans' daring to invent knowledge in service of and through doing on the continent. They in no shape or form constitute authoritative accounts on the two institutions' histories, only aspects relevant to the theme of this chapter and book.

* * *

Best known as the founding president of South Africa's African National Congress (ANC), John Langalibalele Dube (February 22, 1871–February 11, 1946) was educated by US missionaries at the American Board Zulu mission, and in 1887 he was sent to the United States for further education. He enrolled in Ohio's Oberlin College preparatory department and remained in school for two years. Afterward he lived in the United States working at various jobs before returning to Natal in 1892 to preach among the Zulu and then returned to the United States in 1896 for missionary fundraising and to further his theological education at the Union Missionary Training Institute. He lived nearby in Brooklyn, New York, until 1899, when he was ordained as a Congregational minister, and then traveled extensively across the United States before returning to cofound, with his wife Nokutela Dube, the newspaper *Ilanga lase Natal*, where the emerging school- and self-educated African intelligentsia published matters relating to their own people.

On one of these trips, Dube visited Alabama's Tuskegee Institute and met its president, Booker T. Washington, developing an instant connection to the person and his educational model. Washington helped Dube raise adequate funds for his dream and in 1897 gave him a platform to market himself by inviting him to give the commencement speech at Tuskegee and prevailing upon Hampton to invite him to do the same (Marable 1974b, 401–2; Hunt 1975; Dube 1897, 1898). In his speeches Dube strategically deployed flattery to get into white hearts to loosen their pockets, by comparing his people's "heathen" ways with the "wonderful things, such as steamships, wagons, frame houses, furniture, machinery, etc., which the civilized man can make." Provided his own people were similarly trained, Dube promised that they too could make them. But he also performed a double act by expressing concern that Africa's minerals were being extracted for processing in Europe and North America, even while black people were consigned to unskilled, miserly paid jobs (Dube 1898, 438).

Dube was so convinced of an industrial education approach for African advancement that he became known as "the Booker T. Washington of Africa." In 1901 he established the Zulu Christian Industrial School, modeled after Tuskegee, and it became another leading black educational center in South Africa. In 1909, Dube it became known as Ohlange Institute. Its object was to train Africans in industrial, commercial, and entrepreneurial skills. On the one hand, one might very well be correct to see the Hampton-Tuskegee model as white America and South Africa's program to contain black agitation for equality. My interest lies only in the Tuskegee or Ohlange of one whose generational mandate is to build physical infrastructure for solving problems especially in one's community. Hence I emphasize that, on the ground, the institute was an outcome of community (cooperation and solidarity) that defined a role for industrial education, no longer as a Tuskegee or Hampton idea, but vetted by Dube on-site in the United States, deemed by him to answer specific needs back home. That is a possible role for the African diaspora that wishes to see Africa develop on its own terms.

The siting was decided upon by the entire community through traditional forms of democratic consultation and decision making, the local chief raised the capital to purchase the land, and villagers offered their labor and ideas in the construction of the institute. Local elders served as the school's founding trustees. Africans came from as far as three hundred miles to witness "a native, one of their own people, being at the head of the enterprise" (Wilcox 1909; Marable 1974a, 258). That is a role that traditional leaders, councilors, and communities can play—this professional, eager, communal, and anchoring role for a diaspora bringing resources it (not foreigners, NGOs, or government) has deemed appropriate to address local needs and turn them into opportunities.

To his students and staff, Dube conducted himself, and Ohlange, as visible role models of what he expected of them. He wanted students to respect him as "the man who works," not "the idler" who just talks. Business and hard work, not political agitation, was for him the only way toward self-rehumanization (Dube 1901). Like Tuskegee, Ohlange started only with agriculture—thirty acres of maize, potatoes, beans, and ground nuts, with orange, peach, and banana trees between the contours of the fields. In the mornings, students sat in wooden chairs attending lectures in arithmetic, social sciences, and humanities in the middle of the plowed field for lack of built classrooms. In the afternoons, they worked the fields, made furniture in the workshop, practiced masonry through building structures Ohlange (they) needed, and so on. In 1902 when a classroom was completed, students slept on its floor on mats, others on the ground in two unfinished buildings with neither windows nor door. More than one-half of the funds (nearly $6,000 in 1898–1902) came from the American Committee, which also sent frequent consignments of iron bedsteads, school room desks, library books, blacksmithing and carpentry tools, and household utensils (Dube 1901). Dube undertook fundraising tours of the United States, including a 1904–1905 family tour with his brother Charles Dube, a Wilberforce College graduate, when he visited Brooklyn and Chicago, securing $5,000 from Emaroy June Smith, the wife of a Chicago businessman and philanthropist, to build a men's dormitory dedicated in 1907 (Marable 1974a, 259).

Those who have never built anything or been in a position of responsibility may not be able to distinguish strategic deployment of rhetoric from truly held inner principles of those thinker-doers who have to raise money to build something. There is a language intended for public display to create a zone of comfort with investors or funders, so that they do not see you as threatening, or simply to meet their criterion for awarding the funding. When we examine John Dube and Washington's rhetoric, we see a strategic, deliberate, and careful design and deployment of language to dispel suspicion and get into investors' pockets. Regrettably, in our readings, we have been too consumed with the rhetoric and the writing; our judgments are colored by the political struggles of the time, our strong desire to conquer freedom, and create a self-determined path.

Consider this statement by John Dube if, for example, read later by militant political formations in South Africa like the Pan-Africanist Congress: "In Natal, as no doubt elsewhere, we may divide labour and labourers into two classes—the skilled and the unskilled. To enter the field of skilled labour is in our Colony, the prerogative of the White man." That statement, titled "The Zulu's Appeal for Light and England's Duty," if read from the feisty nationalistic mood of the twentieth century, makes Dube a pathetic puppet. So too his elaboration that unskilled labor was the black person's "allotted sphere of activity, and with it we are at present abundantly satisfied. . . . We wish to be trained and be intelligent 'unskilled labourers,' in the house, store, workshop and farm—useful servants and assistants, small jobbers and peasant farmers" (Dube 1910, 22–25). That self-portrait of total black submission massaged the white ego, caressed the white heart into silky softness, and loosened the talons on the purse. White Americans and Britons poured thousands into Ohlange thinking it would produce a mass of tame black laborers.

On the ground however, Ohlange emphasized the humanities and social sciences as the base and agricultural and industrial arts as the tools to turn critical thinkers into doers. The portrait of industrial education as preparing the African for unthinking manual work was a hoodwink; students had to take Latin, English literature, science, history,

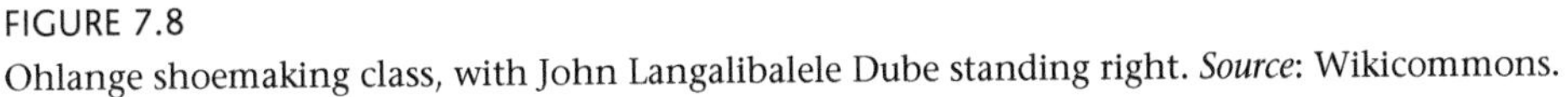

FIGURE 7.8
Ohlange shoemaking class, with John Langalibalele Dube standing right. *Source*: Wikicommons.

geography, bookkeeping, shorthand, Zulu, and Sesotho first. In one prospectus, Dube plays with words in ways that would gratify any white donor while making clear the core idea of black economic empowerment through value addition and entrepreneurship:

> When we have Native Missionaries who will not only preach on Sunday, but teach the people how to live a Christian life of toil six days a week, we shall then see the curtain of darkness lifting. Teach my people trades by which to supply their wants and gratify the desires which their enlightenment creates, and then we shall see a great and grateful people eagerly competing with their brethren in the manufactures and commerce of the world and glorifying the Lord our God. (Houle 2011, 233)

Dube's successor, D. G. S. M'timkulu, would say in 1951 that industrial education trained students to create jobs, not seek employment, hence the phasing out of agricultural courses by the mid-1920s. The industrial portfolio of courses was limited to entrepreneurial trades only, such as carpentry, tailoring, and leatherwork. By "industrial," white people meant unskilled labor trained to take instructions without questioning or reasoning; by industrial education, both Dube and Washington talking about storekeepers, tailors, and craftsmen who engaged in private enterprise, fully conscious of the social, political, economic, and technical (M'timkulu 1950, 22–23).

True—the formation of the ANC in 1912 in the wake of the Bambata Rebellion of 1906–1908 has been popularized as the most significant moment of the early 1900s. It is normal for a century of consciousness preoccupied with "the political." As important as that milestone was, it helps shed light on struggling for a cause, what is needed to struggle, how to "struggle," but does not provide us with ingredients to render fruitful the legitimate struggle the ANC went on to wage. Neither will that moment be ever erased from history, even if it doesn't tell us how to put in place physical and intellectual structures to turn independence into business, jobs, and materially satisfying lives.

Some will jubilate in Dube's rhetoric as founding president of ANC; I place more weight in his contribution as builder of Ohlange and in the students that acted on his inspiration as a critical thinker-builder of institutions to improve material lives of their communities, with themselves as catalysts. The Business League that he created in the 1930s was intended as a business development platform and vehicle along the lines of Washington's National Negro Business League. It played a key role in promoting start-ups and growing private enterprise and capital, with a focus on keeping and multiplying money within the community. The logic being that money black wage earners like teachers, government workers, mineworkers, agricultural workers, and so on earned would stay within the black community. Even within the ANC, he defined his presidency as one of being "up and doing" and of self-help as a pathway to self-emancipation, with Washington as inspiration, dedicated to the education of his people (Walshe 1970, 55). (The ANC has never quite steered away from accommodationism; hence militant elements like Josiah Gumede in the 1930s and Julius Malema post-2000 were ejected. The reasons lie in delicate strategic deployments that go back to the ANC alliance's formation in 1912.) Some scholars bought too easily into Dube's rhetoric and took it for his beliefs rather than rhetoric as strategic deployment; meanwhile the aspiring students had to work on Ohlange's thirty-acre farm to prove themselves and earn money for enrollment. Or they sought labor to pay for their termly fees (Marable 1974b, 403). Accommodationist tactics enabled Dube (like Washington) to be left alone to engineer a confident, proud, and economically empowered black person, and with that would inevitably come a new assertiveness.

The language of accommodation with white paternalists was intentional: Dube, like Washington, wanted white people to fund black economic emancipation; after all, they had enslaved Africans. Even if they did not fund it, Dube was not going to wait on them (Marable 1974b, 401). The fundamental reason why Washington and Dube's industrial education received criticism was because it was easily labeled capitalism, in an age when socialism (and Marxism and Leninism in particular) was seen as a marker of an intellectual and revolutionary. The ANC elite vociferously defended black entrepreneurship while attacking communism; it was an example of black people not talking about what they would do when they at last governed themselves, but deploying entrepreneurship to attain political power, and an example of the shape that economic self-empowerment could take after attaining power.

As tool or method of self-liberation, economic betterment was utterly useless against apartheid, only meeting force with force could compel recalcitrant racists to accede. It is also true that such militant methods, full of rhetoric, don't build sustainable bridges, roads, and other economic infrastructure, or guarantee the very freedom the nationalist leaders promised would flourish after *uhuru* (Marable 1974b, 406). The assumption that the political and guerrilla leadership would make good presidents and governments does not seem to have materialized in the majority of Africa. *Chihondo-hondo* (liberation war-ism) quickly became a weapon to threaten and stifle the freedom of expression and expression of freedom the people fought for, freedom that is the oxygen innovation needs, without which Africa has suffocated talent into leaving for the same European countries that colonized the continent. The liberation struggle was fought to liberate us from the colonizer. The colonizer went home. Now we have followed the colonizer to Europe and submitted ourselves, willingly this time. After independence we had leaders that were critical thinkers, very theoretically savvy, most of them humanists and social scientists, in Savile Row suits, eloquent in French and English, but never doers. For this struggle to sustainably develop Africa, we need people that at least built something tangible—or died trying—and who have distinguished themselves in doing so, not just populists who are sweet on rhetoric but hollow on action, useful puppets of Beijing, Paris, London, and Washington, DC. John Langalibalele Dube was one such thinker-doer leader.

* * *

I have said that white people read too much into Dube's rhetoric that Ohlange needed money and donations to train submissive black laborers that accepted their place under the white man, and who would act as pacifying agents of the colonizer among their people. The government assumed that building an institution it could totally control (unlike Ohlange, which was privately owned) would speed and scale up the process, not simply within KwaZulu-Natal, or South Africa, but across Africa, now that such an institute would become the new Tuskegee of Africa.

The man chosen for the task was the black educationist Davidson Don Tengo Jabavu (popularly known as DDT), graduate of University of London, who went on to be one of South Africa's most prominent black professors of his time. DDT was the son of John Tengo Jabavu, editor of *Imvo Zabantsundu/Black Opinion*, South Africa's first black-owned newspaper. DDT had attended Lovedale back when his father was there, and then Morija,

FIGURE 7.9
Father and son—John Tengo Jabavu (left) with his son Davidson Don Tengo Jabavu, 1903. *Source*: Wikicommons.

before entering a Quaker high school in Wales. In 1906 he earned his BA at the University of London and then enrolled in the MA program at Yale. In 1913, the South African minister of native affairs commissioned Jabavu to go and study the "methods being used" at Tuskegee "with a view of adopting them in connection with the education of the natives of South Africa" (Marable 1974b, 402).

This section draws from his book, *The Black Problem* (which mirrors closely Washington's *The Negro Problem*), in which DDT deliberately addresses the mission he was sent to execute at Tuskegee. Namely: to learn what he can and return to establish a similar institute appropriate for black people in the Union of South Africa. In one particular chapter titled "BOOKER T. WASHINGTON: What he would do if he were in South Africa," DDT chronicles his visit to Tuskegee to study the institute's organization and methods "with special reference as to their applicability to the peculiar circumstances of Natives in South Africa" (Jabavu 1920/1969, 27).

DDT was not interested in what Booker T. Washington had actually accomplished in the United States but rather what he "would have done had he been in South Africa." That was a strategic deployment of Washington, not an intellectual, analytical reading of Tuskegee (Jabavu 1920/1969, 28). Being at Tuskegee in person enabled DDT to see students, teachers, and community at work—the summer school attracting four hundred teachers across states, faculty and students engaged in extension work in the surrounding communities and in industrial and agricultural projects during the vacation (Jabavu 1920/1969, 29).

At Tuskegee, DDT observed "the sense of public responsibility" and practical application in the education curriculum. For instance, students spent three days a week in the classroom, three "at their trade," whereas in the British system in South Africa, the black student learned in the classroom only without the trade part. The student spent one out of two days in masonry or harness making, skills he deployed in "the struggle for life," while the student in the British system recited Shakespearean poetry and learned European history—things totally alienated from the *imbongi* (poets) and *izibongo zake* (totemic praise poetry) and from African history, including that of their own ancestors. The one brought the student in touch with the community, the other created an elitist academic engaged in an academic subject of no everyday necessity to the masses. The aim of the curriculum was to equip students with "a sound, moral, literary and industrial training, so that when they leave the school they may, by example and leadership, help to change and improve the moral and industrial condition of the communities in which they live." This would be accomplished through linking and combining academic studies and industrial training in order to stress the social and moral significance of skilled labor while illustrating through doing "the practical meaning of the more abstract teaching of the class room" (Jabavu 1920/1969, 32).

Tuskegee tuition was free for all students, it being considered unconscionable to turn away a diligent and worthy student due to lack of means; in any case, as young women and men with applied skills, those skills could be applied to holiday work to pay for or in lieu of fees. Night school was also an option for the classroom portion of the learning (Jabavu 1920/1969, 53–54). All students were enrolled in an academic department, divided into day and night school. The program took seven years to complete, three for preparatory work, four for the actual normal course. The Day School Junior curriculum (equivalent to the Standard VI of the British) was composed of reading, grammar, arithmetic, concrete geometry, writing, drawing, geography, and gymnastics (for girls). The Day School B. Middle Class (first-year pupil teachers) consisted of reading, grammar, arithmetic, US history, botany (half year), agriculture (half year), hygiene (half year), and gymnastics (for girls). The Day School A. Middle Class (second-year pupil teachers) was composed of required reading, including spelling grammar; any three of algebra, bookkeeping, ancient history, or chemistry; and any two electives, at least one directly related to the student's trade. The Day School Senior Class (third-year pupil teachers) was required to take English, and two electives from education, economics, modern history, bookkeeping, and any one of either geometry or physics or chemistry (Jabavu 1920/1969, 36–37).

In *Story of My Life*, Washington had stressed the need of the academic program "to teach little and thoroughly, than much and insufficiently." The overall catalogue was quite comprehensive, each course subdivided into further subjects, all oriented toward practical applications: English mathematics (arithmetic, algebra, concrete, and plane geometry); elementary civil engineering; bookkeeping; free-hand industrial drawing; writing; economics; history; geography; industrial history; natural science (chemistry, physics, mechanics, heat, electricity, light, sound); and so on. Conspicuous by their absence were "modern and dead languages," which omission DDT concluded to derive from "the necessities of the heavy industrial counterpart programme than to an underestimation of their cultural value" (Jabavu 1920/1969, 37–38). Needless to say that his comments on studying Greek politics and history instead of language shows Washington's doubts on the urgency of such.

To apply Tuskegee to South Africa, the difference between the experiences and context of the black people involved had to be taken into account, starting with the universities and colleges they could attend. DDT divided the US's black educational institutions into two: The black degree training universities and colleges offering theology, dentistry, pharmacy, medicine, law, humanities, and so on, such as Howard, Wilberforce, Lincoln,

Meharry Medical College, Tennessee, and more than ninety others. Then there were the around four hundred agricultural, mechanical, and industrial institutions, the largest of which was Tuskegee, which took seven times the number of students the colleges and universities trained (Jabavu 1920/1969, 40). South Africa had just one college for blacks that offered a Wilberforce- and Howard-type education: Lovedale. White universities in the Northern states of the United States were open to black students; in South Africa none were open to blacks, prompting black students to seek education in the United States or Britain. Tuskegee drew its faculty from these Northern institutions (Jabavu 1920/1969, 41).

DDT's was a selective reading of theory and experiences elsewhere in search of resources that could be strategically deployed back home to design a black university. In one instance, he found that the hygiene and bookkeeping syllabi of Tuskegee were "highly necessary" in South Africa's black schools. In another, he dismissed a Bible Training School to "not offer much in the way of new suggestions" to add to what was already in practice (Jabavu 1920/1969, 41). DDT particularly appreciated the four principles of Tuskegee's industrial training: cultivating the dignity of labor; vocational training; supplying trained industrial leaders to the nation; and assistance with tuition, including working for tuition (Jabavu 1920/1969, 42).

The genius of Tuskegee was not just in learning through making or making through learning, but learning and making for sale. The business agents department was responsible for buying and selling for the institute, as well as processing or value addition divisions, among them bakery and butchery (Jabavu 1920/1969, 42). In total, DDT counted forty trades and industries being taught at Tuskegee, including architectural and mechanical drawing; blacksmithing; woodturning; brickmaking; brick masonry; plastering and tile setting; carpentry; applied electricity; founding; harness making; printing; carriage trimming; steam engineering; machine shop practice; plumbing and steam fitting; shoemaking; tinsmithing; tailoring; wheelwrighting; and bookkeeping and accounting as applied to the trades. Lovedale also offered apprenticeship in wagon making, printing, bookbinding, shoemaking, carpentry, and so on, but these courses were optional, whereas at Tuskegee all students were required to choose a trade. In the process of training, the instructors ensured that the quality and commercial value of the carpentry, wheelwrighting, repairs, machining, and steam engineering were high, offering business opportunities that no black in South Africa could ever even dream of (Jabavu 1920/1969, 42). DDT wished that for the envisaged institute in South Africa.

Tuskegee also had a Department of Women's Industries, which was in charge of trades such as plain sewing, dressmaking, women's tailoring, millinery, cooking, laundry, soap making, domestic training of girls, mattress making, basketry, broom making, and child nursing and nurturing (Jabavu 1920/1969, 42). The Division of Cooking had two kitchens, three dining rooms, a sitting room, and a bedroom and bathroom, the housekeeping being done by the female students undergoing this four-year course (Jabavu 1920/1969, 43). The admission of women into college struck DDT as a fundamental requirement back home.

But agriculture was, DDT said, "much a practical lesson to [the Union of] South Africa as anything else Tuskegee can offer," and it was not a question of whether or not to adopt, "but how best to introduce it on the model of Tuskegee." The Union could simply "import" a "complete staff" of Tuskegee graduates as the German government had done in Togo and station them at Lovedale on an experimental basis, or select black students and sponsor them to go to Tuskegee, "take the full course" at Tuskegee, and come back to implement it. The US government had done something similar when bringing twenty Puerto Rican students to Tuskegee. A few of the candidates could similarly be trained as farm demonstrators and return to set up demonstration farms along the Tuskegee model (Jabavu 1920/1969, 45).

In DDT's opinion, Washington had shown the most interest in and attention to Tuskegee's Department of Agriculture than any other unit. The department was organized into "divisions": farm crops; gardening; fruit growing; care and management of horses and mules; dairy husbandry; dairying; swine raising; beef production and slaughtering; canning; veterinary science; and poultry raising. The Farm Crops Division raised all foodstuffs and managed the institute's 1,200 head of livestock, providing hands-on farming training in equipment and methods. DDT found the Two Weeks' School for farmers applicable to all black schools in South Africa that offered agriculture courses; through it the school could "exert vital influence around its vicinity" (Jabavu 1920/1969, 47).

The Department of Research, Consulting Chemist, and Experiment Station deployed technical, experimental methods to generate scientific facts to inform all branches of agriculture. The lab work involved physical and chemical analysis of soils, fertilizers, forage plants, milk, butter, cheese, and so on, while the Experiment Station conducted experiments to improve soil, cotton, and corn breeding. DDT also witnessed different types of incubators in operation and feeding, breeding, and other experiments underway related to all things poultry. It was not the science itself that impressed him; only when it was shown that students were not just learning chemistry, but also applying the knowledge of chemistry to enrich the soil and to cooking and dairying and that of geometry and physics to blacksmithing, brickmaking, farming, and so on did white and black, North and South, begin to cooperate (Jabavu 1920/1969, 50). In other words, the possibility of science-at-the-intersection, as activity through which black and white, people of all shades, could be brought together.

It bears reemphasizing that DDT's reading of Washington's writings and lessons of Tuskegee was, as is mine, a selective one intended for strategic deployment in inventing new ways of fusing training and real life, school knowledge and everyday practice, what is useful and what to take home. A case in point is his selection and reading of Washington's *Working with the Hands*, which talks about how Tuskegee experiments in its neighborhood, which had poor agricultural soils, led to the doubling of sweet potato yields (Jabavu 1920/1969, 49). The "pithy, practical, and commonsense character" of the text impressed DDT "enough to suggest its wide circulation among white and black in South Africa"; it had to be taught to every black South African student at the very least (Jabavu 1920/1969, 50).

DDT also took note of the "practical effects" of the Extension Department, which Washington saw as just as critical as the institute's educational work. Its role was to change public opinion and educate people on better methods of farming and inspire the youth to remain on farms instead of leaving to work in cities. *The Farmers' Institute, Demonstration Farming, and the Jesup Agricultural Wagon were all geared toward providing extension services. The wagon, another George Washington Carver invention, was a mobile workshop—a vehicle equipped with tools and instructional aids to go around the countryside transforming fields, pastures, forests, and farmsteads into "transient" farmer education "workspaces."* With these methods Tuskegee reached no less than one hundred thousand farmers in Macon County, Alabama, themselves graduates of the institute's training of black people in hand, head, and heart and entrepreneurship, which had boosted property ownership in the county by 600 percent. DDT was in awe of these milestones and dreamed the same for South African blacks (Jabavu 1920/1969, 52).

DDT deployed Washington's ideas to black education at Fort Hare and outside it. He was a staunch promoter of "better" farming methods, and like his role model emphasized in his public rhetoric the value of manual work and building a society based upon "racial cooperation." By the time of his retirement in 1944, DDT had become not just an academic giant (the preeminent professor in Fort Hare), but touched so many lives beyond the institute and, by extension, that of communities from which his students came, through whom access to his self-help knowledge became possible. Besides students as vehicles for community transformation, DDT also set up the South African Native Farmers' Association dedicated to improving the agricultural practices of Africans and fostering environmental conservation. It was modeled on Washington's phenomenally successful Annual Negro Conference established in February 1891, which brought together black farmers, mechanics, school teachers, and church ministers, but attracted mostly farmers inspired not by theory, but the deliberate solutions to concrete problems they encountered during farming.

FIGURE 7.10
The University of Fort Hare, 1938. *Source*: National Library of Scotland.

FIGURE 7.11
Students eating in the dining hall, Fort Hare, 1938. *Source*: National Library of Scotland.

Founded in 1916, South Africa's biggest "Tuskegee," the National College at Fort Hare, today's University of Fort Hare, would go on to become a strong bridge between Black America and Black Africa, where visiting black intellectuals from Howard, Fisk, Penn School (South Carolina), Hampton, and Tuskegee visited to lecture on and debate issues of black struggle. Its students also went abroad for further education not only at historically black universities, but also at Columbia, Yale, and Cambridge.

Inevitably most of the African nationalist leaders educated at Fort Hare were teachers passionate about farming. The College Diploma in Education was one of the very first curricula offered at the college (in 1916). Core courses included principles, history, and psychology of education, elementary biology and physiology, hygiene, methods of teaching, and school organization. For teaching practice the students were deployed to the countryside. Although most of the students were drawn from South Africa, by 1924 increasing numbers of students continued to come from all the African territories under British rule or protection in Africa, south of the Great Lakes. Some traveled overland by train from Nyasaland, Northern and Southern Rhodesia, and Bechuanaland Protectorate. Others coming from Natal and east Africa landed at the seaport of East London, the shipping lines and state railway offering very generous student concessions. No matter where the student was coming from, those not living at a railhead or port rode there on horseback or bullock, ox-drawn wagon, or walked, the donkey alongside him, carrying the luggage. It was a depressing sight when, having walked tens of miles, the student found the river flooded and missed the train or steamer (Kerr 1968, 150–16, 125–128).

While at Fort Hare, the student also discovered that the government could be challenged through organized political protests. Most of the foreign students joined the ANC. Among those who went to Fort Hare were Nelson Mandela and Albert Luthuli of the ANC, Botswana's Seretse Khama, Zimbabwe's Robert Mugabe, and Desmond Tutu, who was the chaplain, while Z. K. Matthews was one of the African professors.

8 *KUJITEGEMEA ZAIDI:* EDUCATION FOR SELF-RELIANCE

Julius Kambarage Nyerere led his country Tanzania to independence in 1961 and then became its president until his voluntary retirement in 1985. He was born on April 13, 1922, in Mwitongo in the Mara region of Tanzania, one of twenty-five children of Nyerere Burito, chief of the Zanaki people. Nyerere means caterpillar, a name denoting Burito's birth in 1860, the year of the worm caterpillar plague. Julius Kambarage's mother Mugaya Nyang'ombe was born in 1892, marrying Burito in 1907 at age fifteen. She bore the chief four sons and four daughters, including Kambarage.

The British colonial government had a policy of actively encouraging chiefs to have their sons educated to ensure an enlightened (read brainwashed) royal elite that was easier to manipulate. Hence in 1934 the young Nyerere was sent to the Native Administration School in Mwisenge, Musoma, some twenty-five miles away from home, leaving his age mates who could not afford primary education. Having passed with distinction in the 1936 exams, Nyerere was admitted into the elite Tabora Government School, completing his secondary education in 1942 and gaining admission in Makerere College, Uganda, where he graduated with a diploma in education in 1945 and returned to Tabora to a life as a teacher, farmer, and Tanganyika African Association activist. Four years later he was admitted into the University of Edinburgh for a master of arts degree, graduating in 1952, the year he returned to Tanganyika to teach history at St Francis's College, Pugu, near Dar es Salaam. A year later he was elected president of the Tanganyika African Association based on his oratory skills and educational qualifications.

On July 7, 1954, the party formally changed its name and became a political party, the Tanganyika African National Union (TANU). Influenced by Mahatma Gandhi, Nyerere steered TANU toward a nonviolent campaign for independence. In the inaugural election of August 1960, TANU ran the table, sweeping seventy of the seventy-one available seats.

FIGURE 8.1
Julius Kambarage Nyerere, founding president of Tanzania. *Source*: Wikicommons.

On December 9, 1961, Tanganyika gained independence from British rule, with Nyerere as prime minister, and president a year later. With the unification with Zanzibar in 1964, the country's name was formally changed to Tanzania, with Nyerere as president. Julius Nyerere died of terminal leukemia on October 14, 1999, at age seventy-seven.

Just like I have done with the other critical thinker-doers, I resist being enlisted into arguments and judgments that others have made about the former president. We are writing for different reasons. There is no one Nyerere, but many Nyereres, depending on what we look at. Some will privilege the great statesman, and stalk the ire of those who see a ruthless ruler who imprisoned his opponents. Some will draw inspiration from Nyerere the great Pan-Africanist and African liberationist par excellence, others a reckless Tanzanian president who squandered the country's sweat and wealth on Pan-Africanist ventures that benefitted nothing to Tanzania. Some will be allured by Nyerere the theorist of African socialism, drawing fire from those who see a copycat of communist thought that he simply instantiated in Tanzania. And so on.

I have no interest in those interpretations at all, or responses of Tanzanians to any of these specific declarations, positive or negative. That is for Tanzanians to write. As a rule, I do not engage their Nyerere's speeches because he implemented every word he said, or because he was a saint, or because I endorsed his bad deeds. My deference to his work is a selective one. The reason is simply that I am looking for ideas that are relevant to designing and engineering the sustainable futures of Africa and that I can strategically deploy in my own efforts in my community and country in particular.

For that purpose, I have found Nyerere's speeches and lectures on *ujamaa*, self-reliance, education for self-reliance, and the role of the university/higher education indispensable to the critical thinker-doer mode that I embody. I will now discuss each in turn, reflecting Nyerere's thought through my own program and perspective.

UJAMAA AND SELF-RELIANCE

TANU was committed to socialism as a longstanding African tradition, to, with modifications, guide the creation of a new Tanzanian society based on equality and self-reliance. By 1966, it had become obvious that the philosophy and policies of African socialism were not well understood and were being in fact questioned (Nyerere 1968, 1).

The rationale for the Arusha Declaration, delivered on February 5, 1967, in the town of Arusha, on the site that the Uhuru Monument now occupies, was therefore to define socialism and to provide a roadmap toward achieving it. TANU chose to call socialism *ujamaa* (familyhood) to emphasize "African-ness," communality, building on the past as a foundation, "building to our own design," not importing foreign ones into Tanzania, "and trying to smother our distinct social patterns with it." To grow as a society "out of our own roots but in a particular direction and towards a particular kind of objective." Selecting and emphasizing properties of traditional organization and "extending them so that they can embrace the possibilities of modern technology" to solve real-life problems. In short, strategic deployment of our ways of social living to create something "uniquely ours" with methods unique to us (Nyerere 1968, 2). If it was true socialism is universal, then it was able to "encompass us as we are—as our geography and our history have made us" without demanding that we remold ourselves to a singular pattern. In other words, universality in the plural based on different methods, customs, and styles, each equally valid (Nyerere 1968, 3).

Ujamaa was a return to, and illustration of the importance of, history to designing a prosperous Tanzanian future. "The past and present are one," Nyerere had declared when opening the International Congress on African History in 1965. What should drive our inquiry should be "our own desire to understand ourselves and our societies, so that we can build the future on firm foundations," not other people's reasons to understand us. The primary sources of our history are not the vast written archives that exist inside and outside the continent but the "localized unwritten historical knowledge" critical to reconstructing a "really African history." Knowledge of Africa in 1965 was mostly

FIGURE 8.2
Uhuru Monument in Arusha, Tanzania. *Source*: Wikicommons.

written by and for outsiders. It still is. The authors were "fascinated" with the "discovery of Africa," the "great explorers'" travelogues, and European empire, only lately taking interest in African society and culture and the effects of European intrusion. But it is only "when these things are looked at from Africa outwards that an 'African history' will develop," otherwise it remains a history of white people, white things, and white values in Africa. It is a useless knowledge, useless history—what do I, as an African, do with that? What is its use to me? (Nyerere 1965e, 85).

We Africans have a special responsibility. We are "of this continent, and concerned so intimately with its future as well as its past," and the kind of historical knowledge it needs. Our responsibility is to ask serious, deep, and searching questions. Two of Nyerere's questions are important: "What are the African concepts of the past? Of what objective value is the history which was passed from generation to generation in traditional

FIGURE 8.3
The restoration of communality to labor, production process, and benefits was a central part of *ujamaa*, as this 1960s' photo shows. *Source*: Unitedrepublicoftanzania.com.

societies, and how can it most accurately be collated?" (Nyerere 1965e, 83). Not just the accumulation of facts, but also an interpretation of those, how they help us address our African challenges and unleash futures saturated with prosperity and happiness (Nyerere 1965e 1965, 85).

Our growth comes out of our own roots, not grafting onto roots of alien cultures. Our own politics.—Needs.—Perspective.—Ideas.—Experiences.—Experiments. "We start from a full acceptance of our African-ness and a belief that in our own past there is very much which is useful for our future" (Nyerere 1967g, 316). African-ness to Nyerere was not just an attitude but a geographic reality. "Africa is determined that Africa shall be governed by Africa," Nyerere declared emphatically in a lecture at the International Press Club in 1965. The West had to accept the principle that Africa was not free until every particle of African soil was free or else it would be "fighting against Africa" (Nyerere 1965c, 53).

And this is why Nyerere rejected a "theology of socialism," the "true" doctrine according to intellectuals living in the world of writing, whereas leaders had problems to solve, people to serve, and new nations to engineer, on the ground, in ways that defied "the book." Nyerere was a Christian. The Bible to him was the word of God. So was the Qur'an. Books on socialism? "They are written by men; wise and clever men perhaps—but still men. . . . No man is infallible." To think there was one "pure socialism," its recipe already known, was "an insult to human intelligence." Humankind has not found all answers to the problem of living in society, which leaves room for all of us to still make a contribution. We should "continue thinking," continue the discussion. Accepting human equality regardless of gender, color, shape, race, creed, religion, or sex is a good start. Putting humanity's inequalities to the service of equality. Accepting that equality in law is not enough. Even impossible without equality as the basic organizing concept of society. Each person must control their own means of production—a farmer/carpenter their own tools. Access to "at least some of the modern knowledge and modern techniques," definitely yes, provided it was group ownership (Nyerere 1967e, 301–305).

Hence Nyerere's historical and cultural interest in the traditional African family because it lived according to the *ujamaa* principles. "I was the first to use the word *ujamaa*," Nyerere said. It refers to a family way of life, with nobody living rich at the expense of others, no exclusive individual ownership of land, an easy life where wealth belonged to the entire family, and nobody used it to dominate others (Nyerere 1966a, 137). A way of living together).—A way of living together.—Working together.—Togetherness.—Communality. In sickness and in health, in life and in death, in happiness and in grief. Where everybody has enough before I get extra, "our food," "our land," "our cattle," the "I" always within the we." "I am because we are, I am nothing without the we." Identity "in terms of relationships; mother and father of so-and-so; daughter of so-and-so; wife of such-and-such a person" (Nyerere 1967h, 337). In Zimbabwe it is *hunhu*, in South Africa *ubuntu*. At core, love, respect, recognition, rights, place in society. It is property held in common, sharing not individualism. And a culture of work, obligating everybody to work, a dignity in work, including work as communal belonging (Nyerere 1967h, 339).

Everyone takes the dignity and welfare of everyone as their responsibility. Nobody goes hungry while others have food; those with food share with those without. One good turn deserves another; today it is someone else going hungry, tomorrow it may be you. Nobody is an island. No one lives off other people's sweat unless too young, too old, or sick, to work for themselves. "Traditional African society was not called 'socialist'; it was just life. Yet it was socialist in the principles upon which it was based," Nyerere said. Human equality. Mutual responsibility. Community. Work. Welfare of everybody (Nyerere 1967f, 312).

Ujamaa in traditional society had two weaknesses. First, it treated women as unequal and inferior to men. They also did their fair share of work in the home and fields and got less out of it than men. "Although it is wrong to suggest that they have always been an oppressed group," Nyerere said, "it is true than within traditional society ill-treatment and enforced subservience could be their lot." In the new society, therefore, "it is essential that our women live on terms of full equality with their fellow citizens who are men." Second, poverty as a result of ignorance and the economics of a very small scale of crop and other production, hence *ujamaa* villages as a scaling-up strategy while retaining what was best of *ujamaa* (Nyerere 1967h, 340).

Under African socialism as *ujamaa*, Man—the service of Man—is the purpose and justification of all social activity, of society itself, not nation, not production. "All human beings. Male and female; black, white, brown, yellow; long-nosed and short-nosed; educated and uneducated; wise and stupid; strong and weak." Human dignity foremost, everywhere. Democracy. Sovereignty not just of or pronounced by the state, but felt and exercised by the people. A society only of workers, each one contributing, working, to produce wealth and welfare, receiving return proportionate to effort and contribution to the community's well-being. The exceptions: those too young, too old, and sick. For everybody else, work is a duty and a right (Nyerere 1968, 6).

SELF-RELIANCE

Nyerere's ideas about socialism as *ujamaa* were not just fascination with theory, but a search for a formula to create a new society with the rural (and agriculture), not urban (industry), as focal point. People already practiced *ujamaa*, but small-scale, family-based production was grossly inadequate for national economic development unless coordinated into one vast national program. Nyerere had lived *ujamaa* in Zanaki and needed no experiment to get proof of concept, and he had experienced its self-reliance, its creative resilience, in practice, in everyday life. Later we shall see why he preferred self-reliance to donor dependency or foreign aid; here, it is enough to note that he had lived life without always waiting for outsiders to come and "help."

"It means that for our development we have to depend upon ourselves and our own resources," Nyerere said when defining *ujamaa* as self-reliance in 1967 (Nyerere 1967g, 318). The president did not say so, but self-reliance is a function of creative resilience; and what he was doing was to defer to African knowledge and its concepts, not simply through research, but tapping into actual lived experience as an archive for designing and building the development of the independent nation. That is why he could say determination, sacrifice, courage, enthusiasm, and endurance (resilience) are one part of the equation—not the sort of terms one gets from the proverbial paper archive of the Western-and-white-centric academy. There has to be the creative part, populated by intelligence, discipline, problem-solving, creating and seizing opportunities, being methodical and organized, and weighing different methods and systems (Nyerere 1965a, 33).

Nyerere identified four prerequisites for the creation of a new society powered by increased agricultural production. Namely, hard work (people), land, good policies, and leadership. In the cities, people worked 7½ to 8 hours daily for 6 to 6½ days per week, or 45 hours a week per year. "For a country like ours these are really quite short working hours. . . . We are in fact imitating the more developed countries," Nyerere observed. By comparison, women in the village worked 12–14 hours daily, including even Sundays and public holidays—harder than anyone in Tanzania. "But the men who live in villages (and some of the women in towns) are on leave for half of their life" (Nyerere 1967a, 244–245). The rural area was the school of self-reliance. The people and their hard work were the foundation of national development, while money and industries were the product of hard work, not its spur. To curb exploitation, everyone must work maximally and live off their labor. "Nobody should go and stay for a long time with his relative, doing no work, because in doing so he will be exploiting his relative," Nyerere said. "Likewise, nobody should be allowed to loiter in towns or villages without doing work—which would enable him to be self-reliant without exploiting his relatives" (Nyerere 1967a, 247). If everybody is self-reliant, so will the cell, ward, district, region, nation, and continent.

As a tool for engineering a new society, *ujamaa* would start with village communities mobilized into a "socialist nation." That is, villagers meeting to elect officers to a committee and to establish rules for organizing work. A village carpenter's workshop, clinic, or shop was organized by a committee member for everybody. Village officials coordinating with government and NGO partners and neighboring villages in development projects.

Government could then slot in nicely as an enabler, not dictator of work and decisions. To advise them technically—what, how, when. By stationing permanent experts among the people, organizing them, advising on sources of funding or loans, markets, foot soldiers for mechanized production and value addition, testing soils for diseases, as a port of arrival for state-of-the-art technology.—Training some among the community to then teach their kin.—Not doing things for people but with them.—Partnership between the people and government (Nyerere 1967h, 365). All strategic deployments of "traditional African democracy, social security and human dignity" synthesized with "modern knowledge" and stratagems. In time such villages could become "more than simple agricultural communities . . . selling their crops and buying everything manufactured from outside." The state could then only intervene to provide what villages could not—like personnel (teachers, agricultural extension officers, nurses, etc.)—but even then, restricted to "a definite role to play in a society organized on the basis of such communal villages" (Nyerere 1967h, 353–355).

Thus, *ujamaa* agriculture was Tanzania's best leg forward, conducted as cooperatives, by people living, farming, and marketing together. That could be the traditional family group, or non-kin group living and working according to the tenets of *ujamaa* and large enough to strategically deploy modern methods. The watchword would be "our," applied to the land, means of production, crops, shops, workshops, revenue, wealth, and welfare.

Each would have a manager (to task and supervise), a treasurer (financial prudence), and a governing committee (transparency and accountability), all members of the group/community, all investing in their community (Nyerere 1967h, 351–353). In other words, the success of *ujamaa* villages would depend on the people's "self-reliant activities." The people themselves would create and maintain the villages with their own resources and resourcefulness. The goal was to move Tanzania from "a nation of individual peasant producers" gradually taking in aspects of a capitalist system toward "a nation of *ujamaa* villages" cooperating in small groups that form joint enterprises (Nyerere 1967h, 365).

With "a little help and leadership," the people had through their hard work built schools, dispensaries, community centers, and roads, dug and built wells, water channels, animal dips, and small dams without waiting for money first or government (Nyerere 1967a, 246). And one little help government availed was the land—the "basis of human life" (Nyerere 1967a, 247). Another was policy, hence *ujamaa*. The government would not assume that people knew what self-reliance was; it would teach them its meaning and practice—to be self-sufficient in food, clothing, housing, security against internal and external enemies, to be always ready to defend the nation when asked (Nyerere 1967a, 247). Progress as "telescoped" evolution, not revolution (Nyerere 1967g, 317).

That is, to progress as an integrated program based on linked principles, convinced that only Tanzanians were "sufficiently interested to develop Tanzania in the interests of Tanzanians, and only Tanzanians can say what those interests are" (Nyerere 1967g, 318). Clustering scattered villagers into *ujamaa* villages made planning and access to government services easier. For example, teaching agricultural techniques to forty people instead of one family at a time. One water pump for the lot of them, instead of one each to every separate house in a scattered community, and shorter pieces of pipe where hundreds of meters would have been wasted. Moreover, if a village of thirty to forty families approached government together and offered to provide physical labor themselves, the pump and pipes would be cheaper and faster to install on government's part (Nyerere 1967h, 355).

Nyerere talks about persuasion, especially using the language of reality, not theory or jargon. As he says, farmers tend to be "cautious about new ideas however attractive they may sound; only experience will convince them."—Explain.—Encourage.—Participate. Not "Do this" but "This is a good thing to do for the following reasons, and I am myself participating with my friends in doing it." Words, never mind how persuasive, are not enough. Show.—Prove.—Work with them.—Live with them. Young people are more willing to experiment.—Encourage them. It's far better to mix young and old, involve them together, create a balanced community (Nyerere 1967h, 355–357).

Starting with the best that is good about people's values inserts a sense of ownership by them regarding what government or anyone intervening is proposing. That is why

Nyerere emphasized the *ujamaa* principle of sharing according to the work done, with everyone in the family sharing equally regardless of effort. Second, the setting aside of a portion of the total return to assist the sick, crippled, old, and orphaned children. Third, setting aside a portion for expansion or investment, having purchased all inputs like fertilizers and seed, or built essential infrastructures like stores.

Ujamaa would unfold in three phases. First, relocating into a single village near water or designated water points, and planting the next season's crops near their homes. Next, persuading them to a ten-house cell to set up a communal plot or activity to work cooperatively and sharing the harvest. Or parents starting a community farm to work with a school (and school children) and sharing the proceeds. Finally, everybody now confident to invest their entire effort in the community farm and putting their individual gardens for their own vegetable needs. The formula could be varied according to potential, soils, arising and unforeseen constraints, and customs as local context allowed (Nyerere 1967h, 357–358).

Nobody expected livestock farmers to put all their cattle into one herd; it had to be a step-by-step process. First, pool the herd of each into one big herd, taking turns to tend to them, and freeing up production time, per Africa-wide custom. Next, each farmer contributing "a number of cattle into a community herd, assigned reserved pastures, and cared for under modern methods," the milk going to supplementary feeding for school children. Then, maybe deciding to invest profits from the herd to build a dip tank or dam or improve housing (Nyerere 1967h, 363).

No great leap forward—just an evolution in methods and tools. The self-reliant farmer uses a big hoe, not a small one, graduates to using an ox-drawn plough, not a hoe, uses fertilizers and insecticides, grows the right crops for the right soil or season, chooses carefully when to plant, weed, and harvest. Already armed with this "knowledge and intelligence," the farmer now combines with others their hard work and scales up cotton, coffee, cashew nuts, tobacco, and pyrethrum production—as they did in 1964–1967. The goal was not "massive agricultural mechanization and the proletarianization of our rural population," but ox ploughs to eliminate the *jembe* (hoe) and then ox plough by tractor. Rather than give every farmer a lorry imported from Europe, better strategically deploy the ox- or donkey-drawn cart made inside Tanzania, appropriate to local paths, affordable or accessible to each farmer. Aerial spraying of insecticides? Why? Rather use hand-pumps. Think in terms of local availability, low to no cost, needing no specialized technical knowledge, thus no need for training, cheaper to implement and maintain (Nyerere 1967g, 320).

These baby steps in agriculture would create necessity for inputs or value addition at point of production instead of carting raw materials to the cities for processing. Small industries and service centers could thus emerge organically, diversifying the rural economy (Nyerere 1967g, 320). That too could develop as a communal project at the initiative of local cooperatives, not outsiders, and take the form of a poultry processing plant, tannery, woodwork workshop, a truck, an irrigation with a water wheel, or a shirt-making and knitting shop, which benefits the entire group. In other words, rural industrialization as "cottage industries," not "large modern factories." More like a fixed, shared workspace, each specializing in aspects of work according to their knowledge (Nyerere 1967h, 359–362).

At any rate, Tanzania had neither the money nor the skills to undertake urban industrialization, Nyerere warned. "We have neither the money nor the skilled manpower nor in this case the social organization. . . . We have put too much emphasis on industries . . . [that] 'Without money there can be no development' [and that] 'Industries are the basis of development, without industries there is no development.'" Development does not begin with industries, Nyerere argued. If sited in the countryside, Tanzania would have to borrow money from overseas to set up the transport and communications infrastructure such industries required, aid that came with conditions (Nyerere 1967a, 242; Nyerere 1967g, 321).

These necessary changes in rural areas were forms of rural innovation and entrepreneurship, building on what people know and are doing well, or a mode of start-ups from bootstrapping, with minimal to no borrowing from local lenders or grants from governments and NGOs, and making "the widest possible impact." Recognizing that the vast majority lived in rural areas, and that land was "the only basis for Tanzania's development," Nyerere resolved to "build up the countryside" to ensure better living standards (Nyerere 1967g, 321).

The countryside was also a hive of public-facing work involving the communities partnering government and bootstrapping their way to development. Nine months into the Five-Year Development Plan, 316 miles of new feeder roads, 97 small bridges, 664 classrooms, 134 clinics, and 49 cooperative stores had been completed, while 18,542 acres of communal farms were cleared and planted, 517 miles of irrigation channels dug, over 5,000 fishponds set up, and cash crop (cotton, coffee, cashew nut) acreage dramatically increased. But Nyerere was still unhappy about the country's dependence on rain-fed agriculture and lack of irrigation infrastructure (Nyerere 1965b, 40–41). Since private investment in rural development was "a disappointment" despite special tax incentives and investment guarantees, progress would have to be gradual. Everything was "started by the people from their own efforts . . . using their own tools," providing seed from their stocks, or getting an advance from their cooperative or the National Development Credit Agency to buy seed, fertilizers, and other inputs (Nyerere 1967h, 364).

The rationale for nationalization of government's majority ownership in industries strategic to Tanzania was to prevent economic sabotage by multinational corporations and their foreign governments. It had nothing to do with being capitalist, fascist, communist, or socialist—just the desire for economic sovereignty, to own and control the means of production, distribution, and exchange. While political independence had happened in 1961, all industries were still in foreign hands, and even when universities trained engineers, these were glorified mechanics that built, fitted, maintained, and at best renovated plant, parts, and infrastructures designed in Europe and the United States. This was the consequence of independence based on negotiated settlement whereas countries that underwent revolution secured economic control at their founding as nations, such as the United States, the Soviet Union, and China, which all seized their enemy's property. The same European countries that would impose sanctions on African countries for nationalizing were the same ones routinely vetoing privatization of major industries or commercial enterprises likely to externalize their control. "The question," Nyerere pointed out, "is not whether nations control their economy, but how they do so" (Nyerere 1967c, 264). The purpose of nationalization was therefore to have national control, not socialism: to ensure that the elected government of Tanzania had total control of money and credit use, to ensure the profits of banks and industries benefitted Tanzania (Nyerere 1967f, 311).

While promoting private investment in industrial and agricultural activities, Nyerere also accelerated Tanzania's control over its economy through the establishment of the National Development Corporation. Several food processing companies were placed under it, among them Tanzania Millers, Chande Industries, Pure Food Products Ltd., G. R. Jivray, Noormohamed Jessa, Kyela Sattar Mills (Mbeya), Associated Traders Ltd. (Mwanza), and Rajwani Mills (Dodoma). The government also nationalized the National Insurance Corporation Ltd., wherein it had a majority stake. This way, any potential investor would know that these and other industrial and commercial enterprises were reserved for government (Nyerere 1967b, 251–255).

Self-reliance did not mean isolationism politically or economically—just simply depending on ourselves, trading with others (not relying on the charity of others), making everything that can be made locally and importing only those that cannot (Nyerere 1967g, 321–322). The purpose of industry was to move first from import dependency to import substitution as a stepping stone to export-oriented industrial development. It means establishing an industry in every village, to balance out the continued urban-based industrial model and prevent rural-urban migration. The problem was that Tanzania was borrowing foreign loans to develop urban industries, but the foreign currency repayments were coming from rural agricultural production. The people who suffered from foreign debt were not the people who benefitted from the money, the largest chunk of which was spent in the urban areas. All the referral hospitals, the tarmac roads, electric lights, water pipes, hotels, telephone, and other infrastructures were urban (Nyerere 1967a, 243). The city had to be self-reliant, but it was continuing the colonial practice of leeching the countryside to develop itself.

Self-reliance was also not a rejection of international existence. Tanzania would continue seeking aid for specific projects (the Tanzania-Zambia railway and China, for example) and development in general. However, Tanzania did not believe that without aid nothing will be done. The country could do many things for itself "the hard way" and warmly thank those who assisted, but Tanzanians would not build Tanzania by sitting on their hands looking at the road (Nyerere 1967g, 322–323).

Tanzania welcomed foreign knowledge—engineers, doctors, managers; not accepting their directions was "just being stupid," Nyerere said. Policy was the responsibility of Tanzanians, but implementation, being a technical question, required people with qualifications and experience. Nyerere criticized the prejudice or false pride that got in the way of accessing scientific capacity. Even North America and Western Europe also recruited knowledge from India and Pakistan, the United States from Western Europe, "attracting workers on whose education it has not spent one cent." Tanzanians by contrast celebrated when a trained white person left for Europe or North America as "an opportunity for Africanization, or for self-reliance!" That was not what self-reliance meant; for Nyerere it was "childishness" (Nyerere 1967g, 324).

At the end of the day, Nyerere conceded that TANU had "chosen the wrong weapon for our struggle, because we chose money as our weapon"—a weapon that Tanzania did not possess in any significant quantity as a newly independent country. Water, roads, schools, hospitals, development—money! Money! Money! Throw money at everything, increase taxes to get more money (Nyerere 1967a, 237). So-called "assistance" was actually borrowed money, not "gifts or charity" but loans with repayment conditions on due dates and interest rates. Private investment came with insistence to repatriate profits, with little if any reinvested in developing Tanzania. A donation or loan, or even the start of negotiations toward it, was met with a blaze of publicity (Nyerere 1967a, 238). That "money solves everything" attitude was for Nyerere "stupid" because it compromised the country's independence and put it into debt. The whole basis for attracting foreign investment was what? So that foreigners would then take all their profits back to their countries? What "stupid" logic was that? Nyerere could not understand it (Nyerere 1967a, 240).

In any case, more countries were seeking assistance than there were countries able or willing to assist. This selectivity means that time better spent in self-reliance projects was being expended preparing detailed applications for project assistance that often never materialized. Assuming the bid was successful, funding was biased toward projects with a significant "import content" from the lending countries, which shows that they were clearly opening up markets to and subsidizing their own industries. Any project that might compete for a market with lending or donor countries was turned down. The less employment and exports the financing would generate, the better the chances of a successful application. Their priorities, not those of African countries. Even then, usually partial and conditional "aid" was given, for imports from lending countries, with African countries compelled to finance all local costs. As Nyerere concluded, "The lesson again is an obvious one. *Kujitegemea zaidi*. Self-reliance" (Nyerere 1965b, 47).

Nyerere therefore proposed sacrifices from Tanzanians through taxation and savings. A National Savings Bond was established in 1965 for people to invest in the factories or farms and contribute to the nation's development. Taxes were increased to boost revenue. Nyerere introduced some frugal measures within government, including cutting out any spirits at all state receptions, serving only beer, tea, coffee, and soft drinks (Nyerere 1965b, 49). Thankful for, but not expecting, outside assistance, Tanzanians had to rely upon themselves, deciding what comforts to spend on now, and how much to sacrifice and invest in the future. Perhaps forego state-of-the art roads, schools, farm equipment, and industrial machinery for the very basics that can get things done if that saved money for investing in sustainable futures. It made no sense selling abroad and getting foreign currency, then spending it on imported goods when locally manufactured products were available, and then seeking foreign aid that needed to be paid in foreign currency at huge interest. That was why Nyerere launched the "Buy Tanzanian" campaign in 1966 (Nyerere 1966b, 168).

EDUCATION FOR SELF-RELIANCE

"Only when we are clear about the kind of society we are trying to build can we design our educational service to serve our goals," Nyerere declared in a now famous speech in 1967, titled "Education for Self-reliance." In designing the future, Nyerere also planted deep roots in history, drawing attention to African modes of education and knowledge in the long historical trajectory interrupted by colonial occupation and which, having expunged the colonizer, the now independent, self-determined Africa was free to revisit as a critical ingredient in designing its own education system fit for purpose:

> The fact that pre-colonial Africa did not have "schools"—except for short periods of initiation in some tribes—did not mean that the children were not educated. They learned by living and doing. . . . They learned the kind of grasses which were suitable for which purposes, the work

which had to be done on the crops, or the care which had to be given to animals, by joining with their elders in this work. . . . Education was thus "informal"; every adult was a teacher to a greater or lesser degree. But this lack of formality did not mean that there was no education, nor did it affect its importance to the society. Indeed, it may have made education more directly relevant to the society in which the child was growing up. (Nyerere 1967d, 268)

The purpose of colonial education was not to equip youths with knowledge in service of their society and country, but to inculcate colonial values and serve the colonizer. Granted, individuals "worked hard, often under difficult conditions," as teachers and institution administrators. Nor were all values taught in schools bad. Modeled on the British system, the colonial education stressed subservience (vs. initiative and self-confidence), white-collar skills (vs. critical thinker-doers), materialistic individualism (vs. cooperative/communal ethic), human inequality (vs. equality), and alien knowledge values (vs. traditional knowledge values). The purpose of the education was to turn African society into a colonial society (Nyerere 1967d, 269).

Consequently, the pupils coming out of the education system found it difficult to integrate what they had learned or themselves into their societies. They had an "attitude of inequality, intellectual arrogance and intense individualism." The bottleneck elitist education was designed to alienate them from society. The rationale for primary education was to prepare pupils for secondary school; only 13 percent got in, leaving 87 percent "with a sense of failure," referred to as "those who failed" rather than as "those who have finished their primary education." The other 13 percent had a sense of deserved achievement, a prize, high wages, comfort, in the city, status. A sense of superiority in the latter, inferiority in the former (Nyerere 1967d, 275–276). The secondary and primary schools were nearly all boarding schools, taking children from age seven, and drilling in them knowledge alien to society, for 7½ hours daily. The school was not part of society, several miles away from the villages, "an enclave." The skills at school were untranslatable to everyday life. When finally Dar es Salaam University College was established, the same philosophy of isolated, untranslatable knowledge continued. The student lived in "comfortable quarters," fed well, studied hard, expecting they would get a salaried job of no less than £660 per annum; serving the community was secondary, even incidental to that. The early postindependence university could afford to train for employment because skilled Africans were scarce, and demand outstripped supply. All of us went to school, to university, seeing the purpose of education as to get a job, help our parents, live a better life; returning to lift up everyone else, the country, or continent, was not the rationale (Nyerere 1967d, 276–277).

The colonial system drilled in us the mentality that all knowledge worth knowing comes from "educated people," certificates, books, and universities attended. Experiential knowledge is waved away. Certificates equal competence. No interest in other attributes.—Attitude.—Character.—Hard work.—Kindness.—Creativity.—Resilience.—Doer.—Maker. Only the ability to pass (written) exams matters. Opinions of an expert by experience and traditional ways of knowing with no formal education are dismissed. It's all about book learning; not knowledge acquired or made in any other ways, or intelligence measured by any other yardstick. Which is not to say traditional knowledge production and education systems are all virtue and no vice, all wisdom and no fools, or that, as Nyerere said, old means wise (Nyerere 1967d, 278).

Nobody is saying that educational qualifications are unnecessary or worthless or that old age is an automatic sign of wisdom. Nobody can be considered an expert in factories *a priori* simply because of twenty years of experience on the shop floor. Equally, simply because someone has a doctorate, does not mean ability, even though such qualifications are necessary for successful and today's enterprise. Same for agriculture: vastly experienced farmers, producing and applying knowledge in/through practice, their methods always in need of updates, are never to be dismissed simply because they never went to or far with formal schooling. Ox plough for hoe, tractor for ox plough, same task different, faster tools, all improvements but no reason to dismiss the knowledge generated through experience, hence the need for a mixture of both, not always a replacement, only by the other (Nyerere 1967d, 277–278).

The undecolonized education made Tanzania's children despise their parents as primitive and ignorant; the education system said nothing positive about the knowledge of their elders or everyday knowledge that the children could respect as useful knowledge. And the worst caricatures were amplified, the useful knowledge silenced. It did not help that the school extracted the children away from the everyday life of learning as they worked to one where "they simply learn[ed]" while just sitting there! Here Nyerere compared our useless education with that of the United States where "young people work their way through high school and college." Not in Tanzania: "The vast majority do not think of their knowledge or their strength as being related to the needs of the village community. . . . Jobs like digging an irrigation channel or a drainage ditch for a village, or demonstrating the construction and explaining the benefits of deep-pit latrines, and so on." The service ethic, of students deploying their lecture room knowledge in service of improving people's lives, not just for pay, but to make a better world, while acquiring experience for a postgraduation career and life, was anathema (Nyerere 1967d, 279).

There was an education system for white people (now the rich), another for black people (now the poor). Nyerere had abolished the racial segregation in education and expanded facilities and access to primary and secondary education—from 490,000 to 825,000 children attending primary school, and from only 11,832 children in secondary schools (and just 176 advanced-level schools) to 25,000 and 830, respectively, in 1967. But the content and methods of black education were still colonial. True, they were no longer learning British and European history but history of Africa too.—Singing national songs.—Dancing African dances.—Using the national language (Kiswahili) in the curriculum.—Civics classes to de-isolate the school from society. But all these were modifications to an inherited education system. No tangible results had been felt six years into independence.

Therefore, in March 1967, President Nyerere presented his "Education for Self-Reliance" lecture, the first of his policy directives after the Arusha Declaration. After the presentation, several working parties composed of educational administrators and teachers were convened to design an implementation strategy. Meanwhile, the country's schools got busy setting up farms and workshops on their premises and undertaking "nation-building tasks." "We have never really stopped to consider why we want education—what its purpose is," the President said in his seminal speech. Up until the lecture, the government had not questioned or reformed the colonial education system it had inherited. Education was still about training teachers, engineers, administrators, and so on and acquiring skills to earn high salaries. Twenty percent of the budget was devoted to education but what was the product?

Specifically, what was the product's proportionate relevance to the society Tanzania was trying to create? Nyerere quipped:

> The educational systems in different kinds of societies in the world have been, and are, very different in organization and in content. They are different because the societies providing the education are different, and because education, whether it be formal or informal, has a purpose. That purpose is to transmit from one generation to the next the accumulated wisdom and knowledge of the society, and to prepare the young people for their future membership of the society and their active participation in its maintenance or development. (Nyerere 1967d, 268)

The time had come to ask what became, and remains, the most important question ever asked on education not just in Tanzania, or in Africa moreover, but the world and planet: "What is the educational system . . . intended to do—what is its purpose?" The relevance of the system and content being modified itself was under serious interrogation (Nyerere 1967d, 271). "We cannot integrate the pupils and students into the future society simply by theoretical teaching, however well designed it is," Nyerere said when proposing an education for self-reliance. Nobody expected twelve- to thirteen-year-old kids leaving primary school to be useful citizens. The "primary school leaver" problem stemmed from a flawed early childhood education system—or lack thereof. Kids started school aged five to six, finished still too young to assume the responsibility of citizenship, in a society expecting them to assume wage employment. To solve this, Nyerere proposed raising the school starting age so that students were more mature, learned faster, left school older (Nyerere 1967d, 280).

Because not everybody could go on to secondary school, primary school education had to be self-contained, the skills set intended and exams set tailored to make graduates useful to themselves and as citizens on its own terms. Where the British had intended primary school as preparation for secondary school, and secondary school for college or university, Nyerere had to deal with "the economic facts of life in our country . . . the practical meaning of our poverty." Public secondary school spending could only be justifiable as education for "the few for service to the many," to groom those that went on to college to major in public-facing careers—agricultural extension officers, engineers, doctors, teachers, and so on—otherwise it was unjustifiable.

To prepare young people for the realities and needs of their countries, the very idea of education, let alone the education system, needed radical change. Nyerere suggested downgrading written exams (mastery of facts) and assessing instead the reasoning, character, and willingness to serve. To emphasize the class as continuous assessment in thinker-doer skills, not a preparation for the exam from Day 1 of school. What is the sense of setting our exams to international standards with next to nothing directly relevant to our "particular problems and needs"?—At home.—In the community.—Of our country.—Continent. What is the point of teaching all primary school pupils things that a doctor, engineer, teacher, economist, or administrator needs to know when only a tiny few of them will ever qualify for or take that career path? Rather, teach all kids lifelong skills, knowledge, attitudes, and values every citizen—the majority—needs to live happily and well in their specific society. As Nyerere puts it, "schools must, in fact, become communities which practise the precept of self-reliance" (Nyerere 1967d, 282).

Every school and college must contribute to its own upkeep, and infrastructure for self-sustenance—a farm (food production), workshop (manufacturing and repairs), and so on. Not simply for training purposes; "every school should also be a farm" (Nyerere 1967d, 383). "The most effective classroom is an efficient farm. The most effective teachers are the efficient farmers," Nyerere had said when laying the foundation of permanent buildings at the Morogoro Agricultural College on November 18, 1965. "This is as it should be. An agriculturalist should not be a man with clean, soft hands. He must be a practical man, who uses his hands at the same time as he uses his head" (Nyerere 1965f, 105). The students who learn planning, production workflow, record keeping, packaging, and marketing skills through doing could, upon going back to the village, engage their kin and create a synthesis of knowledge systems with college/school and everyday knowledge as ingredients. The culture of dependency on donations and grants from government and NGOs must be replaced by self-reliance, income from sales of farm products, services offered in the workshop, and so on. "It is a recognition that we in Tanzania have to work our way out of poverty," Nyerere says (Nyerere 1967d, 383).

Learning by doing must never be confused with child labor. Every African society brought up its young that way.—Demonstration.—Showing.—Reality.—Practice.—Experiencing.—Experimenting.—All these rolled into one. Hence fertilizer properties are seen operating live through applying a different fertilizer to each plant, assessing effects on each. Different methods of grazing tested out. Effects of terracing on soil conservation, water retention, and crop yield observed. Not expensively assembled, highly mechanized demonstration farms; that would be too artificial, something absent in a people's everyday life. Not capital intensive but making maximum use of resources available locally, and limiting buying or donations to those things that cannot be made locally. Labor and inputs mobilized through traditional cooperative (communal) methods of mobilization and working (Nyerere 1967d, 284).

In short, theory tested and produced in/through actual production. Mistakes will be made; learning by/from mistakes is part of the creativity process. Dealing with the realities of everyday life, engaging with necessary knowledge, renders the knowledge being learned doable; doing produces knowledge. A culture of learning through work with important lessons: the relation of "work to comfort," the meaning of "living together and working together for the good of all, . . . with the local non-school community." How to do that becomes a major strategy question. Nyerere proposed siting the farm in the school premises; this is the standard practice in Africa. The mistake: its purpose is to serve the school curriculum, contrary to his suggestion. These sites typically tend to be prisoners of a curriculum alienated from everyday knowledge. To be useful, therefore, the school farm and the villager's farm (the pupils' homes), academic and everyday knowledge, theory and practice, must be integrated.

I am a product of the rural and the ghetto. Never went to boarding school. Like their father, my children go to public school. As such Nyerere's description of the spoiled boarder is something I only met as an undergraduate at the University of Zimbabwe when I was already mature. Compare this to children sent to primary and secondary boarding school, where cleaners, cooks, and gardeners did all the work—cooked for the students, washed their plates, cleaned their rooms, and grew vegetables for them. Here Nyerere asks some questions: "But is it impossible for these tasks to be incorporated into the total teaching task of the school? . . . Can none of these things be incorporated into classroom teaching so that pupils learn how to do these things for themselves by doing them?" (Nyerere 1967d, 286).

There is a serious danger that today's children in Africa are growing up soft. Those of us who grew up poor, walking miles to and from school, on a "what you grow is what you eat" and "work or starve" diet feel that our children have become victims of our success. At the same time, we have allowed cultures of highly industrialized countries to criminalize learning through doing as child labor, based on values that have no basis in our own societies. How does the same law apply to less-than-a-dollar or no-money-a-day countries and rich countries at the same time? Nyerere articulately prescribes the solution:

> The pupils must remain an integral part of the family (or community) economic unit. The children must be made part of the community by having responsibilities to the community, and having the community involved in school activities. The school work—terms, times, and so on—must be so arranged that the children can participate, as members of the family, in the family farms, or as junior members of the community on community farms. At present children who do not go to school work on the family or community farm, or look after cattle, as a matter of course. (Nyerere 1967d, 287)

Which suggests that the "Future of Work" discussion in Africa must also be a "Future of Learning" conversation: because work is the most important medium of learning and teaching in non-Western—and poor—societies. Western-and-white society faces a different "Future of Work" problem whereby it invented robots to relieve itself of the burden of doing work, and now cries when these bots indeed take the jobs. The issue at stake is neither the history or future of work but the cultures of work and how they have shaped past, are shaping present, and will shape future work. Robots are things; like all things, they become tools or technologies only when societies put their hand on them, not *a priori*.

For Nyerere, there is a tension between the exam (thinking/writing) and practical work (doing/working). When pupils now spend more time doing work that gives them livelihood and develops their community, they have little time for exams. As he says, nobody said the current examination system or format is sacred. Some countries had abandoned exams at lower grades altogether; others had combined written and practical assessments. Nyerere's answer lay elsewhere: revising the school calendar from fixed terms with long holidays to fit in with everyday life. "Animals cannot be left alone for part of the year," the president said. "Nor can a school farm support students if everyone is on holiday when the crops need planting, weeding or harvesting." By staggering school holidays, with some classes ending and some in session, continuous presence was assured, momentum of production sustained at home (so that home becomes school) and on the school farm (so that schooling becomes production). This integration of the school and university into the community (complete, I might add, with the synchronization of the school calendar with the farming seasons) sets us up to produce a public-facing student "unlikely to forget their debt to the community." To sweeten the deal, the work at home and in the school farm earns the student credits counting toward the exam (Nyerere 1967d, 288–290).

As Nyerere said, "*Only when we are clear about the kind of society we are trying to build can we design our educational service to serve our goals*" (Nyerere 1967d, 272; my own emphasis). An education system must mirror the values and objectives of the society its people are trying to create. The rational for centralizing the rural people in Africa is different from the United States or Europe, which are heavily industrialized and therefore have a predominantly urban-centered production (and economic) system. Thus any design must "accept the realities of our present position . . . at present a poor, undeveloped, and agricultural economy. We have very little capital to invest in big factories or modern machines; we are short of people with skill and experience. What we do have is land in abundance and people who are willing to work hard for their own improvement" (Nyerere 1967d, 273). Therefore, in a predominantly rural economy, the country's principal means for developing are the rural people that live and work in the countryside. Urban-based industries and factories can come later.

The priority is the village: to live a good life there, to "be able to find their material well-being and their satisfactions" there, required the designing of a radical educational system that fosters living and "working togetherness" (the spirit-as-skill set, not the practice) for the common good. An education system equipping the youth to be dynamic builders of a new society, sharing the benefits and burdens of community and country fairly. Inculcating within them "a sense of commitment to the total community," embodying "values appropriate to our kind of future, not those appropriate to our colonial past." An education system, therefore, based on a new value system.—Communal (Nyerere uses cooperative) endeavor, not individual advancement.—Equality.—Service.—Humility, not intellectual arrogance (Nyerere 1967d, 273). This "commitment to a particular quality of life" placed the human at the center of education: "There are more important things in life than the amassing of riches. . . . The creation of wealth is a good thing. . . . But it will cease to be good the moment wealth ceases to serve man and begins to be served by man" (Nyerere 1967g, 316).

The new education system must train critical thinker-doers. Nyerere did not use that term, I do, but spoke of a system producing not "passive agricultural workers . . . who simply carry out plans or directions received from above," or "robots . . . who work hard but never question what the leaders . . . are doing and saying." Rather, the education must produce graduates "able to think for themselves," to "interpret," to "make judgments," and to "implement," graduates who have "an enquiring mind," that learn from others and make their own decisions, and that are confident of themselves as free and equal members of society while valuing others as such. As Nyerere said, "Only free people conscious of their worth and their equality can build a free society" (Nyerere 1967d, 274–275).

With a problem-solver graduate, unemployment is not a problem. And Nyerere was not convinced that investment in big, urban-based industries would solve it. One factory set up at a cost of £125,000 would employ just forty workers per shift. In any case, the problem of school leavers not finding formal employment did not mean that "they have no work they can do." The problem was one of perception among parents and the youth of what constituted real employment for someone who had gone to school. That is, working in an office. "It is time that we dropped it, and realized that this is an independent country, with no room for colonial attitudes of mind," the president made clear (Nyerere 1965d, 72). "*Why is it that in highly developed countries like America, Britain, and Germany, you find university graduates working as farmers?*" Nyerere asked. And answered himself: "Because they realize that education is intended to make a person better able to do whatever job is important and profitable, both to himself and his country" (Nyerere 1965d, 72; my own emphasis). An adaptive mode of education, capabilities applicable to many fields, not just one, always a mind in anticipative mode, hands ready to do, and heat committed to serve.

It was no longer just a question of reformulating the curriculum, but also building societally responsive institutions. Here, the university's role in the making of a new society became key, that is: "the whole problem of what a university could, and should, do in a developing society." For Nyerere, "the pursuit of pure learning," or learning "for its own sake," could be a luxury depending on the society's conditions, especially in a society where "people are dying because existing knowledge is not applied." And the pursuit of pure knowledge was a traditional function of the Western-and-white and colonial university—to inquire into things that happen or exist, it being enough and legitimate simply to establish the facts (Nyerere 1966c, 180).

There are usually two types of scientific research: pure and applied. The one usually happens in the universities and is informed by intellectual curiosity, regardless of whether the knowledge produced will be applied to problem-solving or not. The second is usually classified as R&D (research and development), that is, research in service of problem-solving or product development in, say, industrial or agricultural processes. It typically involves industry, government, and research institutions in partnership, assuming that the former are not research entities themselves. The priority of a "developing society," that is "a society in the process of rapid change," is knowledge that addresses the "immediate future" and "immediate present"; our universities cannot pursue "knowledge for its own sake" and applied science at the same time because of limited resources. They must choose between what is absolute necessity and what is luxury (but not useless) and our approaches under our specific conditions, "our particular and urgent problems" as African countries (Nyerere 1966c, 181).

The university that puts its professors and students in blinkers serves neither "the cause of knowledge nor the interests of the society in which it exists," Nyerere argued. It can search for the truth, for facts, yes, but it should also have a "commitment to our society—a desire to serve it." The state university of a developing society should focus on "subjects of immediate moment" to the nation and humanistic goals to justify the heavy public expenditure upon it, including knowledge as a service to community, for free. It must produce a graduate who thinks scientifically, analyses problems, and applies facts learned to solve present and future societal problems (Nyerere 1966c, 183). It must be an anticipative education.

Regrettably, the physical architecture of the university is designed to encourage and reinforce the isolation of the student from society. The "fine buildings" on campus contrast sharply with the living conditions of the masses that paid the taxes for their construction, utility bills, and maintenance. There is a temptation to forget who one is once admitted to these universities, especially the social institutions of communality and interdependence left in the village, in contrast to the individualism of college life. "I am on my own now, on campus, working towards my degree—the key to personal advancement." And with that a culture of arrogant individualism sets in. "It is my knowledge, my degree, my achievement, I worked my socks off for it, and nobody helped me to study." On campus, the individual from the village shifts belonging to fellow students, equally arrogant, and as a group demand "rights without responsibilities." "We want better conditions of study. Larger allowances. Better food. Freedom of speech." And the umbilical link between the quest for learning and the imperatives of the community is broken (Nyerere 1966c, 185).

The university should be remodeled in such a way that the students never have to be alienated from their community, but are instead active agents in its transformation, giving service to it, be socially responsible to it. Not in the arrogant sense of "giving aid to the poor" but as *"an attitude of wanting to work, in whatever work there is to do, alongside and within the rest of the community, until finally there is no more distinction between a graduate and an illiterate than there is between a man who works as a carpenter and his fellow who works as a brickmaker." In which case, Nyerere concluded, "graduates and illiterates would then accept their tasks as distinctive, and as making different demands on them, but as being in both cases but a part of a single whole"* (Nyerere 1966c, 186; my own emphasis).

9 DARE TO INVENT THE FUTURE

I come here to bring you fraternal greetings from a country of 274,000 square kilometers whose seven million children, women, and men refuse to die of ignorance, hunger, and thirst any longer. . . . A people who, in the land of their ancestors, have decided to henceforth assert themselves and accept their history—both its positive and negative aspects—without the slightest complex. . . .

A world in which humanity has been transformed into a circus. . . .

I make no claim to lay out any doctrines here. I am neither a messiah nor a prophet. I possess no truths. . . .

I speak on behalf of the millions of human beings who [live] in ghettos because they have black skin or because they come from different cultures, and who enjoy a status barely above that of an animal.

I suffer on behalf of the Indians who have been massacred, crushed, humiliated, and confined for centuries on reservations in order to prevent them from aspiring to any rights and to prevent them from enriching their culture through joyful union with other cultures, including the culture of the invader.

I cry out on behalf of those thrown out of work by a system that is structurally unjust and periodically unhinged, who are reduced to only glimpsing in life a reflection of the lives of the affluent.

I speak on behalf of women the world over, who suffer from a male-imposed system of exploitation. As far as we're concerned, we are ready to welcome suggestions from anywhere in the world that enable us to achieve the total fulfilment of Burkinabè women. In exchange, we offer to share with all countries the positive experience we have begun, with women now present at every level of the state apparatus and social life in Burkina Faso. Women who struggle and who proclaim with us that the slave who is not able to take charge of his own revolt deserves no pity for his lot. This slave alone will be responsible for his own misfortune if he harbors illusions in the dubious generosity of a master pretending to set him free. Freedom can be won only through struggle, and we call on all our sisters of all races to go on the offensive to conquer their rights.

> I speak on behalf of the mothers of our destitute countries who watch their children die of malaria or diarrhea, unaware that simple means to save them exist. The science of the multinationals does not offer them these means, preferring to invest in cosmetics laboratories and plastic surgery to satisfy the whims of a few women or men whose smart appearance is threatened by too many calories in their overly rich meals, the regularity of which would make you—or rather us from the Sahel—dizzy. We have decided to adopt and popularize these simple means, recommended by the WHO and UNICEF.
>
> I speak, too, on behalf of the child. The child of a poor man who is hungry and who furtively eyes the accumulation of abundance in a store for the rich. The store protected by a thick plate glass window. The window protected by impregnable shutters. The shutters guarded by a policeman with a helmet, gloves, and armed with a billy club. The policeman posted there by the father of another child, who will come and serve himself—or rather be served—because he offers guarantees of representing the capitalistic norms of the system, which he corresponds to. . . .
>
> I am human, nothing that is human is alien to me.
>
> —Thomas Isidore Noël Sankara, "Freedom Must be Conquered" (1984, 154–165)

In this famous "Freedom Must Be Conquered" speech to the UN General Assembly in New York in 1984, Burkina Faso's president, Thomas Isidore Noël Sankara, proclaims the struggle for reasserting the African personality, a new Négritude that could pave way to a new humanity built on what we have seen in Blyden, Césaire, Senghor, Fanon, and Cabral before him.

In it he declared himself the spokesman of the discriminated, oppressed, workers, women, and children. These were, continue to be, for some of us, the realities that necessitate daring. The possibility of a world in which one day my daughter's skin and my sons' kinky hair, will not be a complex upon which the content of their character, their very humanity, is judged. He was no Mohammed nor Christ, and was not claiming that what he was saying was the truth and nothing but the truth. He was not the power but the expression of it; the force was the millions of human beings whose surest citizenship was disadvantage, dealt a bad hand by history, consigned by race to a status no different from animals. Not just black people, but those chewed and spit as vermin by the system of global domination everywhere—Domination not just by capital but also systems where males disadvantaged women for no other valid reason than anatomy. No speaker of empty words, demagoguery without deeds, theory absent deeds and doing but emboldened by the Burkinabè experiment. A theory of the world order derived from daring to invent a new future. More a testimony than a wish, statements the outcomes of doing informed by critical thought. The "we are" and "we have"—Plurality of ownership but singularity of purpose. Laboring together, leading the people, led by the people—the one a visionary, the other belief, claiming

FIGURE 9.1
Thomas Sankara. *Source*: AFP/dpa/picture-alliance.

ownership of the inspiration, and by communality of effort and purpose, turning daring into futures becoming.

Indeed, to clear the way for self-rehumanization to exist, and for the human to freely come in and occupy the world made good by daring. It starts with a refusal to accept mediocrity as a permanent state, to look inside oneself, put oneself on trial, with the future as judge, asking: "Did you do enough?" The excuses are as vast as the ocean: "It is not my business." "I did all I could." "I never discriminated against anyone." "I donate to the Red Cross." "There was nothing I could have done; the government is the problem." "My career needs focus. I can't do too much." "It is a division of labor." "I should." And so forth. For those unsatisfied with these answers Sankara advises:

> You cannot carry out fundamental change without a certain amount of madness. In this case, it comes from nonconformity, the courage to turn your back on the old formulas, the courage to invent the future. Besides, it took the madmen of yesterday for us to be able to act with extreme clarity today. I want to be one of those madmen. . . . We must dare to invent the future. . . . Everything man is capable of imagining, he can create. I'm convinced that's true. (Sankara 1985a, 232)

In a speech delivered at the International Conference on Trees and Forests in Paris, on February 5, 1986, Sankara made flattering reference to his host François Mitterrand's book *L'abeille et l'architecte* (*The Bee and the Architect*) when talking of the revolutionary force driving the madness and turning visionary into builder: "We can win this struggle if we choose to be architects and not simply bees. It will be the victory of consciousness over instinct. The bee and the architect, yes! If the author of these lines will allow me, I will extend this twofold analogy to a threefold one: the bee, the architect, and the revolutionary architect" (Sankara 1986d, 260). It is this revolutionary spirit that made daring possible, gave energy and purpose to creative resilience, even when the outcome might be the ultimate sacrifice. Death.

"Perhaps this will cost us our life, but we are here to take risks. We are here to dare," Thomas Sankara declared in his speech at a mass rally in Ouagadougou on March 26, 1983. He was talking about a culture of sloth and business as usual in the civil service, where employees came into their offices at 9:00 a.m., left at 10:30 a.m., and threatening to dismiss such corrupt civil servants and soldiers wasting the people's scarce resources (Sankara 1983b, 57). In 1985 Sankara added: "You have to dare to look reality in the face and dare to strike hammerblows at some of the long-standing privileges—so long-standing in fact that they seem to have become normal, unquestionable" (Sankara 1985a, 199).

Daring is taking risks, especially those that are presumed ordinarily too dangerous. It is a refusal to accept orthodoxy, the way things have always been, the way things are. Daring is "to risk new paths to achieve greater well-being." To apply new techniques. New ways of organizing ourselves. Liberating our countryside from stagnation or even regression. Democratizing our society. Opening our minds to "a world of collective responsibility in order to dare to invent the future." *And conquer freedom.* To put in place mechanisms where everyone could "at least use his brain and hands to invent and create enough" to have two meals daily and water to drink (Sankara 1984b, 160–161).

When asked by a journalist if he wasn't always afraid that this revolution, his life, could be ended any time, Sankara dismissed the question: "No, that kind of fear I don't have. I've told myself: either I'll finish up an old man somewhere in a library reading books, or I'll meet with a violent end, since we have so many enemies. Once you've accepted that reality, it's just a question of time. It will happen when it happens" (Sankara 1985a, 231).

Sankara always knew that one day he would be killed. At a mass rally in Ouagadougou on March 26, 1983, Sankara called out such people: "As we speak to you, we know that in this crowd are people who would very much like to shoot us right now. . . . We say to them: 'Shoot!' When you shoot, your bullets will turn around and hit you" (Sankara 1983b, 60). In Maputo during the funeral ceremony for Samora Moises Machel, Sankara offered a speech reminiscent of Sekou Touré's at the burial of Amílcar Cabral, imploring revolutionaries to not waste a tear weeping, but instead to "adopt a revolutionary attitude." The purpose of a revolutionary's death was not to put an end to a life (because that single life in and of itself was not important except as a "we"), but to enlighten and strengthen those left to carry on the struggle of the "we":

> To avoid falling into sentimentalism, we must not weep. With sentimentalism one cannot understand death. Sentimentalism belongs to the messianic vision of the world, which, inspires lamentation, discouragement, and despondency as soon as this man disappears . . . to avoid being confused with all the hypocrites here and elsewhere—those crocodiles, those dogs. . . . We do not want to join in the competition among cynics who decree here and there this-and-that many days of mourning, each one trying to establish and advertise his distress with tears that we revolutionaries should recognize for what they are.
>
> Samora Machel is dead. This death must serve to enlighten and strengthen us as revolutionaries, because the enemies of our revolution, the enemies of the peoples of the world, have once again revealed one of their tactics, one of their traps. . . .
>
> We say to imperialism and to all our enemies that every time they carry out such actions, it will be yet another lesson we have learned. . . . They did not reckon with the special force men have within them that makes them say no despite the bullets and the traps. . . . These were the conditions in which Samora Machel dared to pick up the flag carried by Eduardo Mondlane, whose memory is still with us. (Sankara 1986b, 313–316)

One week before his assassination on October 15, Sankara gave a eulogy at an exhibition on the life of Ernesto Che Guevara. In attendance was Camilo Guevara March, Guevara's son, at the head of a Cuban delegation. "We want to tell the whole world today that for us Che Guevara is not dead. Che Guevara was cut down by bullets, imperialist bullets, under Bolivian skies. And we say that for us, Che Guevara is not dead. . . . *While revolutionaries as individuals can be murdered, you cannot kill ideas.* Ideas do not die" (Sankara 1987f, 421; my own emphasis). To be clear, with Sankara ideas are not simply abstract statements; they are at once outcomes of visions put to work. He is an incandescent light because he spoke while wiping sweat off his face or during pauses to straighten his back, to share the Burkinabè story with Africa and the world, and to chastise those who sought to keep the Burkinabè dependent on them by distracting the bent-down citizens, busy at work. Thomas Sankara should not be abused by those who are comfortable, who have forgotten where they come from.

In an interview with the Cameroonian writer Mongo Beti in November that same year, Sankara added: "Everywhere today, in the four corners of the continent, there are Nkrumahs, Lumumbas, Mondlanes, etc. Should Sankara be physically eliminated today, there will be thousands of Sankaras to take up the challenge to imperialism" (Sankara 1985b, 241). Instead of inducing fear, the death of a revolutionary is a site of learning, of accepting the inevitability of death and making life meaningful, an opportunity to struggle to make life worth living and offer important resources for those who shall be left to carry on. Having made peace with and dispensed oneself with the worthless fear of death (which chews one's precious living time where others are, this very moment, barely hanging in there just for one more day to live), the critical thinker-doer soldiers on to make life worth living and death a happy ending. And Sankara could declare: "The *only fear I know is the fear of failure, the fear of not having done enough.* You can fail because of a disagreement, but not because of laziness. . . . I should have, and I didn't. . . . This would be a cross of my own incapacity, of my own escape from responsibility" (Sankara 1985a, 232; my own emphasis).

THE BURKINA'S PLIGHT

Sankara lamented "the terrible price paid by Africa and the Black world for the development of human civilization. A price paid without receiving anything in return." Nobody with any conscience can ever absolve that as a cause of Africa's problems since. "It is our blood that fed the rapid development of capitalism. . . . For every Black person who made it to the plantations, at least five others suffered death or mutilation," to say nothing of the continent's loss (Sankara 1984b, 172).

From the experiential location Sankara explained what inequality felt like, as one who grew up seeing a European principal at the primary school in Gaoua whose children had a bicycle that all other children could only dream about. Having at last failed to persuade these privileged kids to allow him to at least grab a ride, Sankara rode the bike without permission. He was expelled from school and his father—a veteran of the Second World War taken prisoner by the Germans—was jailed for the sin of his son (Sankara 1985a, 193).

His father was not alone, hence for Sankara it was not the personal but the "we" that mattered. To attain greater justice, each had to notice the people's situation and sacrifice for them to have a just existence—50 percent of every one thousand children dying within three months of their birth. Only 16 percent of one hundred able to go to school. Only 18 percent of one hundred able to get a high school degree. A paltry 0.003 percent able to go for postsecondary education, only accessible abroad, and assured of employment upon return. A misery 0.004 percent of Burkina Faso's seven million citizens "touched by even part of this luck." And that percentage, constituting the civil service, consumed 30 billion CFA ($27 million)! This urban elite spoke loudest on human rights, loss of buying power, oblivious to its condemnation of all those children to their premature deaths having protested feverishly against tiny cuts to their salaries to enable the construction of even the tiniest clinic (Sankara 1985a, 199).

Sankara reformed the judicial system to assume an educational rather than punitive purpose. "We are judging one man in order to give rights to millions," he said. The rights of man, yes, not the rights of one man. Not formalist law, "obsessed by procedures and protocol, . . . tricking the people, turning the judge—draped in his robe, decked out in his sash, and sometimes even a wig—into a clown" (Sankara 1984d, 114). How could "ignorance of the law is no excuse" be a dictum of justice when the population was

95 percent illiterate and the law was a written document they could not read? How could it be said that "the law alone may employ force" when that law was enacted to defend and uphold the interests of the minority ruling classes, who alone could buy a lawyer and judges, who alone were educated enough to read the esoteric and elitist language of that law? What justice was there in law when its beneficiaries were those that could afford oratory and riches, in a society with "two systems of justice and two systems of democracy: that of the oppressors and that of the oppressed, that of the exploiters and that of the exploited"? (Sankara 1984d, 115–119).

Burkina Faso was the "distillation of all the natural evils" afflicting humankind.—Hundreds dying of hunger, famine, disease, and ignorance.—Rivers drying up.—The environment deteriorating.—Trees dying.—The desert approaching at the speed of 7 km/year. "Here I am merely the humble spokesperson of a people who, having passively watched their natural environment die, refuse to watch themselves die," Sankara declared to the International Conference on Trees and Forests in Paris on February 5, 1986 (Sankara 1986d, 255). For all the Western-and-white liberal pretensions to outrage at hunger and disease-induced infant mortality, the truth was that these same paragons of human rights were "part of the international complicity of men of good conscience." Meanwhile, half the nation's children were malnourished, had an average caloric intake of 1,875 a day, 21 percent below recommended intake (Sankara 1985a, 199, 212).

In a somber address to the same conference, Sankara gave the context within which the menacingly rapid approach of the Sahara threatened to swallow his country. Namely, how the planetary and global relationship between the "machines that spew fumes [that] spread carnage" and the biosphere that animated the "colonial plunder ha[d] decimated [Burkina's] forests without the slightest thought of replenishing them for [local people's] tomorrows." That plunder of the biosphere went unpunished because the countries involved in these "savage and murderous forays on the land and in the air" were the very same ones with the power to strangle the economies of those whose resources they plundered. As Sankara lamented, "Those who have the technological means to find the culprits have no interest in doing so, and those who have an interest in doing so lack the technological means. They have only their intuition and their innermost conviction" (Sankara 1986d, 258).

"We've become great predators," the president remarked, before launching into a discussion of the annual consumption of firewood at the heart of the Burkina Faso emergency. "If we were to place end to end the carts traditionally used to transport wood here, they would form a convoy the equivalent of 4.5 times the length of Africa from north to south." The dilemma was to allow or to forbid people from cutting down trees—their only source of energy.

The Burkinabè were not just waiting to be swallowed by the Sahara. On January 15, 1986, Sankara had launched the "Popular Harvest of Forest Seeds," a vast operation to create seven thousand village nurseries under a program called "The Three Battles" to defend the trees and forests and cultivate a "deep and sincere love between Burkinabè men and trees" (Sankara 1986d, 255–257). Under "The Three Battles," "unplanned" and "anarchic" cutting of wood and the transport of wood "except in specially white-washed vehicles that are clearly identifiable" were all banned (Sankara 1985a, 215). One had to now produce a lumber merchant's card.—Stay within areas designated for wood cutting.—And cutting in the prescribed way ensuring regrowth. Acts of arson and unauthorized tree cutting were tried in the Popular Courts of Conciliation in the villages, and those found guilty were not sent to jail or punished, but made to plant a specified number of trees (Sankara 1986d, 255–257).

The second battle banned the free roaming of livestock, after human deforestation the second major cause of unmitigated habitat destruction. Any animal found grazing on crops was to be slaughtered on the spot without recourse to the law to "force our livestock raisers to adopt more rational methods." Sankara acknowledged that the method of stock raising was still "purely contemplative." Owners could keep a herd of five thousand cattle, derive pride from their large herds, without regard to how they would feed them. They destroyed neighbors' crops and devastated the forests. Deprived of enough food, these puny animals were lightweight, produced little milk, and hardly pulled (Sankara 1985a, 215).

The third battle (launched in 1983) was to regenerate nature through reforestation. Every village and town was obligated to have a grove of trees. The ancestors had maintained "sacred woods" where rituals like initiation were held, and which "supposedly possessed certain powers that protected them." European Christian churches weakened these practices. The woods, now rendered just ordinary collections of trees, disappeared. The speed of desert encroachment gathered rapid pace. Sankara's government brought back the groves, not with the religious meanings and functions the ancestors had imbued them with, but equally sentimental values, such as planting a tree to mark every happy event (marriages, baptisms, award presentations, and visits by prominent individuals, etc.). Planting and successfully nursing seedlings into trees became a condition for approval of purchase or rental of public buildings. Negligence of—let alone leading to the death of—the trees was grounds for eviction by the Committee of the Defense of the Revolution (CDR) (Sankara 1986d, 256). Sankara declared that fifteen million trees had been planted under the People's Development Program inside fifteen months by 1986. Each family was required to plant one hundred trees in the villages and river valleys per year. Between February 10 and March 20, 1986, thirty-five thousand rural people took part in a bootcamp

FIGURE 9.2
Burkinabe women building a terrace to trap run-off water and prevent spread of the desert, 1986. *Source*: United Nations (John Isaac).

training on economic and environmental organizing and management (Sankara 1986d, 255–257).

Under the same battle, energy efficiency was implemented as a technology to maximize output from wood and reduce the quantities of trees cut down. In 1984 the government launched and heavily subsidized the making of improved cookstoves to promote and popularize their use, producing eighty thousand of them by the end of December 1985, the uptake increasing as wood became scarcer (Sankara 1985a, 216). To welcome the new year in 1986, all the students and schoolchildren in Ouagadougou handmade over 3,500 improved cookstoves and offered them as gifts to their mothers as part of their effort to reduce firewood consumption and protect trees and life (Sankara 1986d, 256).

The problem of vegetation cover was tied to that of water, and Sankara had responded by allocating a dedicated ministry to it "to clearly formulate the problems in order to be able to resolve them." The new ministry of water isolated the solutions to just three: borehole drilling, reservoirs, and dams. But the international lending institutions handed Ouagadougou "one-sided contracts and draconian conditions" that would lead to a traumatizing debt trap. Sounding resigned, the president concluded his speech:

> A handful of forestry engineers and experts getting themselves all worked up in a sterile and costly manner will never accomplish anything! Nor can the worked-up consciences of a multitude of forums and institutions—sincere and praiseworthy though they may be—make the Sahel green again, when we lack the funds to drill wells for drinking water a hundred meters deep, while money abounds to drill oil wells three thousand meters deep! . . . This struggle to defend the trees and forests is above all a struggle against imperialism. Because *imperialism is the arsonist setting fire to our forests and our savannas*. (Sankara 1986d, 259; my own emphasis)

Sankara faced the reality of a rural countryside neglected in development since the advent of French rule, overused soils that were becoming less fertile by the season, rapid population growth, increasingly unpredictable, ever receding rainfall patterns, and landownership and usage held to ransom by speculative white (French) monopoly capital. Green beans that traditionally thrived in the Kougassi region were shipped to Europe, especially France, on the French-owned Union de Transport Aérien and Air Afrique. These airlines could refuse to ship the beans, or ship just twenty of the thirty tons, and all of the rest would start rotting at the airport, which had no cold chain facilities. The best-quality beans arrived to be bought as second grade in Europe, were repackaged, and sold as first grade under another label. Because the beans were perishable they could not be repatriated home (at more cost) so the Burkinabè ended up taking any price (Sankara 1985a, 212). In regions like Orodara, the enterprising farmers grew fruits and vegetables, but they rotted for want of transport to market. That is why people had to be mobilized to build landing strips to fly these perishables posthaste to Ouagadougou and Dori (Sankara 1983b, 62).

And the French could systematically boycott Burkina's exports to "strangle" its economy and cause tensions with growers and, as eventually happened, sow seeds of a coup. Burkina Faso was a major producer and exporter of livestock. In 1985 the French refused to buy Burkinabè cattle and imposed such impossible conditions to shut out the imports completely. They also exerted pressure on cement exporters to stop imports critical to Ouagadougou's ambitious public construction projects, thus rendering the workers at construction sites hostile to the government (Sankara 1985a, 212). The need for cement was especially acute in housing construction projects. Real estate speculation was rife; workers already underpaid suffered through rent gouging. But the solution was to undertake large-scale construction of modern residential homes, which the French now sabotaged (Sankara 1983a, 107).

Establishing markets and collection and distribution points for and an agro-industry to process the crops (citrus fruit, vegetables, livestock) required an injection of modern agricultural techniques in the rural countryside. To diversify agriculture and have regional specialization. And agricultural production to anchor a manufacturing industry (Sankara 1983a, 104–105). Therefore, to construct small workshops needing little training for agro-processing near production zones needed old clunkers, not high-technology machinery, because the latter required lengthy and sophisticated training for an illiterate population and huge capital outlay (Sankara 1983b, 62). "We don't favour big industrial installations since automation eliminates jobs and requires the use of substantial amounts of capital, which we do not have," Sankara said, long before the escalation of robots. In any case, such technology would be expensive and difficult to maintain on spares only made abroad, thus perpetuating a dangerous dependency—as if the country did not have too much of that already. "A single broken part can mean dispatching a plane to Europe because the replacement can be obtained only there" (Sankara 1985a, 210).

Sankara was never embarrassed to be Burkinabè or to show affluent visitors the poverty of his people, by which their lot was measured, where Europe and North America defined theirs according to happiness and development. According to how many pounds of steel per citizen.—How many telephone lines.—How many tons of cement produced. Here in Burkina an incoming foreign ambassador was not squirrelled away quickly to the presidential palace to insulate his nose and eyes from the stench and fly-sights of poverty, but was taken to its epicenter, the rural countryside, driven on the dusty and thirsty roads, and introduced to the people, to whom his country had sent him to represent its own people. At the end of the journey, an informed debrief could follow, based on a concrete experience of Burkinabe everyday life: "Mr. Ambassador, your Excellency, you have just seen Burkina Faso as it really is. *This is the country you must deal with, not those of us who work in comfortable offices*. We have a wise and experienced people capable of shaping a certain way of life. While elsewhere people die from being too well-fed, here we die from not having enough food. Between these two extremes there is a way of life to be discovered if each of us meets the other halfway" (Sankara 1985a, 206; my own emphasis).

"AGAINST THOSE WHO EXPLOIT AND OPPRESS US"

Sankara was one of few leaders of his generation who could take on France's policy of Françafrique, a term derived from France-Afrique coined by Ivory Coast's founding president Félix Houphouët-Boigny in 1955 to describe Africa's relations with France, and extended by François-Xavier Verschave to describe France's control over its former colonies through a combination of finance, treaties, military means, and personal connections. In French-speaking Africa, under the act of coloniality, France demands and receives 85 percent of all the foreign reserves of its fourteen (former) African colonies into its treasury every year to pay for "the benefits of colonization," including infrastructure built with the sweat and blood of Africans forced to give their labor to construct extractive infrastructures to syphon their wealth to France! Even today, France has right of first refusal on all raw materials discovered in those countries; in public procurements and bidding; exclusive right to supply military equipment and train officers (who usually wage the coups when needed); to deploy its troops and defend French interests, and charge the costs to its continuing colonies; that French must be the official language and of education; to use the colonial Coopération Financière en Afrique Centrale (Financial Cooperation in Central Africa) currency; and to send an annual financial balance and reserve report to France, and other don'ts and musts. Those leaders who have dared France have been met with scorched earth, like Guinea in 1958, wherein after Sékou Touré declared independence the departing colonizers packed up and shipped all that could be moved to France, and destroyed all that could not: public buildings, schools, vehicles, medicines, books, horses, warehouses. Or the many coups that have met those who defy Paris, and the lavish palaces and lifestyles for those who have acquiesced. This is Françafrique—good old colonialism.

Meanwhile, in France, there is no black or white. Only citizens and noncitizens; either you are French or you are not. It creates a semblance of civility, of equality, of no discrimination. A category imposed from the top, the category of those who benefit from colonialism, not that of those who feel and experience it. Partnership, yes, but between a rider and a horse, whereby the white is the rider and the nonwhite is the horse, the one enjoying the ride that citizenship allows, the other the burden, the silencing, and nonacknowledgement of race carries. You cannot claim your rights as a black person; only as a citizen. The blindness blights even France's academics, including the most celebrated theorists, who are so confident talking about biopolitics and nonhuman agents, acquiring fame among people naïve about, complicit in, and unaffected by race—mostly white and liberal scholars or the coconuts, who are black outside but live and think white.

Sankara was as direct as he was subtle. And he could do it with such panache and satire. Case in point his welcome speech to French president François Mitterrand in 1986, appropriately titled "Against Those Who Exploit and Oppress Us." The latter had toured the country years before while it was still Upper Volta; and Sankara now welcomed him home to "an entire program" called Burkina Faso, with its own "code of honour and of hospitality," which said: "Cursed is the person at whose door no one ever knocks and

FIGURE 9.3
Sankara with Mitterrand during the Tenth Franco-African Summit, Vittel, France, October 3, 1983.
Source: AFP.

whose home no hungry and thirsty traveller ever visits or enters." And the thirsty traveler was Mitterrand! Stopping at "our home and, regaining strength after a mouthful of refreshing water, has engaged us in conversation in order to get to know us better, to understand us better, and to take back with him, to his home, memories of our home." Here he was, welcoming not simply the president but the person, "welcoming François Mitterrand, . . . the person who came to see and to attest in good faith and in all objectivity to the fact that something is happening somewhere under the African sun, in Burkina Faso." So here was Burkina Faso, "a construction project, one vast construction project" (Sankara 1986f, 325–326).

The sarcastic host was just getting started. "We are far from the myth of the labours of Sisyphus," Sankara veered toward Greek mythology, where Sisyphus is the fruitless laborer that foolishly tries to roll a huge boulder up the hill, which rolls back down to crush him. The Burkinabè had done much with little, more that might appear little to others, but mammoth to the Burkinabè, the garment quite surprisingly fitting given such a small cloth available. Built many dams, small they might be, dwarfs by comparison with "those great works" Europe had built, that "have their merits, and inspire us, legitimately I believe, with pride." A primary healthcare station "in every village of Burkina

Faso." Millions of children vaccinated not just in Burkina, but even in neighboring countries too!

And who better to carry this message to France than François Mitterrand the man—if he was a real man enough "to tell things the way they are." And the man was the perfect choice of interpreter and spokesperson for the Burkinabè because he had had constant battles in his life and political career. "We are familiar with these battles and they also inspire us, those of us from Burkina Faso." First Sankara massages the ego of the colonizer:

> You like to speak, sometimes stubbornly in certain recalcitrant milieus, of the rights of the peoples of the world. . . . In the French region of Berry, I believe, your name, Mitterrand, means "mid-sized field"—or perhaps "grain measurer." In any case, a man of common sense. Common sense that is close to the men who are tied to the land, the land that never tells lies. Whether it is grain or whether it is a field, we think that the constant factor is that you yourself remain tied to the land. (Sankara 1986f, 327–328)

Then the son of an African woman—bless his mother—really sat the colonizer down:

> It is in this context, Mr. François Mitterrand, that we did not understand how bandits like Jonas Savimbi and killers like Pieter Botha, have been allowed to travel up and down France, which is so beautiful and so clean. They have stained it with their blood-covered hands and feet. . . .
>
> Peace . . . also means the Saharawi Arab Democratic Republic.
>
> Peace means Libya; Chad.
>
> Your efforts can only be of considerable aid due to . . . the direct or indirect involvement of your country in those areas. I would like to assure you that, for our part, in Burkina Faso, we are fully ready to lend a hand, to give our assistance to whomever requests it, as long as the fight to be waged is a fight that reminds us of the France of 1789. It is for this reason that I would like to say to you that Burkina Faso is prepared to sign a defence agreement with France to allow all these arms you possess to come and be stationed here, in order to continue on to Pretoria, where peace calls us. (Sankara 1986f, 329–331)

President Sankara continued "to address the man," goading the colonizer in him to dare come out, this time confronting the devil of Françafrique, debts that Mitterrand had criticized, which criticism Burkina was grateful for, but had changed nothing even in France. "Debts of yesterday" the Burkinabè people had never asked for, being asked to compensate France for bringing so-called development they had never asked for. Any that agreed to that stupidity stood condemned in the eyes of the people: "In our 'Song of Victory,' our national anthem, we call those who bear full responsibility here in Africa local lackeys. Because, subject to a master, and without understanding their actions, they executed orders here that went against their people." To make matters worse, before becoming president Mitterrand had admitted who benefitted from the debts. Lest he had forgotten, Thomas Sankara jerked the colonizer's memory:

> Mr. President: You have written somewhere that currently the amount of French aid is declining.
>
> And, unfortunately, you add, the amount of aid evolves according to France's political ambitions and, worst of all—sorry, you said and emphasized, what is worse—it's the capitalists who profit from this. Well, we believe that is also right. You wrote it, I believe, in that book *Ma part de verité* [My share of truth]. This small share of the truth is a truth. It is indeed the capitalists who profit from this, and we are ready to fight together against them. (Sankara 1986f, 332)

Sankara ended his satirical homage to Mitterrand with a contrast of the humanity extended to the French citizens in Burkina, including some rather despicable elements, and African immigrants in France subjected to the most dehumanizing of treatments even while building France:

> We want to think of, we want to address our thoughts to all those over there who are cut to the quick every day, whose souls are wounded, because somewhere a Black or a foreigner in France was the victim of a barbaric act taken without any consideration for his human dignity. . . . As for the immigrants in France, although they are there for their happiness—like any man seeking sunnier horizons and greener pastures—they are also helping and building France for the French. A France, which has, as always, welcomed freedom fighters from all countries on its soil. Here in Burkina Faso, French people are struggling seriously at the side of Burkinabè, often in nongovernmental organizations. Although, it must be said, not all these nongovernmental organizations represent respectable institutions in our view—some of them are purely and simply reprehensible liars—there are some that have great merit. (Sankara 1986f, 333–334)

The concentration and presence of NGOs represented a failure of state-to-state relations. Out of six hundred NGOs, four hundred were French, some of them serving as spy agencies for their government (Sankara 1985a). They built their wells in the English, German, or French style, but the water was drunk Burkinabè style. The NGOs did not share information; one repeated the mistakes of another. Others were just interested in good press coverage in Europe: "You see, my good people, we are over there saving souls. Give us your pennies, God will repay you." Some were not clear about their intentions and got misunderstood by local leaders. Keen to please, they became complicit: "You've come from Europe, very good. You have money and you wish to help the country, bravo, that's what needs to be done because people are starving here. But you're going to need an office, so why not rent mine. You'll need a national director since we very much want to assure some continuity—I have a cousin who is ready to do that. For receptionist, I have a cousin. And as janitor there's my nephew." I scratch your back (good press in Europe, more donations), you scratch mine (I have my donors, if you vote for me you will get free food, clothes, agricultural inputs). The creation of the Office for Overseeing Nongovernmental Organizations was intended to enable the NGOs to retain their flexibility of operations, ensuring "they all learn from the experiences of those who came before them," knew what the priority areas were, and guarantee the effectiveness of their work (Sankara 1985a, 209).

In concluding his speech to the International Conference on Trees and Forests in Paris in 1986, President Thomas Sankara put Burkina Faso's problem in context. Here was a country that knew what the solution to the rapidly encroaching desertification problem was, but the power of the purse resided elsewhere:

> Neither fallacious Malthusian arguments—and I assume that Africa remains an underpopulated continent—nor the vacation resorts pompously and demagogically christened "reforestation operations" provide an answer. We and our misery are spurned like bald and mangy dogs whose lamentations and cries disturb the peace and quiet of the manufacturers and merchants of misery.
>
> That is why Burkina has proposed and continues to propose that at least 1 percent of the colossal sums of money sacrificed to the search for cohabitation with other stars and planets be used, by way of compensation, to finance projects to save trees and lives. We have not abandoned hope that a dialogue with the Martians might lead to the reconquest of Eden. But in the meantime, earthlings that we are, we also have the right to reject a choice limited simply to the alternatives of hell or purgatory. (Sankara 1986d, 259)

Western multinational corporations came in with a commitment to create jobs and contribute to local economic development, and got tax breaks and other waivers, only to announce massive job cuts a few years later. "Today, when you've squeezed the lemon dry, you want to throw the lemon out. No!" (Sankara 1983b, 57). What "a huge drain" on the economy! Literally siphoning out locally created wealth abroad; more money going out than coming in, yet African heads of state lined up to beg the companies to come (Sankara 1983a, 85).

As if that were not enough the West then claimed to be bringing in aid to "help" Africa "develop." But as Jacques Giri had shown Sankara in *Le Sahel demain* (*The Sahel of Tomorrow*), aid was designed only to facilitate "bare survival" and continued dependency on the West, not prosperity, hence it covered only the nonproductive sectors. On one hand, African governments could not deploy such money (usually balance of payment support) to productive sectors to grow it; on the other, these loans were accruing interest. The debts increased. Money intended for development was redirected toward debt installments. Unable to produce more for export, trade deficits increased. The debts continued to grow (Sankara 1984b, 159).

The countries that lent the money were also the colonizers, "the same people who ran our states and our economies," who put Africans in debt to their financiers. The former colonial officials in preindependence Africa were now the technical assistants. "Actually, it would be more accurate to say technical assassins. They're the ones who advised us on sources of financing, on underwriters of loans." African governments were told to sign their people to fifty- to sixty-year debts, even longer—"a cleverly organized reconquest of Africa" turning the people into a "financial slave—or just plain slave" of Europe (Sankara 1987d, 381). The very same people who lent the money were the ones who "often suggested, proposed, organized, and set in place" projects to be done and who must execute them. The process was like a military operation: "They have quite a system. First come the members of the assault squad, who know exactly what they are going to propose. Then they bring out the heavy artillery, and the price keeps going up." The investors did not deposit the money in their banks back home because the yields were low. "They have to create the need for capital elsewhere and make others pay." The president shifted to a tobacco analogy:

> Do we really need to smoke this or that brand of cigarette? They've convinced us, "If you smoke such-and-such brand you'll be the most powerful man on earth, capable of seducing any woman." So we took up smoking, and got cancer as a bonus. The most privileged among us have gone to Europe to be treated. And all to give a second wind to your tobacco market. (Sankara 1985a, 204)

It was an entire organized system, not just one bank. If the country did not pay, its planes would be impounded or critical spare parts would be held up. *"Either we resist collectively and refuse categorically to repay the debt or, if we don't, we'll have to go off to die alone, one by one,"* Sankara implored the OAU heads of state and government in Addis Ababa in November 1984. Unfortunately an individual government went behind others to negotiate deals with the lenders to get cheap kudos. It was branded and extensively publicized as "the best organized, the most modern, the most respectful of written agreements" (Sankara 1985a, 205; my own emphasis). Thus Burkina Faso would only accept aid that "respects our independence and our dignity." Any that offered benefits only to the leaders to "buy off consciences," conditional upon Burkina's purchase of the lender's products or opening bank accounts in that country, would be turned down. Burkina had turned down aid from the Soviet Union that "did not meet our expectations." As he said in 1985, "We have our dignity to protect" (Sankara 1985a, 197, 209).

Sankara laid out his case to refuse to pay the debt at the OAU summit in Addis Ababa on July 29, 1987:

> The debt cannot be repaid, first of all, because if we don't pay, the lenders won't die. Of that you can be sure. On the other hand, if we do pay, we are the ones who will die. Those who led us into debt were gambling, as if they were in a casino. As long as they were winning, there was no problem. Now that they're losing their bets, they demand repayment. There is talk of a crisis. No, Mr. President. They gambled. They lost. Those are the rules of the game. Life goes on. [Applause]

> We cannot repay the debt because we have nothing to pay it with. We cannot repay the debt because it's not our responsibility. We cannot repay the debt because, on the contrary, the others owe us something that the greatest riches can never repay—a debt of blood. It is our blood that was shed.
>
> People talk of the Marshall Plan, which rebuilt the economy of Europe. . . . But they don't mention the African Plan, which enabled Europe to face Hitler's hordes at a time when their economies were under siege, their stability threatened. Who saved Europe? It was Africa. There is very little talk about that. . . .
>
> The rich and the poor don't share the same morals. The Bible and the Koran can't serve in the same way those who exploit the people and those who are exploited. There will have to be two editions of the Bible and two editions of the Koran. [Applause]. . . .
>
> This is not a provocation. . . . I hope our conference sees the necessity of stating clearly that we cannot pay the debt. Not in a warmongering or warlike spirit. This is to avoid our going off to be killed one at a time. *If Burkina Faso alone were to refuse to pay the debt, I wouldn't be at the next conference.* (Sankara 1987d, 381; my own emphasis)

Sankara was a thorn in the flesh of his fellow African leaders, whom he derided for absconding meetings "to discuss Africa, in Africa," yet who "are ready to dash off to Paris, London, or Washington when called to meetings there." He went as far as calling for OAU sanctions against such heads of states and to reward regulars—what he called "a coefficient of Africanness"—in the form of priority in development project funding from the African Development Bank (Sankara 1987d, 374).

What then to do with the money saved? He again warned his colleagues in his last OAU address in Addis:

> When we tell countries we're not going to pay the debt, we can assure them that what is saved won't be spent on prestige projects. We don't want any more of those. What is saved will be used for development. In particular we will avoid going into debt to buy arms. Because an African country that buys arms can only be doing so to use them against an African country. What African country here can arm itself to defend against the nuclear bomb? Every time an African country buys a weapon, it's for use against another African country. It's not for use against a European country. It's not for use against an Asian country. So in preparing the resolution on the debt we must also find a solution to the question of armaments. (Sankara 1987d, 381)

Outsiders, Sankara warned, are not the solvers of Africa's problems. Some of the European aid volunteers were "very sincere" but ignorant about Africa, come as "lesson-givers" with a "patronizing attitude," hence make mistakes, and return home "completely disgusted with Africa" (Sankara 1985a, 191). The job of sacrificing for Burkina was for the Burkinabè; outsiders had their own problems. "We are the ones who are making our revolution . . . We are the ones who should pay the price" (Sankara 1985a, 203). When Sankara came to power in 1983, Burkina's coffers were empty. The deposed regime had agreed to a 3 billion CFA franc ($5.46 million) economic structural adjustment loan.

That would be the last loan Sankara would sign. "We fill the hole by preventing it from appearing," the president said in 1985. "That is, we don't allow a deficit." That required belt tightening. State employees were asked to take cuts in salaries and benefits, adopt more modest lifestyles, and be more responsible in management and using the people's money. Writing only on one side of the paper was prohibited (Sankara 1985a, 201).

When asked if demanding too many sacrifices would not backfire, Sankara replied, "Not if you know how to set an example," and the leaders did not demand sacrifices that they themselves did not make.—By living a modest lifestyle: if a schoolteacher was appointed a minister, they were paid a schoolteacher's salary. The president himself as an army captain got a captain's salary. Sankara's own carpool was ordinary—a far cry from the circus of his peers' motorcades. All ministers travelled economy class and were given a modest 15,000 CFA francs ($27) per diem. The same for the president. This power of example meant that a cabinet minister or head of state could not be "black kings who buy themselves cars and build mansions" while taxpayers were "dying for lack of money to buy a tiny capsule of quinine." At the beginning of 1983 the country had a budget deficit of 695 million CFA francs ($1,264 million) managed by the French. One year later that deficit was just a million that Burkina directed and implemented itself. At the beginning of 1985 it had a budget surplus of 1.0985 billion CFA francs ($2 million) (Sankara 1985a, 201, 203).

Ultimately, the success of a revolution was to be measured by what it produced, not slogans and praise singing. Sankara rebuked those that had composed the song "O CNR, Thomas Sankara, may he forever be president!" as a dangerous and counterproductive form of praise singing that valued form over content. "The revolution will not be measured by the number of slogans and by the number of tenors and basses doing the chanting. It will be measured by . . . the level of production. We must produce, we must produce. That's why I welcome the slogan, 'Two million tons of grain.'" There was no story to write home about when (in 1986) Burkina Faso was still appealing for food aid, unable yet to declare victory over "instincts of beggars and welfare recipients," prey to "he who feeds you also imposes his will" (Sankara 1986a, 289). How could people slaughter a cockerel, turkey, or sheep during Tabaski or Eid al-Adha (the festival of the lamb), Easter, Christmas, or family gatherings with the eagle eyes of the lender fixed on them, seeing this most human thing as wastefulness? How does the one who draws energy from the imperialist's bowl shout "Down with imperialism!" without jeopardizing his next meal? "Well, your stomach knows what's what. . . . Let us consume only what we control!" Sankara said. "There are those who ask, 'But where is imperialism?' Look at your plates when you eat—the imported grains of rice, corn, millet—that is imperialism. Go no further" (Sankara 1986a, 290).

In his last address to the OAU, Sankara implored Africa to use its immense resources to develop itself.—Fertile soils.—Mineral rich.—Hardworking people.—A vast market north to south, east to west. What was required was to turn the African market into "a market for Africans," not the West or East. Sankara made a powerful argument for value addition well ahead of its time, citing Burkina Faso's strides in just four years:

> Let's produce in Africa, transform in Africa, consume in Africa. Produce what we need and consume what we produce, in place of importing it.
>
> Burkina Faso has come to show you the cotton produced in Burkina Faso, woven in Burkina Faso, sewn in Burkina Faso to clothe the Burkinabè. My delegation and I were clothed by our weavers, our peasants. Not a single thread comes from Europe or America. [Applause] I'm not here to put on a fashion show; *I simply want to say that we should undertake to live as Africans. It is the only way to live free and to live in dignity.* (Sankara 1987d, 380–381; my own emphasis)

REENGINEERING THE MIND

Engineering a new society was foremost a reengineering of the mindset. The revolution was a necessary process to create a morally and socially upright and exemplary person, "a new Voltaic man, . . . a new Voltaic society, within which the Voltaic citizen, driven by revolutionary consciousness, [would] be the architect of his own happiness, a happiness equal to the efforts he will have made" (Sankara 1983a, 98–99). That is why Sankara renamed the country the "Land of the Upright Man": Burkina Faso.

Hegemonic powers find it easier to use cultural means rather than military might. It's cheaper, more flexible, more effective. Champagne, lipstick, and nail polish, yes, not fighter jets, artillery batteries, or bombs. "Decolonizing our mentality and achieving happiness within the limits of sacrifices," and "recondition[ing] our people to accept themselves as they are, to not be ashamed of their real situation, to be satisfied with it, to glory in it," was for Sankara a critical weapon against Western-and-white domination (Sankara 1985a, 197).

In 1983, Sankara signaled that anything "antinational, antirevolutionary, and antipopular" must be banned, that musicians were to "put their pens at the service of the revolution," to sing not just of our "glorious past" but also the people's "radiant and promising future." The country would "draw on what is positive from the past" to shape the future, always centering the "popular masses" as the "inexhaustible source for the masses' creative inspiration." The artists had therefore to know "how to live with the masses, becoming involved in the popular movement, sharing the joys and sufferings of the people, and working and struggling with them." And every time before releasing a song, they were to always ask what audience it was mean for: "If we are convinced that we are creating for the people, then we must understand clearly who the people are, what their different components are, and what their deepest aspirations are" (Sankara 1983a, 106). This had echoes in Cabral's call for the bourgeoisie to commit suicide to their identity through a revolutionary commitment to the deepest aspirations of their own people. And Sankara would say: "Our revolution needs a people who are convinced, not conquered" (Sankara 1987b).

In his "Freedom Must Be Conquered" speech, Sankara declared that Burkina Faso would choose which lessons to learn and which friends to make without need for some self-anointed referee. "We wish to be the heirs of all the world's revolutions and all the liberation struggles of the peoples of the Third World." From the United States, James Monroe's 1923 proclamation of "America to the Americans." Hence "Africa to the Africans," "Burkina to the Burkinabè." From the French Revolution of 1789 the rights of people and liberty. From the Bolshevik Revolution in 1917, power to the proletariat, which is what gave birth, after all, even to the Paris Commune. Having learned through self-determined intellection about "the peoples of the world and their revolutions," inclusive of tragic failures and human rights violations, it was the task of the Burkinabè "to retain only the core of purity from each revolution," thereby avoiding "becoming subservient to the realities of others, even when we share common ground because of our ideas" (Sankara 1984b, 165).

We may recall how Cabral had replied to his Western-and-white audience when they asked if he was a Marxist: "The labels are your business." Ours is to struggle for self-liberation. Mongo Beti posed the same question to Sankara. "For now, I am anti-imperialist. Speaking as your comrade president, this is also the case. . . . That's enough for us—to be useful to our people, especially when they don't trouble themselves with labeling their leaders but judge them above all by their revolutionary action. Later we'll see" (Sankara 1985b, 249). A year later, in another interview with *Jeune Afrique,* Sankara conceded familiarity with the classics of Marxism-Leninism. *Das Kapital*? Had read, but "not all of it." Lenin? Yes, he had read all of Lenin's works, and he would carry with him *State and Revolution* if he were to be marooned on a desert island. "This is a book I take refuge in, that I reread often. . . . But on an island, I would also take the Bible and the Koran." Why? Lenin, Jesus, and Muhammed constituted for Sankara "the three most powerful currents of thought in our world, except perhaps for Asia. *State and Revolution* provides an answer to problems that require a revolutionary solution. On the other hand, the Bible and the Koran allow us to synthesize what peoples thought in the past and what they continue to think, in time and space." Asked which of them was the most revolutionary, Sankara gave a relative qualification. Lenin for the 1980s, without ignoring that Muhammed was "a revolutionary who turned a society upside down." He concluded:

> Jesus was too, but his revolution remained unfinished. He ends up being abstract, while Muhammed was able to be more materialist. *We received the word of Christ as a message capable of saving us from the real misery we lived in, as a philosophy of qualitative transformation of the world. But we were disappointed by the use to which it was put. When we had to look for something else, we found the class struggle.* (Sankara 1986e, 263; my own emphasis)

And it was this sense of independence of imagination and thoughtful revolutionary that made Sankara distrustful of outside attempts to prescribe what formula of revolution was good for Burkina Faso absent the invitation nor input of the Burkinabè. There were those who insisted on the formation of a party because that was the norm in revolutions across the outside world. "A revolution without a party has no future," Sankara had been told. And any revolutionaries worth their salt must be members of the Communist International. Not in Burkina. There was nothing wrong with allowing the masses "to struggle without a party and fashion their weapons without a party" (Sankara 1985a, 221). At any rate, class struggle in Burkina operated quite differently from Europe. In Burkina, weak and barely organized, no strong national bourgeoisie to trigger working-class uprising. Rather, a struggle against imperialism, opposed by feudal-type forces and a bureaucratic bourgeoisie operating from hiding (Sankara 1987e, 384).

In a visit to Harlem on the sidelines of the UN General Assembly in 1984, Sankara declared that Harlem had become "a living heart beating to the rhythm of Africa," that "Our White House is in Black Harlem." To him, Black Harlem was "a link between us and our ancestors, us and our children," every art object expressing "the pain of the African. . . . the sources of energy on which we rely in the fight we're waging," the marks concealing "perhaps the same magic that allowed others to have confidence in the future, to explore the heavens, and to send rockets to the moon. We want to be left free, free to give our culture and our magic their full meaning. It is, after all, a magical phenomenon to simply flip a switch and see light appear suddenly. If Jules Verne had been stopped in his tracks, certainly there would not be all these developments in space today" (Sankara 1984a, 145). And for that reason Sankara had set up a research center on black people in Burkina Faso, to study their origins.—History.—Culture.—Music.—Attire.—Culinary arts.—Languages. "In short, everything that enables us to assert our identity" (Sankara 1984a, 143–146). And he exhorted Harlem: "We must be proud to be Black. . . . We must be Black with other Blacks, night and day. . . . Our struggle is a call to build. We don't ask that the world be built for Blacks alone against other men. *As Blacks, we want to teach others how to love each other. Despite their meanness towards us*" (Sankara 1984c, 151; my own emphasis).

Engineering a new society requires taking command of our narratives, telling our own stories. Sankara observed the many things that we say but that are not written down. "The little that has been written has been written by foreigners—students, university professors, and researchers. . . . We haven't developed the instinct of safeguarding our intellectual capital" (Sankara 1986a, 291). This was in 1986. I was in grade seven and fourteen years old. Now I am a university professor aged fifty and still our narrative is owned and controlled by Europe and North America lock, stock, and barrel. Even where the black writes, the concepts and oracles referenced are white.

In an interview with *Jeune Afrique* in 1986, Sankara touched on the language question in the invention of a new Burkinabè society, specifically the need for an everyday language of writing. "I don't like African novels," he said. In fact, he was "disappointed" by the monotonous narrative. "It's always the same story: the young African goes to Paris, suffers, and when he returns he's out of touch with tradition." He was specifically referring to Cheikh Hamidou Kane's *L'aventure ambiguë* (*Ambiguous Adventure*). "In African literature it's not really Blacks who are speaking. You have the impression you're dealing with Blacks who want to speak French at all costs." Rather speak pidgin French. Still unsatisfactory for the critical thinker-doer president: "Anyhow, the African writers I prefer are those who deal with concrete problems, [not] those who seek to write for literary effect." Asked if he read novels: "No, almost never" (Sankara 1986e, 264).

He had no kind words for African academics, whom he dismissed as something of a lost cause, worthless copycats and consumers of Europe. "We would search in vain for genuinely new ideas that have emanated from the minds of our 'great' intellectuals since the emergence of the now-dated concepts of Négritude and African Personality," he declared, in apparent reference to Césaire and Blyden. There was no originality there, nothing from Africa. "The vocabulary and ideas come to us from elsewhere. Our professors, engineers, and economists content themselves with simply adding color—because often the only things they've brought back from the European universities of which they are the products are their degrees and their velvety adjectives and superlatives." He lumped the African academic with the European as "our enemies of yesterday and today," who should never be left with exclusive monopoly over thought, imagination, and creativity. These elites had to return to their societies and misery the colonizer had burdened us with, and, in words reminiscent of Cabral, Sankara urged them to see that "*the battle for a system of thought at the service of the disinherited masses is not in vain.*" The only way these coconuts could gain international credibility was "by being genuinely inventive, that is, by painting a faithful picture of their people" (Sankara 1984b, 158; my own emphasis). Otherwise they were a lost cause.

And Sankara devoted significant quantities of his ire to his nemesis Joseph Ki-Zerbo (the leftist opposition leader), especially in the wake of strikes in early 1984 that the revolution interpreted as an attempt to "make and break governments"—usually a prelude to a coup fomented and backed by France. The government asked the teachers to terminate the job action, in the midst of which (March 24) one French television network broadcast devoted a whole program to Ki-Zerbo. "The maneuver was transparent," Sankara would say a year later. "They were aiming to build this man up, to give him a certain credibility. It was a double manoeuvre aimed both at putting this kind of individual back in the saddle and destabilizing the situation inside the country" (Sankara 1985a, 224).

In an interview with *Jeune Afrique* a year later, Sankara elaborated that he found Ki-Zerbo's studies "very interesting" but tore into him nonetheless.

> He's still an African with a complex—he came to France, he learned, then he returned home to write so that his African brothers might recognize and see in him what wasn't seen or recognized in France. Nothing is more frustrating for an African than to reach the top without having been crowned in France. He says to himself that at home, at least, he will be recognized as one of the greats. (Sankara 1986e, 265)

Setting aside the internal politics of their quarrel, what is obscured in Sankara's critique is that Ki-Zerbo, who had studied African history with the revered Senegalese Cheikh Anta Diop and Malian scholar Amadou Hampâté Bâ, had spent several years stewarding a primary school curriculum reform to reconcile traditional ways of knowing with the Western-and-white pedagogy when Sankara came to power via the 1983 coup. It was at this point that Ki-Zerbo went into self-imposed exile fearing that he might be killed or jailed. We can now join Sankara, talking about this period:

> When the revolution called, he fled. I've asked him to come back twice, but he wants to hide his continual failures. He never succeeded in Burkina, neither by the electoral route nor the putschist route. That's why he left. I met with him twice before he left. We were happy he left because we sensed he was really very scared, and we didn't want him to die from that, to die on us, which would have brought us terrible accusations. Once he left, he went over to active opposition. But he can come back whenever he wants. The door is open for him. (Sankara 1986e, 265)

Judging the two men, supposing one is not swallowed up in the internal politics of their disagreements and their paranoia toward each other, their projects of rehumanizing the African are perhaps their enduring legacy to us. The one a distinguished scholar that not only wrote about, but had begun a revolutionary synthesis of knowledges in curriculum by starting at the primary level, fleeing in fear of death. The other a fearless man, a critical thinker-doer prepared to take huge risks, without fear of dying for a cause worth dying for. Two sons of the soil. One Burkina Faso.

Like Fanon, Sankara was ambivalent about the French language as a strategic deployment for purposes of self-articulation. The French empire had turned Burkina into part of the French-speaking world "even though only 10 percent of Burkinabè speak the language." On one hand, Sankara stresses that French is "simply a means of expressing our reality." On the other, he insists that French must "open itself up to experiencing the sociological and historical realities of its own evolution." What was "the language of the colonizer, the ultimate cultural and ideological vehicle of foreign and imperialist domination," was now, in the hands of the Conseil National de la Revolution (CNR) a "dialectical method" of analyzing imperialism, organizing against it, and fighting to triumph over it. A "vehicle of cultural alienation" had been reengineered into "a means of communication with other peoples," fellow revolutionaries in five continents, *strategically deploying—not using*—French as a medium to scheme and coordinate a common struggle separated by distance. A medium for discovering the richness of European culture. For workers' solidarity. To deploy French in service of the "the ideals of 1789 more than those of the colonial expeditions" (Sankara 1986c, 269).

France too had to "accept other languages as expressions of the sensibilities of other peoples," and "in accepting other peoples, the French language must accept idioms and concepts that the realities of France have not permitted the French to get to know." To seek refuge in cultural chauvinism was to ignore the fact that other languages had accepted French terms untranslatable in their own. The English adopted champagne. The German the French word arrangement. And African languages *impôts* (taxes), *corvée*, and prison. Sankara concluded rather nonchalantly:

> This diversity [*diversité*] brings us together in the French-speaking family. We make it rhyme with friendship [*amitié*] and fraternity [*fraternité*].
>
> To refuse to integrate other languages is to be unaware of the roots and history of one's own. Every language is the product of several others, today more so than in the past, because of the cultural permeability created in these modern times by the powerful means of communication. To reject other languages is to adopt a rigid attitude against progress, and that approach stems from an ideology inspired by reaction. (Sankara 1986c, 269)

When Mongo Beti suggested that the idea of the French-speaking world was "a strategy to control our creativity and even our future," and whether replacing it with Burkinabè languages might address the problem, Sankara offered a qualified answer. No question the neocolonial strategy was there, but he offered remarks that were also in reply to Senghor, who had commended him for the originality of renaming Upper Volta Burkina Faso, but disparaged on everything else. "Unfortunately it's the 'Africans' who defend it more than the French themselves," Sankara said. "It's a paradox, but one that can be thoroughly explained by the acculturation and the complete cultural alienation of those Africans." While what to do with the relation between national languages and French was under discussion in the educational review and remained unsettled, Sankara's position seemed no different from Senghor's and most African leaders for that matter. Namely, the use of French as "the unifying language of all our numerous nationalities" in service of efficiency in public administration and functions. Sankara argued that language was "above classes" (which is simply false); the litmus test was how the revolution strategically deployed it to advance its goals (Sankara 1985b, 247).

Sankara used indigenous idioms generously in his speeches. "This is a pedagogic style," he told one journalist. "The product of our reality." Speeches. Long answers. Symbols. Products of "the oral tradition of African civilization, where speech progresses with many twists and turns. I most often speak to peasants, so I let my spirit flow in this form of dialogue, debate, and exchange of views" (Sankara 1985a, 217).

Instead of being custodians and inspirers of indigenous idioms, Sankara generally saw elders as the face of obscurantism who had to be watched and educated. "These older people have the feeling they're being excluded," Sankara admitted in 1985. "When they were our age, they displayed admirable courage. Today, they're resting on their laurels." The revolution acknowledged their past achievements to tap into the dynamic energy just a simple word from them could inspire, attacked them as a face of feudalism, followed when they spoke, hence for Sankara: "Just as young revolutionaries must combat young reactionaries, old reactionaries will be fought by old revolutionaries" (Sankara 1985a, 194). A year later Sankara branded elders "tortoises with double shells. . . . Owls with shady looks in their eyes . . . , that is, a certain number of fence-sitting chameleons who think and who calculate that, as in a game of checkers, the revolution has just given them a dangerous opening that they will take advantage of to position themselves to resume their favorite sport—intrigues, ploys, settling of scores, defamation, scheming, and I don't know what else." The National Union of Elders of Burkina Faso was established purely "because we know that if we don't mobilize the elders, our enemies will mobilize them against us" (Sankara 1986a, 273).

A new education system was necessary to engineer a revolutionary mindset, to strategically deploy schools as "instruments at the service of the revolution."—To imbue everybody with a Voltaic personality.—Original, never blind mimicry.—Judiciously tweezing what was useful from other peoples.—Thirsty for knowledge, abhorrent of ignorance (Sankara 1983a, 105–106). Statistically, the starting point was an illiteracy rate of 92 percent and only 16.5 percent school attendance (Sankara 1983a, 85).

Sankara's strategy was "to attack both the container and its content." Whereas the colonizer's school sought to produce clerks to enable exploitation of labor to produce and extract resources, the new school sought to "inject new values" and produce "a new man who understands ideas, who absorbs them, and who functions in total harmony with the dynamic evolution of his people." An education system no longer just for the privileged, but a democratized education with classrooms built and accessible everywhere. Sankara was under no illusion that every child would be educated even if the country's entire budget was committed to education. Other forms of education were necessary beside classical teaching models, hence the program under which "everyone who knows how to read will have the duty of teaching a certain number of others. Those who don't participate will lose the possibility of continuing themselves" (Sankara 1985a, 223, 225). By 1986 Sankara could declare that the literacy rate had risen to 22 percent within two years (Sankara 1986d, 256).

THE PEOPLE

In the engineering of a new society, people were not simply bystanders while their benevolent government did development for them. As Sankara said in 1985, "There exists a force, called the people, and that we must fight for and with the people," an attitude that starts as a sacrifice but will tomorrow be "normal and simple activities" (Sankara 1985a, 228). These "wretched of the earth, . . . expropriated, robbed, mistreated, imprisoned, scoffed at, and humiliated every day, and . . . whose labor creates wealth" had to be partners to government because they "suffer[ed] most from the lack of buildings, of road infrastructure, and from the lack of health care facilities and personnel" (Sankara 1983a, 83). They had to be allies and at one with government in constructing small dams to solve the water problem rather than watch the little that rained wash off to sea. And the result was over 150 wells sunk in about twenty districts to assure drinking water by 1986 (Sankara 1986d, 256).

The problem-solving leader must "immerse themselves in the daily lives of the people" who nourish them with "necessary reserves of energy." The good leader's grasp of the question derives from building their office and tenure in the hearts and minds of the people, feeling their living realities and ideas, never the "I," at all times the "we," true to the best values of *ubuntu/hunhu/ujamaa*, the communal good sacrosanct over the individual.—Being as belonging, never done alone.—A leader because of and only through and from the people, because over the people is the territory of the ruler. When asked what led him to be a listening-hearing leader, Sankara said he did not want to be like a man sent by the villagers to Ouagadougou during a great drought, who waited in queues in vain, had the village's money and his bike stolen, and committed suicide. To the city he was just another dead body, buried without ceremony or consequence, but out in the village, the people continued to wait patiently for his return, unaware he was long dead (Sankara 1985a, 192). Many go to the city and diaspora and are consumed by it, never returning, yet Africa has this expectation toward us.

"Democracy is the people," Sankara declared on Radio Havana during a visit to Cuba in 1987. Democracy is not ballot boxes and elections without fail, to keep up appearances, politicians only visiting their constituents at election time. Democracy is when people can speak their mind every day, when as a leader you can trust the people, and they you. Power in the hands of the people—Economic power.—Military.—Political.—Social.—Cultural. Power to be (Sankara 1987e, 385).

In "Eight Million Burkinabè, Eight Million Revolutionaries," Sankara's speech marking the Fourth Anniversary of the Political Orientation Speech, October 2, 1987, Sankara could declare: "We are not celebrating the backward peasant, who is resigned to his fate, naïve, a slave to obscurantism, and ferociously conservative. We are celebrating the birth of the new peasant, serious and aware of his responsibilities, a man who turns to the future by arming himself with new technologies." *Produce and consume Burkinabè* no longer a slogan but a state of mind actuated in production.—'Peasant' no longer a derogatory term but synonym of respect.—Proud and worthy combatant in the struggle for self-reliance.—Fruits not of machines but the actions of men.—Triumph of human will not steel—Blossom of daring not military might.—The overcoming of inferiority complex as fear—Guided courage, not blind rage. "Yesterday these men were resigned, silent, fatalistic, and passive," Sankara recalled as if in subconscious valedictory. "Today they're standing tall, involved in concrete revolutionary struggle on different projects. . . . We're not saluting them just to be nice. The results we have achieved can be explained scientifically. Power, whether it comes from muscles or is produced by machines, can be measured and compared and is therefore substitutable" (Sankara 1987a, 391).

Battles were being won. Big ones. But the war to "achieve the new society" was far from won. The revolution would be measured as a success when "a new people" emerged with their own identity, assertive, and with clarity of self-determined objectives and means to achieve them, "a convinced people, not a conquered people" (Sankara 1987a, 393).

The army was the engineering arm of revolution and nation building.—Engineering *with*, not *for* them. "The new soldier must live and suffer among the people to which he belongs."—To work in the field and raise cattle, sheep, and poultry with, not like them.—Build schools.—Clinics.—Maintain roads. A soldier "at the service of the people, . . . [not one] who looks down on, scorns, or brutalizes his people" (Sankara 1983a, 100–102). The hierarchical relationship defined by stripes denoting rank created "a caste sitting on top of others" that Sankara wanted changed "so the army fuses with the people" (Sankara 1985a, 226).

> Our soldiers must constantly experience what the people experience. It's not right for military men to be paid regularly while the civilian population as a whole doesn't have the same possibility. So to bring military personnel into contact with reality, we put them in touch with the needs of the day. We've decided that in addition to their purely military, professional, and tactical activities, they should participate in economic life. We've instructed them to build chicken coops and start working in stock raising. . . .
>
> One-quarter of a chicken per soldier per week. This way, not only will the quality of food improve, but in addition, this particular layer of people with regular salaries will not be in the business of buying chickens, and this will surely lower the price for the civilian population. With this kind of training, the soldier who has gotten into the habit of acting like this, either under orders from his officer or on his own initiative, will do the same at home. (Sankara 1985a, 227–228)

The army's role was that of combating internal and external enemies, yes, but with the people. That meant every soldier mastering the lifelong skills or creative resilience of the people, and the people undergoing military training. After all, "the defence of a people is the task of the people themselves." Instead of eighteen months, military service was increased to two years—one-quarter dedicated to weapons and tactics, the rest to production. Sankara elaborated on the first: "The people must defend themselves. They must decide to make peace when they cannot—or don't wish to—pursue a war. They must decide, too, what the army should be." And concerning the second: some national service persons would work in agriculture, others healthcare. That put truth in the "service" in national service; each trainee would "discover or rediscover the Burkinabè people." National service would be for everybody, not just the poor, high-ranking academic and rural people alike (Sankara 1985a, 227).

For Sankara, communication was the most important instrument of leadership: "*All problems between men are problems of communication*," he said (Sankara 1985a, 226; my own emphasis). Fewer instances illustrate this more than this instance in 1985 as he wrapped up an interview with French journalist Jean-Philippe Rapp:

> It's 10:00 p.m. now. Once we've finished this interview, around midnight, I'll be leaving for a small village, where I'll stay until 5:00 a.m. You have to take the time to listen to people and make a real effort to enter into every milieu, including those with little to recommend to them. You have to maintain relations of all kinds—with the young, the elderly, athletes, workers, the great intellectuals, and the illiterate. In this way, you get a mountain of information and ideas.
>
> When a leader addresses an audience, I think he should do it in a way that makes every single person feel included. When congratulations are in order, everyone should have the feeling that he, personally, is being congratulated. When it's a question of criticism, everyone must recognize that his own actions are being judged as well—everyone must know that he has done such a thing himself, have the feeling of standing naked, of being ashamed, and determined to not make the same mistakes again in the future. In this way, we can become aware

> of our errors collectively and retrace our steps together. I must take steps to inform myself. I must break with protocol and everything that boxes us in. At times, too, I must say what I've discovered and denounce specific situations. This shakes things up.
>
> Of course, I'm not informed about everything, especially since there are those who are hesitant to speak to me, who believe I'm not accessible. Efforts should be constantly made to bring us closer together. Every week I answer fifty private letters, at the very least, asking me the most unimaginable and unanswerable questions. But we keep the lines of communication open. I'm extremely pleased when people present their proposals to me in response to the problems I've laid out, even if we don't always accept their particular solutions. (Sankara 1985a, 219–220)

Given the choice between violence and peaceful means, Sankara declared he would choose peace, but "the logic of some situations at times leaves you no choice." He had had to use violence against those who thought they could get away with anything, against which violence had to be used with restraint. "I'm a military man," Sankara declared his bona fides. "I can be called to the battlefield at any moment. On the battlefield, I hope to be able to help my enemy and spare him senseless suffering, even though the logic of the battlefield demands that I use my weapon against him and kill him as quickly as possible in order not to be killed myself" (Sankara 1985a, 231).

He was a very well-educated man, a critical thinker, yet who owned no personal library. "A library is dangerous," he said. "It betrays." He refused to say what he read, never made notes or underlined passages in a book. "Because that's where you reveal the most about yourself. It can be a true personal diary." The book he would have wanted to write was *"A book on organizing and building for the happiness of the people."* He never read to pass time or for love of fine narrative or to unravel the writer's plot. As he summed it: *"I like to look ahead to new men, new situations"* (Sankara 1986e, 264; my own emphasis).

WOMEN

"I still remain a man who sees in every one of you a mother, a sister, or a wife," Thomas Sankara began his famous International Women's Day address in 1987, aptly titled "The Revolution cannot Triumph without the Emancipation of Women" and delivered in French. He implored his sisters who had come from Kadiogo province who did not speak French to "be patient with us, as they always have been," having borne "the task of carrying us for nine months without complaint." The act of giving birth, though painful to the sisters, had ensured that "happiness became accessible." As much as revolution against the previous regime on August 4, 1983, had ushered in the revolution, this "selfish happiness," he said, was an illusion because something important was missing. Women.

Sankara launched his attack on the aspect of Burkinabè culture that confined women to "a depersonalizing darkness" while the men alone marched toward and "reached the edges of this great garden that is the revolution." For women, the revolutionary euphoria torching Burkina Faso was "merely a rumble in the distance, . . . merely a rumor" (Sankara 1987c, 337). Sankara paid glowing tribute to women in this moving eulogy:

> Revolution is the most natural of all relations between one human being and another. . . . This human being, this vast and complex combination of pain and joy; solitary and forsaken, yet creator of all humanity; suffering, frustrated, and humiliated, and yet endless source of happiness for each one of us; this source of affection beyond compare, inspiring the most unexpected courage; this being called weak, but possessing untold ability to inspire us to take the road of honor; this being of flesh and blood and of spiritual conviction—this being, women, is you! You are our source of comfort and life companions, our comrades in struggle who, because of this fact, should by right assert yourselves as equal partners in the joyful victory feasts of the revolutionary action. (Sankara 1987c, 338)

Sankara likened the position of women in Burkinabè and African society to slavery. Here and there a few female warriors or priestesses "broke out of their oppressive chains," but that should never be used to sanitize history (Sankara 1987c, 342). Women, he said, could no longer be seen as mere sexual beings and good only for their biological functions but as an active social force. Retreating into Frederick Engels's dialectical materialism, Sankara attributed the position of women worldwide not simply to physical features (musculature, capacity to give birth) and technological progress, but also societal mores that consigned them to spaces and standards set by men. Enslavement of women, he said, citing Engels, was an outcome of the advent of private property, the man becoming master of slave and woman alike, turning slave into concubine. Private property. Some women took revenge in infidelity; adultery became the dirty opposite to marriage, but the woman's sole defense against domestic slavery (Sankara 1987c, 341).

Sankara cautioned that, to fight and win, women should refrain from making it "a war of the sexes." They needed a large base of the "oppressed layers" of society (workers, peasants, etc.) and not women only because even among themselves they were not united, to say nothing of men. Regardless of how oppressed the man was, he had "another human being to oppress: his wife." Vanity. What vanity? The same man who consorted with prostitutes and mistresses would never hesitate to murder his own wife at the slightest suspicious of infidelity. Prostitutes. Who goes to see prostitutes? Whose demand does she supply—who is the market? One finger of accusation pointed to her, three toward the man (Sankara 1987c, 345–349).

A boy is born; it is a "gift from God." A girl is born; it is fate. The boy is taught "to want and get, to speak up and be served, to desire and take, to decide things on his own." The girl? "To seek the supervision of a protector or negotiations for a marriage, . . . to serve and be useful." To submit. The boy-child plays until they drop dead from exhaustion; but she enters production from a tender age: as assistant housewife. The job of a housewife is deemed as doing nothing. It has no income, it is not a job—according to men, who don't do it. Sankara painted the position of women in rural Burkina in somber terms that diagnosed with the objective of solving:

> Woman—source of life, yet object
> Mother, yet servile domestic
> Nurturer, yet trophy
> Exploited in the fields and at home, yet playing the role of a faceless, voiceless extra
> The pivot, the link, yet in chains.
> Female shadow of the male shadow.
>
> Pillar of family well-being, the midwife, washerwoman, cleaner, cook, errand-runner, matron, farmer, healer, gardener, grinder, salesman, worker
> Labor power working with obsolete tools, putting in hundreds of thousands of hours for an appalling level of production.
>
> First to work and last to rest
> First to go for water and wood
> First at the fire, yet last to quench her thirst.
>
> She may eat only if there is food left and only after the man
> She is the keystone of the family,
> Carrying both family and society on her shoulders,
> In her hands,
> In her belly.
>
> In return, she is paid with oppressive, pro-birth ideology, food taboos and restrictions, overwork, malnutrition, dangerous pregnancies, depersonalization, and innumerable other evils.
> (Sankara 1987c, 353)

In Sankara's depiction, women are the embodiment and spine of Africa's creative resilience. Because only half of them are likely to go to school compared to men, 99 percent are illiterate. They have barely any professional trades training.—Are discriminated against in employment.—Get the worst jobs.—Are harassed and fired without recourse.—Choke under "a hundred traditions and thousand excuses." And yet "women have continued to rise to meet challenge after challenge . . . to keep going, whatever the cost, for the sake of their children, their family, and for society in general. . . . To work for change, to fight to win, to fall down repeatedly, but to get back on their feet each time and go forward without retreating" (Sankara 1987c, 357–358). Not just resilient but creative.

The revolution sought to cultivate a "new consciousness" among Burkinabè women. In every project they were not only involved, but *took up*—were *not just given*—the initiative.—The Sourou (valley irrigation project).—Reforestation.—Vaccination brigades.—The "clean town" operations.—The Battle for the Railroad.—Solving the water problem everywhere.—Helping to install mills in the villages.—Popularizing improved cookstoves.—Creating child day care centers. In 1983, the hospital-bed-to-patient ratio was 1:1,200 citizens, while the doctor-to-patient ratio was 1:48,000 (Sankara 1983a, 86). The goal of engineering a healthcare system for a new society involved making healthcare available to everyone, but to lessen the burden on those who carried nine months of life, gave birth, and then had to continue. Establishing maternity and pediatric facilities.—Mass immunization against communicable diseases through vaccination campaigns.—Mass hygiene campaigns.—Some 2.5 million children aged nine to fourteen vaccinated against measles, meningitis, and yellow fever in two weeks (Sankara 1983a, 107; 1986d, 256). Sankara delivered a powerful, definitive statement on single women to walk tall and proud without letting societal norms of marriage weigh them down:

> We must say again to our sisters that marriage, if it brings society nothing positive and does not bring them happiness, is not indispensable and should be avoided. To the contrary, let's show them our many examples of bold, fearless pioneers, single women with or without children, who are radiant and blossoming, overflowing with richness and availability for others—even envied by unhappily married women, because of the warmth they generate and the happiness they draw from their freedom, dignity, and willingness to help others. . . .
>
> Our society is surely sufficiently advanced to put an end to this banishment of the single woman. As revolutionaries, we should see to it that marriage is a choice that adds something positive, and not some kind of lottery where we know what the ticket costs us, but have no idea what we'll end up winning. Human feelings are too noble to be subjected to such games. (Sankara 1987c, 365)

A new society was one that abolished age-old traditions, that treated women as beasts of burden, who had suffered under colonial rule first as black people just like men, and second at the hands of men. Women's liberation was neither an act of compassion nor charity, but a basic necessity for the revolution to triumph. It was certainly not "a mechanical equality between men and women, acquiring habits organized as male—drinking, smoking, and wearing pants." Or diplomas. The revolution was building "a new mentality" and creating the conditions for women to take charge alongside men for their country's future.—Creating conditions for unleashing women's fighting initiative at all levels.—In conceiving projects.—Making decisions.—Implementing them. Like freedom, women's liberation was not a gift; it had to be conquered (Sankara 1983a, 103).

CONCLUSION: YOU CANNOT KILL IDEAS

What Thomas Sankara had known would happen occurred on the afternoon of October 15, 1987. At around 2:15 p.m. the president and his small group of advisers convened at the old Conseil de l'Entente headquarters. At around 4:30 p.m. gunfire erupted out in the courtyard. The president's two bodyguards and his driver fell first. Sankara and his advisers dashed for cover inside the meeting room. He then got up, ordered everyone to remain inside: "It's me they want." He started walking toward the door with arms raised to meet his assailants. They opened fire and cut him down, and then entered the room to complete their slaughter. Only one man was spared: Alouna Traoré. As it would later become clear, Sankara was murdered by his second in command and comrade in arms, Blaise Compaoré.

The revolutionary as an individual was dead, but the incandescent ideas he championed amidst much loneliness within the revolutionary leadership live as an inspiration to those who dare to invent the future. Tangible things—no, *touched lives*—memorialize one who builds his office in the hearts, minds, and lived realities of a people. The most fitting description of the immediate aftermath of Sankara's assassination is best left to Mwalimu Joshua Njenga, writing on October 12, 2021:

> It is the afternoon of 15 October 1987, and the Burkinabe national radio interrupts its regular programming to play military music.
>
> At about 6 pm, someone announces that "patriotic forces" have overthrown the "autocratic government of Thomas Sankara" . . . .The speaker accuses Sankara of betraying the revolution and of running the country as a "one-man-show." He describes him as "a traitor of the revolution," "a fascist," and "paranoid." He says Sankara is a "Petite bourgeoisie" who pretends to care about poor people but has abandoned the revolution's ideals and is now a close friend of "imperialists."
>
> He then announces that the "Popular Front", headed by a new president, Captain Blaise Compaoré, is now in charge and has dissolved the National Council of the Revolution (CNR) and the government.
>
> He does not reveal, however, that in addition to ousting Sankara, the Popular Front also assassinated him and 12 of his associates at about 4:30 pm. At dusk, a group of 30 prisoners throw their bodies at the back of a military truck and ferry them to the Dagnoën Cemetery in the outskirts of Ouagadougou. . . .
>
> On arrival, the prisoners dig shallow graves under the supervision of heavily-armed soldiers, who use vehicle spotlights to illuminate the corner of the cemetery where this is taking place.
>
> The bodies are bloody and dirty, but they are all identifiable. Sankara is wearing a red tracksuit. His fists are closed and his mouth slightly open.
>
> At about 9 pm, the prisoners lower the bodies into the shallow graves without caskets. They then cover them and then plant sticks with pieces of paper identifying each body on the mound of soil above each grave.

As news of Sankara's execution and burial spread, there is confusion and disbelief.
In the countryside, people gather in shopping centres to discuss what has happened. Most have in their villages a development they can associate with Sankara's stint as president: a peasants' union, a school, a cereals bank for surplus grain, a water reservoir, thousands of newly planted trees, improved harvests of millet and sorghum, or adult literacy classes. They fear all these will be lost. (Njenga 2021)

CONCLUSION: WHY WE MUST DARE

THE STATE OF THE AFRICAN UNIVERSITY AS A STARTING POINT

The persistence of colonial infrastructures of collaboration established during the European occupation presents us with several questions that guide this conclusion.[1] First, what kind of education do we want our students to have in order to meet the opportunities and challenges facing Africa? Second, what kind of ingredients and tools does our education require to be responsive to the needs of all of Africa's people? Finally, how do we go about setting up and sustaining that kind of education and knowing, bearing in mind that over 70 percent of Africa's employment is currently within the informal sector, not in research and development (R&D)?

These questions are not new, but have never been more urgent. In 1997, an expert group composed of deans of engineering schools met at a summit convened by the United Nations Educational, Scientific, and Cultural Organization and the African Network of Scientific and Technological Institutions. On its agenda was a review of quality assurance and the relevance of engineering programs in Africa's higher education institutions (HEIs).

The diagnosis started where it should start: with historical analysis as a tool, in this case Africa's enduring colonial legacy and Africans' efforts that have both sought to escape this legacy and further entrenched it. Many of the engineering schools and curricula at that time were for, not by, Africans. They still follow the disciplinary structures European colonizers set for us, consistent with the economic exploitation they were meant to effect. Most schools of engineering started with agricultural, civil, electrical, electronic, and mechanical engineering, as well as surveying, with other programs being added after independence to cater for postindependence exigencies. Today, many engineering programs are not yet true to their local economies and have not yet crafted degree programs in specific response to requests from government and industry (Kumapley 1997; Kunje

1997; Markwardt 2014; Massaquoi and Luti 1997). In general, the gathering noted that the engineering education was imported from the Global North and therefore designed for other societies (Simbi and Chinyamakobvu 1997, 44). The buildings, campuses, degree programs, and even courses were new, but the "universities of science and technology" continued to be subjected to the traditional lectures that funneled "content knowledge" into students' heads without developing and stimulating "any spirit of inquiry or initiative in the student." The students' duty was that of "memorizing lecture notes for the sake of passing examinations only" (Simbi and Chinyamakobvu 1997, 48; Kunje 1997; Senzanje, Moyo, and Samakande 2006). African engineering content and syllabi are generally both continuations of colonial traditions of engineering (pompous titles, little or no tangible and visible product) and models borrowed from and imitating those of the West. The curricula are still too theoretical and of little relevance to their contexts (Matthews et al. 2012). Our engineering model operates in exclusion of the society, one that it engineers for rather than with. It's engineering without social responsibility, engineering without any creativity.

At the end of its deliberations, the expert group called for a curriculum with "more social sciences, computer courses and industrial attachment"; dissertations reflecting "real life situations" and graduates capable of "solving regional problems" (Massaquoi and Luti 1997, 8). The deans spoke against an imitative model that failed to "address African needs," but simply put "an African complexion to imported copies," thus continuing "a cycle of dependence which makes us lie back and await changes in foreign systems" and then react with minor adjustment to suit our needs (Massaquoi and Luti 1997, 8). The professors called for "committed scholars with creative minds" to critically engage with global ideas and instruments to generate new technologies and provide indigenous oversight on decisions pertaining to foreign things that locals assign technological value (Massaquoi and Luti 1997, 8; Masu 1997). Two decades later, that call remains unanswered; the high rate of unemployment among engineering graduates confirms that Africa's HEIs are churning out graduates with unemployable skills (EARC 2018).

The deans' call predated a current debate among engineering educators in the West. Engineering education generally passes on disciplinary, well-understood, and already-existing formulae for problem-solving; seldom does it try out new methodologies or take creative risks (Beer, Johnston, and DeWolf 2006; Bucciarelli 1994, 2003; Seely 2005; Sheppard et al. 2008). The calls for engineering education reform in the United States, for example, boil down to one question: "How can one teach engineering science courses so that students come to understand what they are not learning?" (Downey 2005, 592).

To answer this question means that engineering has to be opened up more aggressively to the humanities, arts, and social sciences so that engineers better understand the social and political context within which they do engineering (Grasso and Burkins 2010). Top engineering institutions like the Massachusetts Institute of Technology require their students to take a significant number of social sciences and humanities subjects to graduate (MIT 2017). Of course what they do with those insights, how seriously or not they take them, is quite another matter; as a requirement for graduation, it is difficult to tell if students are not just looking to course off, or simply to understand social, cultural, and political context to apply their "real" skills, as opposed to social sciences and humanities as engineering tools in their own right. Reform is needed so that technical skill is only one among many other skill sets an engineer requires to turn the complex social, political, cultural, technical, environmental, and ethical challenges of the profession into opportunities (Faler 1981; Noble 1977). The idea of "holistic engineering," emphasizing context specificity, teamwork, interdisciplinary communication, and lifelong learning, has generally emphasized collaboration between different branches of engineering without actually extending to social sciences and humanities (Duderstadt 2010; Grasso and Burkins 2010; Ramadi, Ramadi, and Nasr 2016).

By 2007, US universities had begun focusing on engineering science (high-tech subjects) at the expense of the traditional engineering disciplines (mechanical, civil, electrical, chemical, and aeronautical), with a resulting critical shortage of engineers of physical infrastructure. An acute dependence on international students and workers followed (Frankel 2008). Today, the antiquated US road, rail, and electricity infrastructure needs upgrading.

Africa should not blindly follow the colonial , US, or Chinese models for that matter, but learn from everywhere it can. In China, for instance, engineers are not in short supply: engineering is the country's largest discipline, with 2,222 (or 92.2 percent) of its 2,409 institutions running an undergraduate program in 2011—and counting. That same year, 8.689 million undergraduate and 0.588 million graduate students (a third of China's enrollment) were engineering majors. This is understandable—China has more than 1.3 billion people and is the world's second largest economy; it is the factory of the world! But like most of Africa's HEIs, China's curriculum prioritizes knowledge accumulation and dissemination and building knowledge systems, not knowledge mastery and practical ability. And it is obsessed with rankings vis-à-vis its competitors as opposed to meeting the needs of industry (Bai et al. 2009; Rutto 2015).

The last thing Africans can afford is to replace Western imports with Eastern ones. The argument advanced is that science and engineering should be brought into interdisciplinary conversation with the social sciences and humanities to forge a new covenant for solving Africa's problems and generating made-by-Africans products and opportunities. It is not enough when training an engineer from Africa to simply make engineering sciences, laboratory experiments, and design legitimate topics for the social sciences, humanities, and arts, or to help engineers "get it" (i.e., better understand the social and political context within which they do engineering). One key obstacle inherited is the colonial mentality that the engineer designs for, not with, society. It reduces society to a spectator when it should be a comrade in arms in research and problem-solving.

By adopting an inclusive, multioptic approach to conception, not implementation or use, and by identifying, conceptualizing, and solving problems together, solutions cease to be imposed from the top down by governments, by foreign countries using "donations" or by "donor agencies" using "soft power" to dominate Africa. Solutions then emerge organically from and with the people affected by the problem. This communality of research and knowledge production is the embodiment of *umoja*, *ujamaa*, *hunhu*, and *ubuntu*. Picture an engineer, a physician, a lawyer, and a specialist in investment finance working with a historian, a sociologist, a political scientist, an environmentalist, a philosopher, a linguist, an informal trader, a blacksmith, a pottery maker, a healer, youths, and an elderly custodian of indigenous knowledge all working within one team, each bringing their skills to bear upon one problem.

This chapter argues for an education that fosters within its students and faculty a culture of inclusive, multioptic problematizing and problem-solving, that is, one that deploys multiple skills sets and sees issues from many angles. To accomplish this, we must invest in programs that synergize and even synthesize the science and engineering curricula with the humanities, arts, and social sciences in order to generate opportunities, solve problems, and create physical and intellectual infrastructures for that purpose.

It is argued further that the solution ought not to be an end in itself, but also a platform for staging completely new innovations. This multiplier effect takes valuable lessons from the value addition Africans are contributing to mobile technology. The mobile money transfer app M-pesa, for example, can be interpreted as value addition to the cellphone and an innovation with multiplier effects. To acquire this research, problematizing and problem-solving capability, the African research university must:

- Have research capability, not just capacity;
- Have financial means to do research;
- Engage with the informal economy;
- Solve problems that the society and continent face;
- Be entrepreneurial (in an innovative and market sense);
- Break received disciplinary silos, cross cultural barriers to knowledge sharing and learning, and anticipate societal needs before they arise; and
- Define, design, and build viable futures.

The argument is not simply that we have no such university that embodies who we are as Africans and what we could be. I am much more worried that we are not even thinking about it with our eye on the realities that define Africa and the futures we may not live to see but which our (grand)children will have to face. As a discipline that makes and builds things, engineering occupies an important space it should open up and share in order to achieve the immense power it potentially has to help African societies build positive, happy futures.

RESEARCH CAPABILITY

> I teach . . . two first-year tutorials per week, nine first-year tutorials per week, four second-year seminars per week, three third-year lectures per week, two third-year seminars per week. Add up and then add four PhD students to supervise. That should give you twenty lectures. Those are the lectures I was giving from July to the end of this week (October 14). And every second semester. So, I hope you have softened your judgment of a brother after looking at the stats.[2]

These are the words of a friend, a faculty member at the University of the Witwatersrand (Wits), explaining to me why it was impossible for him to join me in a workshop I was trying to organize as a visiting professor in July 2016. It drove home a reality I had witnessed when I taught at the University of Zimbabwe from 2000 to 2002—that generally, the African university continues to be a teaching university, with big classes, heavy workloads, poor to nonexistent research funding, and little time off for faculty to conduct research. This is called the "massification of higher education," where students are empty containers whose job it is for the lecturer to fill up. Students' job is to open their ears, imbibe, memorize for and take an exam, pass, graduate, and look for a job. Students approach research as just another exam, and in many instances lecturers' own publication records are razor thin (Kanyandago 2010; Openjuru 2010; Zeelen 2012).

Capability is not to be confused with ability or capacity. Capability refers to talent, skill, or proficiency; the friend cited above, for example, could walk into any Ivy League university and thrive as a research professor as he possesses the necessary capability. Capacity refers to being in a position to do research if one has the ability. All the constraints my friend referred to above impede his capacity to do so. Usually our solutions target one and leave out the other.

The research figures speak for themselves. Based on 2011 figures, the highest-performing African country, South Africa, had 818 researchers per one million people. Compare that to South Korea's 4,627 per million. South Korea produces 3,124.6 science and engineering articles per year; the United States produces 212,394.2; Brazil, 13,148.1; and India, 22,480.5 (Cyranoski et al. 2011; UNECA 2013). Fifteen percent of the world population live in Africa, yet the continent has just 1.1 percent of the world's scientific researchers (one scientist or engineer per ten thousand people) compared with twenty to fifty per ten thousand in more industrialized nations. Africa owns just 0.1 percent of global patents (UNESCO 2015). Institutional rankings put pressure on faculty to publish, and promotion and salary scales are based on them. Individualism, which is detrimental to research collaboration, creeps in (Soudiena and Grippera 2016).

Interestingly, one of the major causes of the problem is beginning to be a potential solution. Especially in the past two decades, Africa has seen its most skilled human resource, graduate students, either drained or draining itself out to greener pastures owing to poor salaries and conditions of service. Graduate students educated on taxpayer-funded subsidies have studied for PhDs abroad, found employment there, and never returned to plow back their skills into the homeland. The statistics are staggering: 43 percent of Zimbabwe's highly educated population live in Organization for Economic Cooperation and Development (OECD) countries, with Mauritius (41 percent) and the Congo Republic (38 percent) close behind. About twenty thousand medical doctors, engineers, professors, and other professionals leave Africa each year. Some thirty thousand of the estimated three hundred thousand Africans who live abroad have PhDs, the vacancies they leave in their homelands being filled by expatriates at a cost of US$4 billion annually. Europe and North America benefit from skills acquired at great cost; for example, in Kenya it costs US$40,000 to train a medical doctor and US$10,000 to $15,000 to educate a university student for four years (Mills et al. 2011). The money used to train these students comes from a budget that includes loans from the International Monetary Fund and the World Bank that African countries must pay back, but the graduates they expended it on now work in the very countries that lent the money.

Africa no longer talks about the brain drain as Lalla Ben Barka of the UN Economic Commission for Africa did in 2014 when she said, "In 25 years, Africa will be empty of brains" (Tebeje 2014). Outmigration has depleted university faculties, and most remaining lecturers have master's degrees rather than PhDs (Chinyemba 2011). Africa is now embracing its capacity to be present throughout the world, to see, learn, master, internalize, and bring back skills to develop the continent—hence the emphasis on the skills developmental diaspora (Plaza and Ratha 2011).

Two programs funded by the Carnegie Corporation of New York are proving just how wrong Barka was by offering the African diaspora and Africa at large a wonderful opportunity to return even while and because of staying where they are. One is the Council for the Development of Social Science Research in Africa's African Diaspora Support to

African Universities program dedicated to social sciences and humanities; the other is the Carnegie African Diaspora Fellowship program, which has a much broader remit. On the one hand, the programs are helping African intellectuals based in North American universities to forge links with African universities. On the other, they are providing financial resources to African universities to identify and host the African diaspora intellectuals they want, through whom they create interuniversity partnerships. Both programs have been mobilizing African academics in the diaspora to contribute to "the strengthening of PhD programs and the curricula," "the filling of gaps and dealing with shortages in teaching," mentoring of young scholars in Africa, and "strengthening relations between African academics in the diaspora and the institutions where they are based and African universities" (CODESRIA 2014; Foulds and Zeleza 2014). The author of this chapter is one of these diaspora intellectuals, and this chapter is an outcome of these collaborations to not only forge overseas partnerships, but also create and strengthen intra-African interuniversity connections.

FUNDING

However, such brain circulation will not solve a perennial research capability problem: funding. How does the university remain financially viable? The students are poor, and the university needs a budget to maintain its operations. The often state-funded universities have no money; research requires money. What is to be done?

The channels through which Africans ended in North America and Europe reveal our education system's enduring colonial ties to and financial dependence on the West and our struggle to evade colonial legacies and be institutionally independent. Our universities have expanded, but funding remains inadequate and susceptible to government budget cuts. This affects research and salaries, discouraging prospective talent and leading to the loss of staff to private-sector and overseas competition. At 0.5 percent of gross domestic product (GDP), African investment in R&D is the lowest in the world. There have been individual country improvements, for instance, Kenya's and Botswana's pledge to commit 2 percent of GDP to research since 2015. Ethiopia already commits 1.06 percent. The bulk of Africa, however, commits much less to research, instead prioritizing primary, secondary, and undergraduate education (APLU 2014; Divala 2016; HESA 2014; Trayler-Smith 2014; UNESCO 2015).

Genuine international partnerships with African institutions have acted as capacity-building vehicles for universities and individual faculty, bringing in much-needed funding, equipment, and staff development, with overseas partners also benefiting from the collaboration (Rampedi 2003). But there are also deceitful, neocolonial partnerships that continue colonial infrastructures of dependency and that reduce and use Africa-based faculty and institutions as the equivalent of data-mining offshore rigs (Ishengoma 2016; Kot 2016). Most of these partnerships are initiated by universities, foundations, and donor agencies in the Global North (Samoff and Carroll 2004). What is seldom highlighted is that most university initiatives start with well-meaning individual faculty and students, with institutions getting involved only later.

Some donor-funded programs continue to serve US, British, and European interests. For example, donor agency partnerships involving the US Agency for International Development are inextricable from US "soft power"—the use of aid and diplomacy in the national interest. Other programs, such as those by the Centers for Disease Control and the National Institutes of Health, are aimed at containing and preventing deadly diseases and their agents from coming to the United States. The United Kingdom and Europe have similar "soft power" and "containment" partnerships (CDC 2015; Kot 2016).

Living in the United States has given those of us in diaspora an appreciation of the pragmatic national interests that drive these host countries' interventions in our homelands. They have interests in Africa; Africans have interests in the United States. That is common ground for building solid bridges and mutually beneficial postcolonial relationships rather than privileging populist but empty political rhetoric that scuttles innovation opportunities. Every North American, European, Chinese, and Australian institution will now have to rethink its African partnership strategy around the African faculty in their employ. In turn, African intellectuals will have to strategically position themselves as bridges facilitating mutual benefit between their host institutions and Africa.

Our universities are still young; most depend on annual central government budgets for their operations and for faculty and other staff salaries. Endowments are the exception; where they exist, they are very small. For example, as of 2010, endowments for some African HEIs were as follows: the University of South Africa, US$300 million; University of Pretoria, US$165 million; University of Cape Town, US$150 million; and Wits, US$100 million (UNISA 2011; UCT 2010; UP 2010; Wits 2011). Comparatively, Harvard's as of 2015 was US$36 billion; Yale, US$25 billion; the University of Texas system, US$24 billion; Stanford and Princeton, US$22 billion; and MIT, US$13 billion. The endowment total of US universities was US$394.94 billion, up from US$219.37 billion in 2005, composed of gifts from alumni and other well-wishers, as well as investment portfolios (Commonfund Institute 2016). Africa's rich and famous tend to build themselves mansions and buy expensive vehicles rather than investing in Africa's education systems. In 2015, the richest person in Africa was Nigerian Aliko Dangote (net worth US$12.6 billion). Twenty-six of the top fifty richest people in Africa are each worth US$1 billion or more

(Forbes 2016). Commendably, Dangote has established a foundation called the Dangote Foundation, "the main objective of . . . [which] is to reduce the number of lives lost to malnutrition and disease."[3] Tony Elumelu does the same with his foundation,[4] while Strive Masiyiwa, chairman of telecommunications group ECONET, and his wife Tsitsi, sponsor talented African students to attend prestigious universities overseas under the Yale Young African Scholars Program (Office of Public Affairs & Communications 2016) and to send thousands of poor and underprivileged children in Zimbabwe to school through their Higher Life Foundation.[5] That is how it should be. What is still needed is to fund research targeting problem-solving at local universities and to create spaces where diasporic talent can come home, walk tall on the African soil, and "do their thing." It does not have to be for free; it can also be, quite simply, business and the diaspora as an investor.

TOWARD INFORMAL SECTOR AND UNIVERSITY PARTNERSHIPS

In Mozambique, only 11.1 percent of the population is employed in the formal sector, 4.1 percent of whom are government employees. Of the 10.1 million labor force, 52.3 percent are self-employed (Robb, Valerio, and Parton 2014). In neighboring Zimbabwe, some 50 percent (5.7 million) of Zimbabweans are employed in agriculture; 42 percent of them are communal farmers or farmworkers. In 2012, 67 percent of Zimbabweans were economically active. The employment rate was 89 percent. About 60 percent of the economy is informal; that is where 20 percent of the country's GDP comes from. Without counting those (self-)employed in the informal economy, the unemployment rate is 80 to 90 percent (ZimStat 2012).

Youth unemployment throughout Africa is increasing rapidly and employment creation programs have had little impact (Hilson and Osei 2014). Fifty percent of graduates on the continent are unemployed (ACET 2016). Simply put, Africa's problem is that it trains for employment when it should be training employers and problem-solvers. Universities' yardstick for successful training is the employability and performance of graduates internationally and their admission into MSc and PhD programs inside and outside the country. Industries require employees with practical skills, since they are subsidiaries of overseas firms and thus do not do R&D locally (Simbi and Chinyamakobvu 1997).

It is a cliché that our universities are not producing graduates who meet the needs of industry (Matthews et al. 2012; McCowan 2014). Our higher education's lack of applicable value to the economy and society explains the high rates of unemployment among graduates (Zavale and Macamo 2016). In the streets of Harare, for instance, unemployed graduates have driven this point home by playing football in the streets in their academic gowns, complete with mortarboard. Any university that brags about the number of (unemployed) graduates it has produced instead of the patents and technology prototypes in its name is not worthy of being called a *research university*.

Examples of courses that produce employable graduates include the University of Zimbabwe's applied engineering and science programs, which began in 1992. This included a shift from the BSc general degree that trained school teachers to an honors program geared to industrial applications as Zimbabwe placed itself on an IMF/World Bank–funded market economy footing. The applied physics program offers courses in industrial, medical, laser and plasma, and environmental physics. Most students choose industrial physics, with courses in workshop practice, computer applications software, theory of devices, computer interfacing, instrumentation physics, quality control, digital signal processing and data communications and networks, and industrial applications of laser and plasma physics, as well as biomedical instrumentation. Upon graduating, students have not struggled to find jobs in industry (Carelse 2002). The applied geology program was a response to expansion in the mining industry, and includes a vacation placement for students doing basic geological jobs like core logging and sampling (Walsh 1999).

These initiatives are geared toward supplying industry with employees. However, if, hypothetically, somebody removed the jobs that these graduates occupy, the initiatives would cease to be effective or relevant. In that sense, our university system is apocalyptic. That is what the protesting students were attempting to convey: that they would rather acquire skills that prepare them for a world in which the job that they end up performing is the one that they go out there and create.

Here we come face to face with the street and the village as (possible and actual) workplaces. What in Africa we call the informal sector in the United States is called small businesses, including home businesses. People in these businesses are self-employed, not unemployed; contrast that with Africa, where only formal employment counts. A paid cattle herder, a street vendor, a farmer, a welder, or somebody who rears livestock in their rural home does not count as employed. Billions of dollars circulate informally, seldom entering the formal banking system—hence Zimbabwe's unending cash crisis (Murwira 2014). Still, not employment.

Deindustrialization threw experienced Zimbabwean workers onto the streets, where they created employment for themselves and others—underneath trees, on pavements, at shopping centers in urban and rural townships, at road intersections, in backyards, on rural homesteads, in wetland gardens, in the fields. Mechanics at Gazaland (Highfield) and Chikwanha, carpenters and leather upholsterers in Glen View, and steelworkers and boilermakers at Makoni—these small entrepreneurs have used their artisanship to dominate manufacturing in the country.[6]

Critics rightly say their record keeping and customer service are poor, and government enforcement of standards impossible because there are too many of them. Few workers have formal contracts and their rights get violated daily. With no registration, most informal entrepreneurs pay no taxes. "Instead of celebrating mediocrity and hiding behind the fallacy of empowerment," one observer notes, "perhaps Zimbabwe should be looking for ways to grow formal industry and get the manufacturing sector working again" (Rudzuna 2014). Small- and medium-sized enterprises (SMEs) in Glen View Area 8 face challenges like capital availability, difficulties in procuring raw materials, low technological capabilities, and difficulties in securing permits and licenses, with the result that SMEs are neither growing nor surviving. Policy frameworks, including the SMEs Policy and Strategy Framework (2002–2007) and the Industrial Development Policy for 2012–2016 are weak on informal-sector participation (Mbizi et al. 2013).

Traditionally, an employee is "somebody who has got a pay slip and can get certain privileges like accessing credit"; therefore, the strategy has been to formalize the informal sector and tax individual workers' monthly salaries (Munanga 2013; Oxford Analytica 2010). In Mauritius, hawkers are licensed and registered with the registrar of companies, the Small and Medium Enterprise Development Authority or municipalities, and taxed 15 percent of all profits. They are only allowed to sell at designated points. Such measures have faced resistance in Zimbabwe. Vendors say they make very little, that banks cannot lend to them, charge exorbitant fees and interest rates, and their enterprises collapsing at any time (Ndebele 2015). Between US$3 billion and US$7 billion circulates in the informal sector (ZEPARU and BAZ 2014). Government says it will "follow where the money now is . . . in the informal sector." It wants informal entrepreneurs to keep books, even if "very simply, very elementary[,] and [to] show the taxman" (Business Writer 2015).

The informal sector, the mainstay of most African economies, is not properly accounted for in the curricula of Africa. The reason is simple: there is no place for community as knowledge producer or partner, comparable to industry/university and transcontinental interuniversity and funder/university partnerships. At most, universities engage in "community outreach"—they send students for service attachments and "allow" people from the community to participate in university activities as part of the "developmental university" (Pitlane Magazine 2017). The closest example of a society-responsive university in Africa to date was Tanzanian Julius Nyerere's notion of "education for self-reliance," a mutually beneficial university/community partnership wherein students acquired real-life experience and the community benefited from academic knowledge, thus creating "a sense of commitment to the total community" (Nyerere 1968, 239). However, Nyerere's revolutionary project lacked entrepreneurship and the capacity to be self-sustaining and profitable. What Nyerere—and all our governments—have done right, we should consolidate and build upon. Whatever errors and weaknesses there are, we should analyze and correct. What tools we can make, we should make. What we do not have, we should import, adapt, and use.

Research has demonstrated the urgency of escalating the technical efficiency of informal-sector entrepreneurship: farming, metal manufacturing, transportation, and marketing are still excessively labor intensive—75 percent of their gross added value is labor (Mujeyi et al. 2016). The case for mechanization of land preparation, weeding, and harvesting is obvious (Thebe and Koza 2012). Very interesting grassroots innovation and entrepreneurship are taking place in the valley wetland gardens of Chihota (Zimbabwe) as farmers import and deploy petrol- and diesel-powered water pumps to draw water from shovel-dug ditches. A traditional method of irrigation, these shallow wells are now many times the size they used to be as farmers replace handheld cans with pumps to scale up their operations. Where they used to grow collard greens, tomatoes, and onions on small areas of a few yards, they now plant hectares of winter cash crops traditionally monopolized by white commercial farmers—potatoes and early maize, for example[7] (Wuta et al. 2016). Research shows that artisan-craftsmen are critical suppliers of agricultural and other tools used daily (Bennell 1993; Mupinga, Burnett, and Redmann 2005), and that rural areas are a potential site of grassroots-driven beneficiation of crops, milk, fruits, and so forth (Bertelsmann Stiftung 2014; Popov and Manuel 2016). Specific examples include fruits, vegetables, and grains that could be processed into juices, dried products, and extracts like oil, as well as organic waste (like cattle, goat, and chicken manure) that could be processed into fertilizer and fuel (Mvumi, Matsikira, and Mutambara 2016; Rusinamhodzi et al. 2013). We have to start reimagining the homestead, the village, as laboratory and factory.

To do this, programs must be initiated to make value-adding tools available to rural and urban sites of informal economic activity and to turn them into venues of vocational-entrepreneurial education. Nonpedagogical ingredients are already present in some countries. For example, the leading German company Bosch Group supplies artisans with hand and machine tools (and user training) in Ghana and Nigeria under its Bosch Power Box program of value addition through improving product quality (Agbugah 2016). Another example is Hello Tractor,[8] an app-based tractor rental for the poor, started by Jehiel Oliver, an African American. The social enterprise is currently operating in Nigeria. Marketing, too, is increasingly being linked via information and communication technology (ICT)–based platforms, which build upon and respond to the needs of farming and add value to their activities. Platforms like eSoko, iCow, Rural eMarket, and M-Shamba (Fripp 2013) were offering early services like start-up market information, weather forecasts,

FIGURE 10.1
The valley wetland gardens of Chihota communal lands, Zimbabwe's horticultural hub, 2018. *Source*: Research || Design || Build.

farming tips, business strategies, market monitoring, supplying, and sourcing. Studies of ICT use often stress how digital tools could be used to improve the lives of the poor, especially by governments and nongovernmental organizations. They talk of computers, printers, telephones, television, the internet, and fax machines (Mugwisi, Mostert, and Ocholla 2015), yet ordinary people use the cheapest cellphones as long as they have one function: WhatsApp. Thus, such studies miss, for instance, how villagers in Chihota strategically deploy WhatsApp to sell their crops, inquire about prices, and arrange pickup of their commodities for transport to city markets after ascertaining that they are not flooded with the same products. Our higher education system is still "too academic and distant from the developmental challenges of African local communities" to capture and collaborate with innovator-entrepreneurs like these (Kaya and Seleti 2013, 30).

The language of research, engineering, science, innovation, and entrepreneurship has no space for the real-life problem-solving, value-generating activities happening at the grassroots level. The usual colonial languages (English, French, Portuguese, and German) are still the official academic and research languages, except in Tanzania, which returned to kiSwahili, only to rescind and ban it in 2022. Scholars who see this as undermining the serious development of research and theory based on indigenous conceptual frameworks and paradigms are right. Our failure to develop indigenous modes of theory and to meet the needs of Africans has robbed us of the opportunity to engage African people as partners in, not recipients of, solutions. Languages die if they are not used (Divala 2016; Hountondji 2002; Gudhlanga and Makaudze 2012).

THE ROLE OF ENTREPRENEURSHIP

Research has shown that about 60 percent of Zimbabwe's start-ups (called SMEs locally) fail in the first year, 25 percent fail within three years, and just 15 percent survive. This translates into an 85 percent start-up failure rate (Mudavanhu et al. 2011).

Africa has already embraced entrepreneurship education (EE), but not entrepreneurship. On paper, the mandate of EE is to educate entrepreneurs who are also innovators, to instill "an entrepreneurial attitude" or "spirit" and expunge "risk-averseness." EE is supposed to equip students with techniques to analyze and synthesize and create risk-takers who initiate innovative start-ups and see them to success (Fayolle and Gailly 2008; Griffiths et al. 2012; Woollard, Zhang, and Jones 2007). Skeptics, however, differ: a certificate does not make one an entrepreneur, and entrepreneurship does not exist without innovation (Walt and Walt 2008).

EE is expanding in Africa at a rapid pace; the demand is "overwhelming" (Robb, Valerio, and Parton 2014). Since 1997, entrepreneurship has been a compulsory subject in Kenya's technical vocational education and training (Farstard 2002), even though at Jomo Kenyatta University of Agriculture and Technology (JKUAT) engineering students were "encouraged" to "audit or attend" entrepreneurship courses, but they were not a requirement for graduation (Marangu 1997). JKUAT offer entrepreneurship specialization at the doctoral level; other universities offer undergraduate programs (Robb, Valerio, and Parton 2014). Since 2008, EE programs have been established at Mozambique's three public and two private HEIs: Universidade Eduardo Mondlane, Universidade Pedagógica, Instituto Superior Politecnico, Universidade Católica de Moçambique, and Instituto Superior de Gestão, Comercio, e Finanças. Under the National Agenda to Combat Poverty, these HEIs are the nation's vehicles for driving the economy forward through entrepreneurial education, start-up incubators, and leveraging overseas partnerships (Libombo and Dinis 2015).

Thus far success stories are scarce (Libombo and Dinis 2015). In general, EE curricula focus more on theory and business plans rather than exposing students to real-life business situations. Entrepreneurship is about taking risks, yet students graduate without ever having taken any (Robb, Valerio, and Parton 2014). Their instructors are themselves risk averse; few have ever been entrepreneurs (Kirby 2006). The institutions that train them have no support structures for start-ups or ties to, let alone collaboration with, industry or the informal sector (Shambare 2013). EE slavishly teaches the Schumpeterian principles of a linear correlation between entrepreneurship and economic development. African entrepreneurship is highly informal, creative, irregular, and often hardship driven, with no access to lines of credit (Libombo and Dinis 2015; Robb, Valerio, and Parton 2014; Sautet 2013). Despite supporting the majority in a continent of limited formal jobs, the informal sector or the everyday does not feature as a space for students to acquire practical skills.

This is where vocational education becomes key to any research university: to not just research, but turn our findings into products. Vocational education is supposed to train people in hands-on, practical, and basic reading and mathematical skills. Empirical research shows that the courses are quite poorly developed, offer limited practical training, and depend on donors for funding and equipment. Usually the programs do not build on predominant activities and local resources that sustain the informal sector. For example, in Mozambique, despite loud political declarations about nonformal vocational education, few programs are devoted to agriculture, which supports 75 percent of Mozambican livelihoods. Furthermore, small-scale farmers contribute 95 percent of agricultural production and 70 percent of the population lives in the countryside. There are similar problems in Botswana and South Africa (Mayombe 2016; Moswela and Chiparo 2015; Oladiran et al. 2013).

Our science does not usher in anything tangible due to the specific circumstances in which it originated, and the notion that technology is an outcome of scientific research and that white men determine what is considered scientific. Since our independence, we have voluntarily chained ourselves to the Haldane principle that emerged in the United Kingdom in 1904, which states that researchers, not politicians, should make decisions about research funding allocations. In 1918, Richard Haldane recommended that government-supported research be placed in a special department and more general research in autonomous research councils. Classical political science designated technology as residual to factors of production (land, labor, and capital); everything starts with research in basic sciences, is applied by engineers, which ushers in technological application, innovation, and diffusion. Thomas Kuhn (1962) further cemented the Haldane principle in *The Structure of Scientific Revolutions*.

The African Union's Science, Technology, and Innovation Strategy for Africa (STISA-2024) is the latest iteration of the Haldane principle, on a continent where many innovations are "neither based on nor the result of basic science research" (Marjoram 2010, 173), but in informal activity. STISA-2024 derived from the "Frascati family" of manuals that OECD National Experts on Science and Technology Indicators have developed since 1960: the Frascati Manual in 1963 (on R&D), the Oslo Manual in 1991 (innovation), and the Canberra Manual in 1995 (human resources in science and technology) (OECD 2002, 2005).

The argument is not that R&D is not important; the issue is what ingredients ought to constitute it so that it works for us. Everything else—who, where, with what—depends on critically addressing that question. Marjoram (2010) points out that promoting the development and application of science, engineering, and innovation must take precedence over education, capacity building, and infrastructure, which the UN Conference on Trade and Development (UNCTAD 2007) emphasizes. Yet both are further downstream of establishing an identity for science and engineering in Africa defined by and for African priorities, as Latin American science, technology, and innovation strategists did when crafting their own Bogota Manual (RICYT/OAS/CYTED 2001). Instead of top-down (science-intensive) R&D, these scholars emphasize the role of "social innovation," "inclusive innovation," "innovation at the bottom of the pyramid," "grassroots innovation," "innovation for development," "jugaad innovation," "reverse innovation," and "community innovation" (Globelics 2012). Our designs "must reflect local conditions, use local resources in response to local problems. Anything from the outside must be complementary to this" (Mamdani 2010, 495).

AFRICA'S GLOBAL OMNIPRESENCE AS AN INNOVATION RESOURCE

With travel and residential experience across the world, Africa's children have learned, seen, and created, insights that only come with a global omnipresence (whereby a country is everywhere through the dispersal of its citizens or descendants all over the world). The last time Europe dominated the world, its diaspora was all over the world. It had colonies. It had a global empire. Africa now has a global diaspora, what one would call a global omnipresence. Africans are everywhere; the term Global Africa is no longer cliché. So why is Africa not as prosperous in the world as imperial Europe?

Today the convergence of the African diaspora and the African digital is the perfect storm for creating a vast global African market. This can happen in one of two ways. While Africa waits to industrialize and add value to its mineral, plant, and agricultural products, all its children could create spaces in localities outside the continent to import and process goods from home. They would create and own the entire value chain from point of production through transportation chains to clearing and value addition in the

United States, Europe, and Australia to marketing. Indians, Chinese, and Brazilians are doing this with not just restaurants, but also food markets and movies. Similarly, Africa's children could and must take the lead in not only investing in Africa, but also identifying and bringing to the continent the latest technologies they adjudge, working closely with the motherland, best suited to positively transform the economic development of the continent. It is their market, yes, but also contributes to the countries where some of them are now citizens. Africa's children must occupy the space of carriageway and bridge between their motherland and their home abroad. The capital generated all round would rapidly provide investment funding for Africa-centered initiatives so that, eventually, capacity is built for value addition in Africa itself, ensuring the continent exports only finished goods, and its children are contracted for specialist services (consultancy, for example) and set up needle-moving design and manufacturing on the continent or in black communities abroad, viable TV stations and digital platforms, airlines, healthcare systems, and so forth. Africa's children can build on the success of Nollywood and African music, the geographies of its market, and its appeal beyond African shores.

As inventors and designers of rich futures, who are globally omnipresent, Africans have eyes, ears, hands, and brains everywhere, and thus the capacity to identify the raw materials that constitute what they adjudge to be the best ingredients for sculpting self-determined futures. This is where self-determined definitions of technology, innovation, and science "from" Africa matter. Nothing ever comes to Africa as technology *a priori*. It only becomes so in view of Africans touching it with their hand, head, and heart. Africans are in the business of very deliberately tweezing out from this global omnipresence what works for them and deliberately bringing it into their communities or homes. No other continent has ever been in the powerful position that Africa finds itself in, where its children are in Silicon Valley, the Cambridge, Massachusetts, area, the Research Triangle of North Carolina, are dotted all around Europe, in Australia and China, not only doing the small time jobs, the menial ones, and the hands-on, but also excelling at some of the best institutions in the world. Africa's children are part of the best of the best.

CONCLUSION: AN INTERDISCIPLINARY, CROSS-CULTURAL, AND ANTICIPATIVE UNIVERSITY

In a world that demands interdisciplinary, cross-cultural, and without-the-box thinking, we have ministers of higher education and vice-chancellors who are taking us where we should be fleeing from. Disciplinary rigidity and the separation of the engineering sciences into electrical, civil, mechanical, or agricultural engineering impedes an integrated approach and leaves no room for productive floor crossing and collaboration.

The physical architectures of the university are such that the humanities, arts, and social sciences are aloof from the science and engineering departments. Internal engagement across lines is nonexistent, to say nothing of interdisciplinary research and teaching. These structural and pedagogic rigidities are a serious obstacle to a multioptic, problem-solving research university. But we are building more of these.

For countries like Nigeria, the path lies in modeling new science and engineering institutions around very specific services and products—energy, materials, chemical and leather technology, industrial research, various types of incubators, biotechnology, remote sensing, and so on (AUST 2016). The danger is that STEM will create a vast pool of monoskilled technicians (what I call "glorified mechanics" of bodies, cars, and the soil, with no historical and identity consciousness), whereas the informal economies that dominate Africa thrive on multiskilled competence. For Ethiopia, the route to a developed nation lies in quintupling the current public universities to thirty-four (Rayner and Ashcroft 2011). For Rwanda, it lies not in numbers, but in merged universities with concentrated researchers and resources (Iizuka, Mawoko, and Gault 2015). Private universities and colleges are sprouting in every African country, absorbing high school graduates in large numbers. There is much money to be made. For example, by 2012, Uganda had twenty-seven private universities compared to just seven public; Ethiopia had thirty private and twenty-two public; Nigeria, forty-five private, thirty-seven state, and thirty-six federal; while South Africa had a whopping eighty-seven private compared to just twenty-three public (Mashininga 2012). The number of PhDs every country is producing is also increasing—for example, Burkina Faso has been lauded for having one PhD for every twenty graduates (UNESCO 2015). However, Africa's problem is no longer one of quantity, but rather the quality of degrees. The professors need to be doers, not just by-the-book "lecturers"; then, students will also be doers.

I truly believe that *criticism, never mind how constructive, is no longer enough*; it is time for criticizing by showing what the right way is, through action. In Africa, people brag about, and hero-worship, those with many university degrees; we did this a lot in Zimbabwe with Robert Mugabe. We lionize those who have, or brag when we have, published many books and journals. It is one thing whether or not anyone reads what they have written; I worry that scholarship is not measured by how many lives we have touched or positively changed, but by how many citations our publications have received. We seldom ask ourselves: What have you ever done for your own community? What minds and lives have you changed? Or what big problem has your thesis helped solve? It would be nice to retire one day having presided over the training of a new generation of graduate who does not simply rant about the unfulfilled promise the president, member of parliament, or mayor made to fix the economy, provide jobs, or stop raw sewage from flowing through the streets, but who seizes the problem as an opportunity to innovate solutions. After all, without problems there is no solution. That is not to say criticism is not important or is not itself a form of action; when it focuses on criticizing without offering solutions, or simply offering speculative solutions, while one just *sits there!*, or if it is simply fault-finding, it always leaves the job of building to someone else. That is what the African education system is producing by and large. Why is that?

Because the production of knowledge and education system is discipline based and designed to produce monoskilled, not multiskilled, graduates, who lack anticipative skill sets. Education experts, university presidents, and industry thinkers say that we are training students for jobs that will be obsolete in ten to fifteen years, because of fear that an automation apocalypse is upon us, that digital technologies will render human labor redundant. I do not see that as a problem, but as an open invitation to training innovation—to produce an anticipative, risk-taking, and multiskilled capability graduate to ride this wave, not get swept away by it. The ingredients must not simply be Western-and-white cultures, values, and systems of work; indigenous cultures of work like *nhimbe* (communal work party) and transient workspaces (Mavhunga 2014) could be mobilized to rehumanize the workspace through values of *hunhu* (*ubuntu*), to establish balance between human (and humane) and artificial intelligence. Initially this may mean an education system that is interdisciplinary, every discipline complementing another, producing a multiskilled, all-weather graduate adaptable to the twenty-first century's ever-changing, digitized economy at the very least. Later, the new knowledge in the service of and through problem-solving will generate its own orders of knowing and learning in which the survival of disciplines as silos may not be guaranteed.

Eventually, it may also mean a dissolution of disciplinary thinking, and the existence of the university or school as enclave. When we get out of our ivory tower and come to the people, it may mean we become intellectuals (the knower and thinker unbound by the academy), not academics (the creature of the book). Even then, the intellectual is not enough; Africa needs an intellectual who is a doer, who deploys knowledge in the service of problem-solving, and for whom the process and space of problem-solving is to produce knowledge. Africa needs thinker-doers, and currently our professoriate and graduates are neither because of a funnel-tunnel system of education where students open their ears, the information is pumped in, students memorize, accept without questioning, take the exam, and pass based on accurately representing, not questioning, dogma. They pass, graduate, look for a job, to replicate dogma. We need to engineer an education that rewires our students to question, to think for themselves, to run toward, not away from, problems, to solve them.

To do that, we must remove the colonial mindset that said true knowledge only comes from the classroom or lab, to expand the knowledge spectrum beyond academic knowing. The school lab is not enough; it is as theoretical as the books that guide it, that have proved inadequate already. We need to engage people and knowledge in everyday life—the village, the street, the marketplace—as an open laboratory for engaged researchers and thinker-doers, where value addition happens. I want to train the new African who refuses to be trapped in the lab, lecture room, and bookish mode, to take the university into the community. In those very spaces our ancestors invented grinding mills, pestles, and mortars to process foods, turned clays into earthenware, plant and insect matter into medicines, mined and processed ore, built Great Zimbabwe in dry-bonded granite rock, and Timbuktu in clay. What have we actually produced? We still export our products raw.

Academics are letting Africa down. We are teaching and producing useless knowledge. If it was useful, why are our graduates asking the municipality to clean up waste? Our governments to create jobs? To fix our roads? While we just sit there, criticize, write books analyzing the African crisis? With all that we know, what do we have to show for it? Whose lives has been changed for the better by it? What have we done for and with our communities lately?

As Amílcar Cabral says about the petty bourgeoisie committing suicide to their identity as a class and identifying with the deepest aspirations of their own people, by the same token, the highly learned African of today will have to commit suicide to the ivory tower if they are to produce knowledge useful to everyday Africans. We need to speak the language of the everyday, of the streets, the village, and the slums if we are to be active agents in lifting Africa.

In other words, the academic generation of the 1950s-1990s could point to independence as the fruit of their knowledge in the service of solving the problem of the colonizer and his colonialism. What can our generation of academicians point to besides our many publications, doctorates, professorships, and conference papers? As theorized by my dzimbahwe ancestors, *Shiri yatakatenda ndinyenganyenga, ndiyo isakambopomerwa kudya kana shanga imwe zvayo mumunda*/The bird we believed is the swallow, for it is the one never to be accused of eating even one grain in the field.

NOTES

CHAPTER 1

1. I would like to express my gratitude to Yolanda Covington, who was my senior at the University of Michigan, for inviting me to present this paper, titled "Black Science and Technology Studies (STS): An African in a White (Man's) Discipline," in the Department of Africana Studies lecture series on "Race, Science, and Technology in the Global African World," University of Pittsburg, December 3, 2018.

CHAPTER 2

1. I owe eternal gratitude to Mamadou Diouf for introducing me to Blyden's work, via the seminal work of George Shepperson.

CHAPTER 3

1. An earlier version of this chapter was published in *History and Technology* (Mavhunga 2015).

CHAPTER 4

1. A draft of this paper was presented to the Department III colloquium at the Max Planck Institute for the History of Science. I would like to thank Dagmar Schäefer for hosting me in 2019–2020 as a visiting scholar.

CHAPTER 6

1. The word "Shona" is an invention of the European colonist, who corrupted the Ndebele's reference to their eastern neighbors as *abatshona* (the ones who go under [retreat into caves and fight therefrom]). I prefer *dzimbahwe* people (those who built their houses in stone), drawing on Mufuka,

Mandizvidza, and Nemerai. *Dzimbahwe,* and Pikirayi, *The Zimbabwe Culture.* Hence my preference for *vedzimbahwe* (those of *dzimbahwe*).

CONCLUSION

1. Elements of this paper appear in the *Journal of Higher Education Research in Africa* (2018).

2. Personal Communication with Chikoko Nyamayemusoro, WhatsApp Chat, October 16, 2016.

3. https://www.dangote.com/foundation/

4. https://www.tonyelumelufoundation.org

5. https://www.higherlifefoundation.com

6. Mavhunga Field Notes, Mbare, Mupedzanhamo, Gazaland (Harare), Chikwanha and Makoni (Chitungwiza), Zimbabwe, January 20–30, 2017.

7. Mavhunga Field Notes, Mbare, Mupedzanhamo, Gazaland (Harare), Chikwanha and Makoni (Chitungwiza), Zimbabwe, January 20–30, 2017.

8. https://www.hellotractor.com

REFERENCES

ACET (African Center for Economic Transformation). 2016. *Unemployment in Africa: No Jobs for 50% of Graduates*. Available online at http://acetforafrica.org/highlights/unemployment-in-africa-no-jobs-for-50-of-graduates.

ACIL Tasman. 2009. *The Economic Impact of the Bushranger Project*. Melbourne: ACIL Tasman.

"African Independence Leader Honored by Lincoln (Pa) U." *Afro-American* (US). October 24, 1972.

Agbugah, F. 2016. *Bosch Is Set to Bring Nigerian Artisans into the Formal Sector of the Economy*. Available online at http://venturesafrica.com/bosch-is-set-to-bring-nigerian-artisans-into-the-formal-sector-of-the-economy.

Alexander, Jocelyn, Jo Ann McGregor, and Terence Ranger. 2000. *Violence and Memory: One Hundred Years in the "Dark Forests" of Matabeleland, Zimbabwe*. Oxford: James Currey.

APLU (Association for Public and Land-grant Universities). 2014. *African Higher Education: Opportunities for Transformative Change for Sustainable Development*. Available online at http://www.aplu.org/library/african-higher-education-opportunities-for-transformative-change-for-sustainable-development/file.

Armah, Ayi Kwei. 1968. *The Beautyful Ones are Not Yet Born*. Boston: Houghton Mifflin.

Armah, Ayi Kwei. 2010. *Remembering the Dismembered Continent*. Popenguine: Per Ankh.

AUST (African University of Science and Technology). 2016. "Science, Technology and Innovation (STI): Case Study: Nigeria." Abuja, Nigeria.

Bai, J., X. Zhang, J. Chen, M. Sun, F. Yu, and Z. Zhou. 2009. "Comparative Study on Engineering Education in China and USA." Unpublished BSc, Interactive Qualifying Project, Worcester Polytechnic Institute.

Beer, F. P., E. R. Johnston, and J. T. DeWolf. 2006. *Mechanics of Materials*. New York: McGraw-Hill.

Bennell, P. 1993. "The Cost-Effectiveness of Alternative Training Modes: Engineering Artisans in Zimbabwe." *Comparative Education Review* 37, no. 4: 434–453.

Bertelsmann Stiftung. 2014. *Mozambique Country Report*. Gütersloh: Bertelsmann Stiftung. Available online at https://www.bti-project.org/fileadmin/files/BTI/Downloads/Reports/2016/pdf/BTI_2016_Mozambique.pdf.

Bhebe, Ngwabi. 1999. *The ZAPU and ZANU Guerrilla Warfare and the Evangelical Lutherahn Church in Zimbabwe*. Gweru: Mambo Press.

Bhebe, Ngwabi, and Terence Ranger, eds. 1995. *Soldiers in Zimbabwe's Liberation War*. Harare: University of Zimbabwe Publications.

Beach, D. N.1989. *Mapondera: Heroism and History in Northern Zimbabwe, 1840–1904*. Gweru: Mambo Press.

Blyden, Edward Wilmot. 1862. *Liberia's Offering: Being Addresses, Sermons, etc.* New York: John A. Gray.

Blyden, Edward Wilmot. 1887. *Christianity, Islam and the Negro Race*. London, W. B. Whittingham & Co.

Blyden, Edward Wilmot. 1888. *Christianity, Islam and the Negro Race*, 2nd edition, with an introduction by Samuel Lewis. London: W. B. Whittingham & Co.

Blyden. Edward Wilmot. 1895. "The African Problem." *The North American Review* 161, no. 466: 327–339.

Blyden, Edward Wilmot. 1902. "Islam in Western Soudan." *Journal of the Royal African Society* 2, no. 5: 11–37.

Blyden, Edward Wilmot. 1905. "The Koran in Africa." *Journal of the Royal African Society* 4, no. 14: 157–171.

Brickhill, Jeremy. 1995. "Daring to Storm the Heavens: The Military Strategy of ZAPU 1976–1979." In *Soldiers in Zimbabwe's Liberation War*, edited by Ngwabi Bhebe and Terence Ranger, 48–86. London: James Currey.

Bucciarelli, Larry. 1994. *Designing Engineers*. Cambridge, MA: MIT Press.

Bucciarelli, Larry. 2003. *Engineering Philosophy*. Delft: Delft University.

Business Writer. 2015. "Chinamasa Defends Presumptive Tax." *The Daily News*, May 28.

Cabral, Amílcar. 1960. "The Facts about Portugal's African Colonies." Text written under the pseudonym of Abel Djassi and translated into English, published as a pamphlet by the Union of Democratic Control, with an introduction by Basil Davidson, London July 1960. In *Unity and Struggle*, 17–27.

Cabral, Amílcar. 1961. "Guiné and Cabo Verde against Portuguese Colonialism." Speech made at the 3rd Conference of the African Peoples held in Cairo, March 25–31, 1961, 10–19.

Cabral, Amílcar. 1962. "At the United Nations." Extracts front a statement made in Conakry in June 1962 to the United Nations Special Committee on territories under Portuguese Administration, 20–40.

Cabral, Amílcar. 1964. "Brief Analysis of the Social Structure in Guiné." Condensed text of a seminar held in the Frantz Fanon Centre in Treviglio, Milan, from May 1 to 3, 1964, 46–61.

Cabral, Amílcar. 1966. "The Weapon of Theory." Address delivered to the first Tricontinental Conference of the Peoples of Asia, Africa and Latin America held in Havana in January, 1966. London: Stage 1. In *Revolution in Guinea*, 73–90.

Cabral, Amílcar. 1968. "The Development of the Struggle." Extracts from a Declaration Made to the OSPAAAL General Secretariat 1968. London: Stage 1, 1974. In *Revolution in Guinea*, 91–102.

Cabral, Amílcar. 1969. "Fidelity to Party Principles." One of nine lectures delivered by Cabral in crioulo (a Guinéan language), during the "Party Principles and Political Practice" seminar for PAIGC cadres held from November 19 to 24, 1969. The lectures were tape-recorded and later transcribed into Portuguese. In *Unity and Struggle*, 98–99.

Cabral, Amílcar. 1969. "For the Improvement of our Political Work." One of nine lectures delivered by Cabral in crioulo (a Guinéan language), during the "Party Principles and Political Practice" seminar for PAIGC cadres held from November 19 to 24, 1969. The lectures were tape-recorded and later transcribed into Portuguese. In *Unity and Struggle*, 99–113.

Cabral, Amílcar. 1969. "Not Everyone Is of the Party." One of nine lectures delivered by Cabral in crioulo (a Guinéan language), during the "Party Principles and Political Practice" seminar for PAIGC cadres held from November, 19 to 24 1969. The lectures were tape-recorded and later transcribed into Portuguese. In *Unity and Struggle*, 83–93.

Cabral, Amílcar. 1969. "Our Party and the Struggle Must Be Led by the Best Sons and Daughters of our People." One of nine lectures delivered by Cabral in crioulo (a Guinéan language), during the "Party Principles and Political Practice" seminar for PAIGC cadres held from November, 19 to 24 1969. The lectures were tape-recorded and later transcribed into Portuguese. In *Unity and Struggle*, 64–75.

Cabral, Amílcar. 1969. "Revolutionary Democracy." One of nine lectures delivered by Cabral in crioulo (a Guinéan language), during the "Party Principles and Political Practice" seminar for PAIGC cadres held from November 19 to 24, 1969. The lectures were tape-recorded and later transcribed into Portuguese. In *Unity and Struggle*, 93–98.

Cabral, Amílcar. 1969. "Struggle of the People, By the People, For the People." One of nine lectures delivered by Cabral in crioulo (a Guinéan language), during the "Party Principles and Political Practice" seminar for PAIGC cadres held from November 19 to 24, 1969. The lectures were tape-recorded and later transcribed into Portuguese. In *Unity and Struggle*, 75–79.

Cabral, Amílcar. 1969. "To Start Out from the Reality of Our Land—To be Realists." One of nine lectures delivered by Cabral in crioulo (a Guinéan language), during the "Party Principles and Political Practice" seminar for PAIGC cadres held from November 19 to 24, 1969. The lectures were tape-recorded and later transcribed into Portuguese. In *Cabral, Unity and Struggle*, 44–75.

Cabral, Amílcar. 1969. "Unity and Struggle." One of nine lectures delivered by Cabral in crioulo (a Guinéan language), during the "Party Principles and Political Practice" seminar for PAIGC cadres held from November 19 to 24, 1969. The lectures were tape-recorded and later transcribed into Portuguese. In Cabral, *Unity and Struggle*, 28–44.

Cabral, Amílcar. 1972. *Our People Are Our Mountains: Amilcar Cabral on the Guinean Revolution*. London: Committee for Freedom in Mozambique, Angola & Guiné.

Cabral, Amílcar. 1973. "The Role of Culture in the Battle for Independence." *UNESCO Courier*.

Cabral, Amílcar. 1974. "The Nationalist Movements of the Portuguese Colonies." Opening address at the CONCP Conference held in Dar Es-Salaam, 1965. London: Stage 1, 1974. In *Revolution in Guinea*, 62–69.

Cabral, Amílcar. 1974. "Practical Problems and Tactics." Text of an interview given to Tricontinental magazine, published in issue no. 8 in September 1968. London: Stage 1, 1974. In *Revolution in Guinea*, 108–122.

Cabral, Amílcar. 1974. *Revolution in Guinea*. London: Stage 1. The following speeches and lectures are taken from this book.

Cabral, Amílcar. 1974. "Tell No Lies, Claim No Easy Victories. . . ." Extracts from Party directive 1965. London: Stage 1, 1974. In *Revolution in Guinea*, 70–72.

Cabral, Amílcar. 1974. "Towards Final Victory." Declaration to Voz de Condensed Vesrion of an Interview Recorded at the Khartoum Conference in January 1969, published in *Tricontinental* no. 12. London: Stage 1, 1974. In *Revolution in Guinea*, 126–132.

Carelse, X. F. 2002. "The Training of Industrial Physicists in Zimbabwe: A Success Story." *Physica Scripta* T97: 28–31.

CDC (Center for Disease Control and Prevention). 2015. *African Union and U.S. CDC Partner to Launch African CDC*. Press Release. Available online at https://www.cdc.gov/media/releases/2015/p0413-african-union.html.

Césaire, Aimé. 1939/2001. *Notebook of a Return to the Native Land*. Translated by C. Eshleman and A. Smith. Middletown, CT: Wesleyan University Press.

Césaire, Aimé. 1955/2000. *Discourse on Colonialism*. Translated by Joan Pinkham. New York: Monthly Review Press.

Césaire, Aimé. 1977. "The Essential and The Fundamental: In 'Tell us . . . Aimé Césaire.'" Interviewed by E. J. Maunick. *Africa Development / Afrique et Développement* 2, no. 4: 43–54.

Césaire, Aimé. 1982c. "Knives of Noon." Translated by A. E. Stringer *The American Poetry Review* 11, no. 5: 7.

Césaire, Aimé. 1985. *A Tempest*. Translated by Richard Miller. New York: Ubu Repertory Publications.

Césaire, Aimé. 1989. "It Is Through Poetry That One Copes With Solitude: An Interview with Aimé Césaire." Interviewed by Charles H. Rowell. *Callaloo* 38: 49, 51, 53, 55, 57, 59, 61, 63, 65, 67.

Césaire, Aimé. 1983/2001. "Wifredo Lam . . ." Translated by Clayton Eshleman and Annette Smith. *Callaloo* 24, no. 3: 712.

Chavunduka, Gordon L. 1978. *Traditional Healers and the Shona Patient*. Harare: Mambo Press.

Chinyemba, F. 2011. "Mobility of Engineering and Technology Professionals and its Impact on the Quality of Engineering and Technology Education: The Case of Chinhoyi University of Technology, Zimbabwe." *International Journal of Quality Assurance in Engineering and Technology Education* 1, no. 2: 35–49.

Chipamaunga, Edmund. 1983. *A Fighter for Freedom*. Gweru: Mambo Press. 1983.

Chung, Fay. 2006. *Re-Living the Second Chimurenga: Memories from Zimbabwe's Liberation Struggle*. Uppsala: Nordic Africa Institute.

CODESRIA (Council for the Development of Social Science Research in Africa). 2014. *CODESRIA Launches a New Programme: African Diaspora Support to African Universities*. Available online at http://www.codesria.org/spip.php?article2175.

Commonfund Institute. 2016. *NACUBO-Commonfund Study of Endowments Results Released*. Available online at https://www.commonfund.org/2016/01/27/2015-nacubo-commonfund-study-of-endowments-2.

C-U Coalition against Apartheid flyer. 1977. African Activist Archive.

Cyranoski, D., N. Gilbert, H. Ledford, A. Nayar, and M. Yahia. 2011. "Education: The PhD Factory." *Nature* 472, no. 7343: 276–279.

Damasane, Paul. 2014. "Spirituality in the Struggle for Independence." *The Sunday Mail.*

Daneel, Martinus. 1995. *Guerrilla Snuff.* Harare: Baobab Books.

Davidson, Basil. 1973. "Cabral's Murder: The Facts." *West Africa* 2913: 459–461.

Davidson, Basil. 1964/1974. "Living Better Isn't Only Eating Better." *African Development* (Br).

Davidson, Basil. 1969. *The Liberation of Guiné.* London: Penguin.

Davidson, Basil. 1973. "Crisis for the Colonialists." *The Guardian* (Januatry 23).

Davidson, Basil. 1974. "The Heroic Republic." *West Africa* (Nigeria).

Davies, J. C. A. 1982. "A Major Epidemic of Anthrax in Zimbabwe, Part 1." *Central African Journal of Medicine* 28 (1982): 291–298.

Davies, J. C. A. 1983. "A Major Epidemic of Anthrax in Zimbabwe, Part 2." *Central African Journal of Medicine* 29 (1983): 8–12.

Davis, R. Hunt. 1975. "John L. Dube: A South African Exponent of Booker T. Washington." *Journal of African Studies* 2 (December).

Devy, Ganesh. 1995. *After Amnesia: Tradition and Change in Indian Literary Criticism.* Hyderbad: Orient Longman.

Dhliwayo, Jabulani. 2012. *The Endless Journey: From a Liberation Struggle to Driving Emerging Technologies in Africa.* CreativeSpace: Jabulani Simbini Dhliwayo (self-published).

Dickinson, Margaret. 1972. *When Bullets Begin to Flower.* Nairobi: East Africa Publishing House.

Diouf, Mamadou. 2019. "Black Atlantic Horizons and the Practices of History: Histories of 'Dismembering' and 'Remembering'" Edinburgh: ECAS2019. Available at https://www.youtube.com/watch?v=U4WLRLS-VF8.

Divala, J. J. 2016, "Re-Imaging a Conception of Ubuntu that Can Recreate Relevant Knowledge Cultures in Africa and African Universities." *Knowledge Cultures* 4, 4: 90–103.

Downey, G. 2005. "Are Engineers Losing Control of Technology? From 'Problem Solving' to 'Problem Definition and Solution' in Engineering Education." *Chemical Engineering Research and Design* 83, no. A6: 583–595.

Dube, John Langalibalele. 1897. "Need of Industrial Education in Africa." *Southern Workman* 27: 141–142.

Dube, John Langalibalele. 1898. "Zululand and the Zulus." *Missionary Review of the World* 21: 435–443.

Dube, John Langalibalele. 1901. "A Native View of Christianity in South Africa." *Missionary Review of the World* 24: 421–426.

Dube, John Langalibalele. 1910. "The Zulu's Appeal for Light and England's Duty." Copy in Missionary Research Library, Union Theological Seminary, New York.

Du Bois, W. E. B. 1903. *The Souls of Black Folk.* Chicago: A. C. McClurg and Co.

Duderstadt, J. J. 2010. "Engineering for a Changing World." In Domenico Grasso and Melody Burkins, eds. *Holistic Engineering Education: Beyond Technology*, 17–35. New York: Springer.

EARC (East Africa Regional Resource Center). 2018. *Kenya: Country Strategic Paper 2014–2018*. African Development Bank Group, Nairobi. Available online at https://www.afdb.org/fileadmin/uploads/afdb/Documents/Project-and-Operations/2014-2018_-_Kenya_Country_Strategy_Paper.pdf.

Faler, P. G. 1981. *Mechanics and Manufacturers in the Early Industrial Revolution: Lynn, Massachusetts 1780–1860*. New York: SUNY Press.

Fanon, Frantz. 1952/1962. *Black Skin, White Masks*. Translated by Charles Lam Markmann. New York: Grove Press.

Fanon, Frantz. 1959/1965. *A Dying Colonialism*. New York: Grove Press.

Fanon, Frantz. 1961/2004. *The Wretched of the Earth*. New York: Grove Press.

Fanon, Frantz. 1964. *Toward the African Revolution*. New York: Grove Press.

Farstard, H. 2002. *Integrated Entrepreneurship Education in Botswana, Uganda and Kenya*. Oslo: National Institute of Technology.

Fayolle, A., and B. Gailly. 2008. "From Craft to Science: Teaching Models and Learning Processes in Entrepreneurship Education." *Journal of European Industrial Training* 32, no. 7: 569–593.

Fernandez-Armesto, Felipe. 1987. *Before Columbus: Exploration and Colonization from the Mediterranean to the Atlantic, 1229–1492*. Philadelphia: University of Pennsylvania Press.

Firmin, Anténor. 1885. *De l'égalité des races humaines*. Paris: Cotillon.

Firmin, Anténor. 2000. *The Equality of the Human Races*. Translated by Charles Asselin. Introduction by Carolyn Fluehr-Lobban. New York: Garland Press.

Forbes. 2016. *The World's Billionaires*. Available at http://www.forbes.com/profile/aliko-dangote/.

Foulds, K., and P. T. Zeleza. 2014. "The African Academic Diaspora and African Higher Education." *International Higher Education* 76: 16–17.

Frankel, E. 2008. *Changes in Engineering Education: Massachusetts Institute of Technology*. Available at http://web.mit.edu/fnl/volume/205/fnl205.pdf.

"Freedom Fighters Declare Week of Mourning." 1973. Zambia Daily Mail (January 26).

Fripp, C. 2013. "Top 10 Mobile Agriculture Applications." *IT News Africa* (November 7). Available at http://www.itnewsafrica.com/2013/11/top-10-mobile-agriculture-applications/.

"Gaddafi Sends Arms and Men to Aid PAIGC." 1973. *Daily News* (Tanzania) (January 26).

Globelics. 2012. *Learning Innovation and Inclusive Development: A Globelics Thematic Report*. Globelics Secretariat.

Gobineau, Arthur. 1853. *Essai sur l'inégalité des races humaines*. Paris: Paris 4.

Gobineau, Arthur. 1915. *The Inequality of Human Races*. Translated by Adrian Collins. London: Heinemann.

Godin, Benoit. 2008. "Innovation: The History of a Category." *Project on the Intellectual History of Innovation* (Working Paper # 1). Montreal.

Grasso, D., and M. Burkins, eds. 2010. *Holistic Engineering Education: Beyond Technology*. Rio de Janeiro: Springer.

Griffiths, M., J. Kickul, S. Bacq, and S. Terjesen. 2012. "A Dialogue with William J. Baumol: Insights on Entrepreneurship Theory and Education." *Entrepreneurship Theory and Practice* 36, no. 4: 611–625.

Gross, Daniel A. 2018. "The Troubling Origins of the Skeletons in a New York Museum." *The New Yorker*.

Gudhlanga, E. S., and G. Makaudze. 2012. "Promoting the Use of an African Language as a Medium of Instruction in Institutions of Higher Learning in Zimbabwe: The Case of Great Zimbabwe's Department of Languages, Literature." *Prime Journal of Social Science* 1, no. 3: 51–56.

Guha, Ranajit. 2003. *History at the Limit of World History*. New York: Columbia University Press.

Gwaunza, Musah. 2014. "Sakupwanya: 'Doctor of the War' Par excellence." *The Manica Post*.

Henriksen, Thomas H. 1975. "African Intellectual Influences on Black Americans: The Role of Edward W. Blyden." *Phylon* 36, no. 3: 279–290.

Herald Reporter. 2011. "VP Nkomo Mourns Bango." *The Herald*.

Herald Reporter. 2014. "Pswarayi Hero Status Clarified." *The Herald*.

HESA (Higher Education South Africa). 2014. *Remuneration of Academic Staff at South African Universities: A Summary Report of the HESA Statistical Study of Academic Remuneration*. Available online at http://www.justice.gov.za/commissions/FeesHET/docs/2014-HESA-SummaryReport-RemunerationOfAcademicStaff.pdf.

Hilson, G., and L. Osei. 2014. "Tackling Youth Unemployment in Sub-Saharan Africa: Is There a Role for Artisanal and Small-Scale Mining?" *Futures* 62: 83–94.

Hoagland, Jim. 1973. "Cabral: Murder of a Key African." *International Herald Tribune* (January 25).

Hoffman, Bruce, Jennifer M. Taw, and David Arnold. 1991. *Lessons for Contemporary Counterinsurgencies: The Rhodesian Experience*. Santa Monica, CA: RAND.

Hossain, Shah Aashna. 2008. "'Scientific Racism' in Enlightened Europe: Linnaeus, Darwin, and Galton." *Biology* 103. http://serendip.brynmawr.edu/exchange/node/1852.

Houle, Robert J. 2011. *Making African Christianity: Africans Reimagining Their Faith in Colonial South Africa*. Bethlehem, PA: Lehigh University Press.

Hountondji, P. 2002. *The Struggle for Meaning: Reflections on Philosophy, Culture and Democracy in Africa*. Athens, OH: Ohio University Press.

Huni, Munyaradzi. 2012a. "A Humble and Bitter War Commander" (Interview with Retired Brigadier Ambrose Mutinhiri). *The Sunday Mail* (August 2).

Huni, Munyaradzi. 2012b. "Gabarinocheka: He Rose from the 'Dead.'" (Interview with Cde Chinodakufa). *The Sunday Mail* (August 25).

Huni, Munyaradzi. 2012c. "Chitepo's Death: Survivor's Horror Account." (Interview with Bensen Kadzinga, War Name Sadat Kufamazuva). *The Sunday Mail* (December 1).

Huni, Munyaradzi. 2012d. "Rutanhire Relives Horrors of the War." (Interview with Jackson Musanhu, War Name Cde George Rutanhire). *The Sunday Mail* (December 15).

Huni, Munyaradzi. 2013a. "Face-to-Face with the Real 'Gandangas.'" (Interview Onias Garikai Bhosha, War Name Cde George Gabarinocheka). *The Sunday Mail* (January 6).

Huni, Munyaradzi. 2013b. "Commander Who Captured the First White Man." (Interview with Luke Mushore War Name Dick Joboringo). *The Sunday Mail* (February 17).

Huni, Munyaradzi. 2013c. "Chimoio Massacre: A Doctor's Nightmare." (Interview with Felix Muchemwa). *The Sunday Mail* (March 10).

Huni, Munyaradzi. 2013d. "Give Me My Gun, My Grenade: I Want To Die." (Interview with Akwino Aquous Muwoni, War Name Guerrilla Texen Chidhakwa). *The Sunday Mail* (April 27).

Huni, Munyaradzi. 2013e. "Cde Chinx: Music was My Gun." (Interview with Dick Chingaira Makoni, War Name Cde Chinx). *The Sunday Mail* (May 19).

Huni, Munyaradzi. 2013f. "The Gun was a Prison for Sell-outs." (Interview with Wereki Sandiani, War Name Cde Philip Gabella). *The Sunday Mail* (May 26).

Huni, Munyaradzi. 2013g. "The Rear was a Big War Front." (Interview with Tendai Kuzvidza, War Name Cde Hondo Mushati). *The Sunday Mail* (May 26).

Huni, Munyaradzi. 2013h. "The Rear was a Big War Front." (Interview with Tendai Kuzvidza, War Name Cde Hondo Mushati). *The Sunday Mail* (June 8).

Hunt, James. 1864. "The Negro's Place in Nature." *Journal of the Anthropological Society of London* 2: 15–56.

Ibekwe, Chinweizu. 1987. *Decolonising the African Mind*. London: Sundoor.

Interview. 1998. "Dr. Timothy Stamps, Minister of Health, Zimbabwe." *PBS*.

Iizuka, M., P. Mawoko, and F. Gault. 2015. *Innovation for Development in Southern & Eastern Africa: Challenges for Promoting ST&I Policy*. United Nations University Policy Brief, no. 1. Tokyo: United Nations University.

Ishengoma, J. M. 2016. "Strengthening Higher Education Space in Tanzania through North-South Partnerships and Links: Experiences from the University of Dar es Salaam." *Comparative & International Education* 45, no. 1: 1–17.

Jabavu, Davidson Don Tengo. 1920/1969. *The Black Problem; Papers and Addresses on Various Native Problems*. New York: Negro Universities Press.

Jahn, J. 1968. *A History of neo-African literature*. London: Faber.

Johnson, Thomas A. 1973. "Cabral buried in Guiné amid Cry of Revolution." *The New York Times*.

Jules-Rosette, Bennetta. 1991. "Speaking about Hidden Times: The Anthropology of V. Y. Mudimbe." *Callaloo* 14, no. 4: 944–960.

Kadungure, Samuel. 2014. "Cde Bandera: Selfless, Gallant Fighter." *The Manica Post* (September 5).

Kagame, Paul. 2014. "Africa Leaders Must Solve Africa's Problems in Africa Not In Paris," Africa Web TV. https://www.youtube.com/watch?v=Nh-fN8QLFqo.

Karsholm, Preben. 2006. "Memoirs of a Dutiful Revolutionary: Fay Chung and the Legacies of the Zimbabwean Liberation War." Introduction to Fay Chung, 7–25. *Reliving Chimurenga*.

Kanyandago, P. 2010. "Revaluing the African Endogenous Education System for Community-Based Learning: An Approach to Early School Leaving." In *The Burden of Educational Exclusion: Understanding and Challenging Early School Leaving in Africa*, edited by J. Zeelen, J. van der Linden, D. Nampota, and M. Ngabirano, 101–116. Rotterdam, Boston and Taipei: Sense Publishers.

Kaya, H. O., and Y. N. Seleti. 2013. "African Indigenous Knowledge Systems and Relevance of Higher Education in South Africa." *The International Education Journal: Comparative Perspectives* 12, no. 1: 30–44.

Kerr, Alexander. 1968. *Fort Hare 1915–48: The Evolution of an African College*. New York: Humanities Press.

Kirby, D. 2006. "Creating Entrepreneurial Universities in the UK: Applying Entrepreneurial Theory to Practice." *Journal of Technology Transfer* 31, no. 5: 599–603.

Kot, F. C. 2016. "The Perceived Benefits of International Partnerships in Africa: A Case Study of Two Public Universities in Tanzania and the Democratic Republic of Congo." *Higher Education Policy* 29: 41–62.

Kriger, Norma. 1992. *Zimbabwe's Guerrilla War: Peasant Voices*. Cambridge: Cambridge University Press.

Kuhn, T. S. 1962. *The Structure of Scientific Revolutions*. Chicago: University of Chicago Press.

Kumapley, N. K. 1997. "Some Issues Relating to Relevance and Quality Assurance in Engineering Education at Kwame Nkrumah University of Science and Technology, Kumasi, Ghana." In *Quality Assurance and Relevance of Engineering Education in Africa*, edited by J. G. M. Massaquoi and F. M. Luti, 36–41. Paris: ANSTI and UNESCO.

Kunje, W. A. B. 1997. "A Look at Issues Affecting Quality and Relevance of Engineering Education." In *Quality Assurance and Relevance of Engineering Education in Africa*, edited by J. G. M. Massaquoi and F. M. Luti, 30–35. Paris: ANSTI and UNESCO.

Lamont, Bishop Donal. 1977. *Speech from the Dock*. Leigh-on-Sea: Kevin Mayhew.

Lan, David. 1985. *Guns and Rain: Guerrillas and Spirit Mediums in Zimbabwe*. London: James Currey.

Lawrence, J. A., C. M. Foggin, and R. A. Norval. 1980. "The Effects of War on the Control of Diseases of Livestock in Rhodesia (Zimbabwe)." *Veterinary Record* 107: 82–85.

Le Baron, Bentley. 1966. "Négritude: A Pan-African Ideal?" *Ethics* 76, 4: 267–276.

Lester, Craig. 1996. *Protection of Light Skinned Vehicles against Landmines—A Review*. Melbourne: DSTO Aeronautical and Maritime Research Laboratory.

Libombo, D. B., and A. Dinis. 2015. "Entrepreneurship Education in the Context of Developing Countries: Study of the Status and the Main Barriers in Mozambican Higher Education Institutions." *Journal of Developmental Entrepreneurship* 20, no. 3: 1550020-1-1550020-26.

Lynch, Holis. 1965. "The Attitude of Edward W. Blyden to European Imperialism in Africa." *Journal of the Historical Society of Nigeria* 3, no. 2: 249–259.

Mamdani, Mahmood. 2010. "Commercialization is Killing Makerere: Interview with Moses Mulondo." *Pambazuka* News 495.

Marable, W. Manning. 1974a. "A Black School in South Africa." *Negro History Bulletin* 37, no. 4: 258–261.

Marable, W. Manning. 1974b. "Booker T. Washington and African Nationalism." Phylon 35: 398–406.

Marangu, S. M. 1997. "Training of Engineering Graduates in Response to Present and Future Needs for Kenya: The JKUAT Experience." In *Quality Assurance and Relevance of Engineering Education in Africa*, edited by J. G. M. Massaquoi and F. M. Luti, 25–29. Paris: ANSTI and UNESCO. Available at http://unesdoc.unesco.org/images/0012/001206/120628Eo.pdf.

Marjoram, T. 2010. "Engineering, Science and Technology Policy." In *UNESCO Report: Engineering: Issues, Challenges, and Opportunities for Development*, 171–175. Paris: UNESCO.

Markwardt, C. E. 2014. *Mozambique Creates a Star of Africa—A World Class Petroleum Engineering Program at the University of Eduardo Mondlane (UEM)*. Maputo: Society of Petroleum Engineers.

Mashininga, K. 2012. "Private Universities Set to Overtake Public Institutions." *University World News* 211. Available at http://www.universityworldnews.com/article.php?story=20120302141207184.

Massaquoi, J. G. M., and E. M. Luti, eds. 1997. *Quality Assurance and Relevance of Engineering Education in Africa*. Paris: ANSTI and UNESCO.

Masolo, D. A. 1991. "An Archaeology of African Knowledge: A Discussion of V. Y. Mudimbe." *Callaloo* 14, no. 4: 990–1011.

Masolo, D. A. 2017. "The Place of Science and Technology in Our Lives: Making Sense of Possibilities." In *What Do Science, Technology, and Innovation Mean from Africa?*, edited by Chakanetsa Mavhunga, 29–44. Cambridge: MIT Press.

Masu, L. M. 1997. "Quality Assurance and Relevance of Engineering Education in Africa." In *Quality Assurance and Relevance of Engineering Education in Africa*, edited by J. G. M. Massaquoi and F. M. Luti, 13–19. Paris: ANSTI and UNESCO. Available at http://unesdoc.unesco.org/images/0012/001206/120628Eo.pdf.

Matthews, P., L. Ryan-Collins, J. Wells, H. Sillem, and H. Wright. 2012. *Identifying Engineering Capacity Needs in Sub-Saharan Africa*. London: Royal Academy of Engineering. Available at http://www.raeng.org.uk/publications/reports/engineers-for-africa.

Mavhunga, Chakanetsa. 2011. "Vermin Beings: On Pestiferous Animals and Human Game." *Social Text* 29, no. 1: 151–176.

Mavhunga, Chakanetsa. 2014. *Transient Workspaces: Technologies of Everyday Innovation in Zimbabwe*. Cambridge, MA: MIT Press.

Mavhunga, Chakanetsa. 2015. "Guerrilla Healthcare Innovation: Creative Resilience in Zimbabwe's Chimurenga, 1971 1980." *History and Technology* 31, 3: 295–32.

Mavhunga, Chakanetsa, ed. 2017. *What Do Science, Technology, and Innovation Mean from Africa?* Cambridge, MA: MIT Press.

Mavhunga, Chakanetsa. 2018. "Modelling an African Research University: Notes towards an Interdisciplinary, Cross-Cultural and Anticipative Curriculum." *Journal of Higher Education Research in Africa* 16, 1/2: 25–50.

Mavhunga, Chakanetsa. 2018. *The Mobile Workshop: The Tsetse Fly and African Knowledge Production*. Cambridge, MA: MIT Press.

Mayombe, C. 2016. "Enabling Labour Market Entry for Adults through Non-Formal Education and Training for Employment in South Africa." *International Journal of Lifelong Education* 35, no. 4: 376–395.

Mbizi, R., L. Hove, A. Thondhlana, and N. Kakava. 2013. "Innovation in SMEs: A Review of its Role to Organisational Performance and SMEs Operations Sustainability." *Interdisciplinary Journal of Contemporary Research in Business* 4, no. 11: 370–389.

McCollester, Charles. 1973. "The Political Thought of Amilcar Cabral." *Monthly Review*.

McCowan, T. 2014. *Can Higher Education Solve Africa's Job Crisis? Understanding Graduate Employability in Sub-Saharan Africa*. London: University of London.

McLaughlin, Janice. 1996. *On the Frontline: Catholic Missions in Zimbabwe's Liberation War*. Harare: Baobab.

Mhanda, Wilfred. 2011. *Dzino: Memories of a Freedom Fighter*. Harare: Weaver Press.

Mills, E. J., S. Kanters, A. Hagopian, N. Bansback, J. Nachega, M. Alberton, C. G. Au-Yeung, A. Mtambo, I. L. Bourgeault, S. Luboga, R. S. Hogg, and N. Ford. 2011. "The Financial Cost of Doctors Emigrating from Sub-Saharan Africa: Human Capital Analysis." *British Medical Journal* 343. Available at http://www.bmj.com/content/bmj/343/bmj.d7031.full.pdf.

MIT (Massachusetts Institute of Technology). 2017. *General Institute Requirements*. Available at http://catalog.mit.edu/mit/undergraduate-education/general-institute-requirements.

Moore, Gerald, and Ulli Beier. 1963. "Introduction." In *Modern Poetry from Africa*, ed. Gerald Moore & Ulli Beier, 15–32. Harmondsworth: Penguin.

M'timkulu, G. S. 1950. "Ohlange Institute." *Native Teachers' Journal*.

Moswela, B., and U. Chiparo. 2015. "An Evaluation of Botswana Technical Colleges' Curriculum and Its Enhancement of Graduate Employability." *Journal of Higher Education Theory and Practice* 15, no. 7: 105–120.

Mudavanhu, V., S. Bindu, L. Chigusiwa, and L. Muchabaiwa. 2011. "Determinants of Small and Medium Enterprises Failure in Zimbabwe: A Case Study of Bindura." *International Journal of Economic Research* 2, no. 5: 82–89.

Mufuka, Ken, K. Mandizvidza, and J. Nemerai. 1983. *Dzimbahwe: Life and Politics in the Golden Age*. Harare: Harare Publishing House.

Mugwisi, T., J. Mostert, and D. N. Ocholla. 2015. "Access to and Utilization of Information and Communication Technologies by Agricultural Researchers and Extension Workers in Zimbabwe." *Information Technology for Development* 21, no. 1: 67–84.

Mujeyi, K., S. Siziba, W. Z. Sadomba, and J. Mutambara. 2016. "Technical Efficiency of Informal Manufacturing Sector Enterprises: Evidence from the Informal Metal Industry of Zimbabwe." *African Journal of Science, Technology, Innovation & Development* 8, no. 1: 12–17.

Mukwenje, Sydney. 2013. "My Liberation War Experience: Part Four . . . Captured, But Released." *The Patriot*.

Munanga, E. 2013. "Financial Challenges Faced by Retail Smes Operating in a Multi-Currency Environment: A Case of Gweru Urban, Zimbabwe." *Asian Economic and Financial Review* 3, no. 3: 377–388.

Mupinga, D. M., M. F. Burnett, and D. H. Redmann. 2005. "Examining the Purpose of Technical Education in Zimbabwe's High Schools." *International Education Journal* 6, no. 1: 75–83.

Murwira, Z. 2014. "$7bn Circulates in Informal Sector." *The Herald*.

Mutambara, Aggrippah. 2014. *The Rebel in Me: A ZANLA Guerrilla Commander in the Rhodesian Bush War, 1975–1980*. Solihull: Helion.

Mutanda, Freedom. 2015a. "Muumbe, the Mujibha of Mujibhas." *The Manica Post* (March 20).

Mutanda, Freedom. 2015b. "It Was No Walk In The Park: Mapuranga." *The Manica Post* (May 8).

Mutanda, Freedom. 2015c. "A Combatant's Reflections on War." *The Manica Post* (May 15).

Mutanda, Freedom. 2015d. "Age Couldn't Dissuade His War Effort." *The Manica Post*. (July 31).

Mvumi, B., L. T. Matsikira, and J. Mutambara. 2016. "The Banana Postharvest Value Chain Analysis in Zimbabwe." *British Food Journal* 118, no. 2: 272–285.

Mwale, Emergency. 2014. "Stark Reminder of the War on the ZIPRA Front." *The Patriot.*

Nass, Meryl. 1992. "Anthrax Epizootic in Zimbabwe, 1978–1980: Due to Deliberate Spread?" *PSR Quarterly* 2: 198–209.

Nass, Meryl. "Zimbabwe's Anthrax Epizootic." *Covert Action* 43 (1992–93): 12–18.

Ndebele, H. 2015. "Street Vendors: Lessons from Mauritius." *Zimbabwe Independent*, 5 June.

Ndlovu, Mary. 2011a. "Interview with Regina Ndlovu." Bulawayo, May 6. Zenzo Nkobi Photographic Archive, SAHA and Mafela Trust.

Ndlovu, Mary. 2011b. "Interview with Dr. Benjamin Dube." Bulawayo, October 14. Zenzo Nkobi Photographic Archive, SAHA and Mafela Trust.

Ndlovu, Mary, and Zephaniah Nkomo. 2010. "Interview with Thadeus Parks Ndlovu." Bulawayo, December 11, 2010. Zenzo Nkobi Photographic Archive, SAHA and Mafela Trust.

Ndlovu, Mary, and Zephaniah Nkomo. 2011a. "Interview with Longman Ndebele." Bulawayo, March 1. Zenzo Nkobi Photographic Archive, SAHA and Mafela Trust.

Ndlovu, Mary, and Zephaniah Nkomo. 2011b. "Interview with Ellingworth Poli." Bulawayo, November 14. Zenzo Nkobi Photographic Archive. SAHA and Mafela Trust.

Nhongo-Simbanegavi, Josephine. 2000. *For Better or Worse? Women and ZANLA in Zimbabwe's Liberation Struggle*. Harare: Weaver.

Njenga, Joshua. 2021. "Thomas Sankara: Rise, Politics, and Assassination," www.joshuanjenga.com. Accessed July 6, 2022.

"Nkomo, John, ZAPU—Secretary of Administration, Minister of Local Government and Rural Development." Interview, Harare, July 21, 1995. Liberation in Southern Africa—Regional and Swedish Voices, Nordic Africa Institute, Aluka.

Noble, David. 1977. *America by Design: Science, Technology and the Rise of Corporate Capitalism*. New York: Alfred A. Knopf.

Nyerere, Julius Kambarage. 1968. *Freedom and Socialism/Uhuru naUjamaa: A Selection from Writings and Speeches, 1965–1967*. Oxford: Oxford University Press. (All of Nyerere's speeches below are from this book.)

Nyerere, Julius Kambarage. 1965a. "Tanzania's Long March is Economic." Speech at a banquet in honour of Chinese Premier Chou En Lai, June 4.

Nyerere, Julius Kambarage. 1965b. "Dissolving the Independence Parliament." Address in kiSwahili to National Assembly, June 8.

Nyerere, Julius Kambarage. 1965c. "Relations with the West." Address to International Press Club luncheon during Commonwealth Conference, London, June 23.

Nyerere, Julius Kambarage. 1965d. "Unemployment is No Problem." Speech on the Laying Stone of the Kibo Match Factory, Moshi, August 16.

Nyerere, Julius Kambarage. 1965e. "Congress on African History." Opening Speech at the International Congress on African History, Dar es Salaam, University College, September 26.

Nyerere, Julius Kambarage. 1965f. "Agriculture is the Basis of Development." Speech when laying foundation of permanent buildings at the Morogoro Agricultural College, November 18.

Nyerere, Julius Kambarage. 1966a. "Leaders Must Not Be Masters." Extempore public speech in kiSwahili during a tour of Mafia Island, February.

Nyerere, Julius Kambarage. 1966b. "The Tanzanian Economy." Address to the National Assembly at the commencement of its Budget Session, June 13.

Nyerere, Julius Kambarage. 1966c. "The Role of Universities" (original: "The University's Role in the Development of the New Countries"). Opening address to the World University Service General Assembly, Dar es Salaam University College, June 27.

Nyerere, Julius Kambarage. 1967a. "The Arusha Declaration: Socialism and Self-Reliance." Arusha, January 29.

Nyerere, Julius Kambarage. 1967b. "Public Ownership in Tanzania." Article published in the Sunday News, February 12.

Nyerere, Julius Kambarage. 1967c. "Economic Nationalism." Speech on the occasion of the extension of the Tanzania Breweries plant, Dar es Salaam, February 28.

Nyerere, Julius Kambarage. 1967d. "Education for Self-Reliance." First of his 'post-Arusha' policy directives, focusing on education, March.

Nyerere, Julius Kambarage. 1967e. "The Varied Paths to Socialism." Address to faculty and students at the University of Cairo, Egypt, on the occasion of the conferment upon him of an honorary degree, April 10.

Nyerere, Julius Kambarage. 1967f. "Address to the Trade Unions." Address to the biennial Conference of the National Union of Tanganyika Workers, July.

Nyerere, Julius Kambarage. 1967g. "The Purpose is Man." Speech delivered at the Teach-In on the Arusha Declaration organized by the Dar es Salaam University College branch of the TANU Youth League, August 5.

Nyerere, Julius Kambarage. 1967h. "Socialism and Rural Development." Second 'post-Arusha' policy paper, September.

Odera-Oruka, Henry. 1983. "Sagacity in African Philosophy." *International Philosophy Quarterly* 23, no. 4: 383–393.

OECD (Organisation for Economic Cooperation and Development). 2002. *Frascati Manual: Proposed Standard Practice for Surveys on Research and Experimental Development.* Paris: OECD. Available at http://www.oecd.org/sti/inno/frascatimanualproposedstandardpracticeforsurveysonresearchandexperimentaldevelopment6thedition.htm.

OECD (Organisation for Economic Cooperation and Development). 2005. *The Oslo Manual: Guidelines for Collecting and Interpreting Innovation Data.* 3rd edition. Paris: OECD and Eurostat. Available at http://www.oecd.org/sti/inno/oslomanualguidelinesforcollectingandinterpretinginnovationdata3rdedition.htm.

Office of Public Affairs & Communications. 2016. "Yale Welcomes Zimbabwean Philanthropists and New Partners to Campus." *Yale News.* Available at http://news.yale.edu/2016/05/12/yale-welcomes-zimbabwean-philanthropists-and-new-partners-campus.

Oladiran, M. T., G. Pezzotta, J. Uziak, and M. Gizejowski. 2013. "Aligning an Engineering Education Program to the Washington Accord Requirements: Example of the University of Botswana." *International Journal of Engineering Education* 29, no. 6: 1591–1603.

Openjuru, G. 2010. "Government Education Policies and the Problem of Early School Leaving: The Case of Uganda." In *The Burden of Educational Exclusion: Understanding and Challenging Early School Leaving in Africa*, edited by J. Zeelen, J. van der Linden, D. Nampota, and M. Ngabirano, 22–47. Rotterdam, Boston, and Taipei: Sense Publishers.

Oxford Analytica. 2010. *Shadow Economies Resist Taxation Efforts*. Oxford Analytica Daily Brief Service. Available at https://dailybrief.oxan.com/Analysis/DB158317/AFRICA-Shadow-economies-resist-taxation-efforts.

Padgen, Gary. 1982. *The Fall of Natural Man: The American Indian and the Origins of Comparative Ethnology*. Cambridge: Cambridge University Press.

Parker, Jim. 2006. *Assignment Selous Scouts: Inside Story of a Rhodesian Special Branch Officer*. Alberton: Galago.

Pikirayi, Innocent. 2001. *The Zimbabwe Culture. Origins and Decline of Southern Zambezian States*. Walnut Creek: Altamira Press.

Pitlane Magazine. 2017. "Community Outreach Activities of African Universities." Available at http://www.pitlanemagazine.com/cultures/community-outreach-activities-of-african-universities.html.

Plaza, S., and D. Ratha. 2011. *Diaspora for Development in Africa*. Washington, DC: World Bank.

Popov, O., and A. Manuel. 2016. "Vocational Literacy in Mozambique: Historical Development. Current Challenges and Contradictions." *Literacy & Numeracy Studies* 24, no. 1: 23–42.

Ramadi, E., S. Ramadi, and K. Nasr. 2016. "Engineering Graduates' Skill Sets in the MENA Region: A Gap Analysis of Industry Expectations and Satisfaction." *European Journal of Engineering Education* 41, no. 1: 34–52.

Rampedi, M. A. 2003. *Implementing Adult Education Policy in the Limpopo Province of South Africa: Ideals, Challenges and Opportunities*. Groningen: Centre for Development Studies.

Ranger, Terence. 1967. *Revolt in Southern Rhodesia*. London: Heinemann.

Ranger, Terence. 1985. *The Invention of tribalism in Zimbabwe*. Harare : Mambo Press.

Ranger, Terence. 1999. *Voices from the Rocks: Nature, Culture and History in the Matopos Hills of Zimbabwe*. Oxford: James Currey.

Rayner, P., and K. Ashcroft. 2011. "Ashcroft Ethiopian Higher Education: Expansion, Dilemmas and Quality." *World Education News & Reviews*. Available at http://wenr.wes.org/2011/06/wenr-june-2011-feature.

Rhodesian Army G Branch. 1975. *Soldier's Handbook of Shona Customs*. Salisbury: Rhodesian Army G Branch.

Richards, Audrey. 1988. "Wifredo Lam: A Sketch." *Callaloo* 34: 91–92.

RICYT/OAS/CYTED. 2001. *Bogota Manual: Standardization of Indicators of Technological Innovation in Latin American and Caribbean Countries*. Bogota: RICYT / OAS / CYTED COLCIENCIAS/OCYT.

Robb, A., A. Valerio, and B. Parton. 2014. *Entrepreneurship and Education and Training: Insights from Ghana, Kenya and Mozambique*. Washington, DC: World Bank.

Rodney, Walter. 1972. *How Europe Underdeveloped Africa*. Dar-es-Salaam: Tanzania Publishing House.

Rudzuna, C. 2014. "Limited Benefits from Informal Sector." *Zimbabwe Independent*.

Rusinamhodzi, L., M. Corbeels, S. Zingore, J. Nyamangara, and K. E. Giller. 2013. "Pushing the Envelope? Maize Production Intensification and the Role of Cattle Manure in Recovery of Degraded Soils in Smallholder Farming Areas of Zimbabwe." *Field Crops Research* 147: 40–53.

Rutto, D. 2015. "Industry Demands and Future of Engineering Education in Kenya." *International Journal of Engineering Pedagogy* 5, no. 2: 31–36.

Ruvando, Vitalis. 2015. "Indigenous Spirits: Enduring Vicars of Revolutions." *The Patriot*.

Sadomba, Zvakanyorwa. 2012. *War Veterans in Zimbabwe's Revolution*. London: James Currey.

Samoff, J., and B. Carroll. 2004. "The Promise of Partnerships and Continuities of Dependence: External Support to Higher Education in Africa." *African Studies Review* 47, no. 1: 67–199.

Sankara, Thomas. 2007g. *Thomas Sankara Speaks: The Burkina Faso Revolution*. Atlanta: Pathfinder Press. (The speeches and interviews below are all derived from this book.)

Sankara, Thomas. 1983a. "Building a New Society, Rid of Social Injustice and Imperialist Domination." Political Orientation Speech, October 2, 1983.

Sankara, Thomas. 1983b. "Who are the Enemies of the People?" At Mass Rally in Ouagadougou March 26.

Sankara, Thomas. 1984a. "Asserting Our Identity, Asserting Our Culture." At Burkinabè Art Exhibit in Harlem, October 2.

Sankara, Thomas. 1984b. "Freedom Must Be Conquered." At United Nations General Assembly, October 4.

Sankara, Thomas. 1984c. "Our White House is in Black Harlem." At Rally in Harlem, October 3.

Sankara, Thomas. 1984d. "The People's Revolutionary Courts." Speech to Inaugural Session, January 3.

Sankara, Thomas. 1985a. "Dare To Invent the Future." Interview with Jean-Philippe Rapp.

Sankara, Thomas. 1985b. "On Africa." Interview with Mongo Beti, November 3.

Sankara, Thomas. 1986a. "The CDRs' Job is to Raise Consciousness, Act, Produce." At First National Conference of CDRs, April 4.

Sankara, Thomas. 1986b. "A Death That Must Enlighten and Strengthen Us." Speech on Death of Samora Machel, delivered in Ouagadougou, October.

Sankara, Thomas. 1986c. "French Enables Us to Communicate with other Peoples in Struggle." Message to First Francophone Summit, February 17.

Sankara, Thomas. 1986d. "Imperialism is the Arsonist of our Forests and Savannas." At International Conference on Trees and Forests, Paris, February 5.

Sankara, Thomas. 1986e. "On Books and Reading." Interview with *Jeune Afrique*, February.

Sankara, Thomas. 1986f. "Against Those Who Exploit and Oppress Us—Here and in France." November 17.

Sankara, Thomas. 1987a. "Eight Million Burkinabè, Eight Million Revolutionaries." On the Fourth Anniversary of the Political Orientation Speech, October 2.

Sankara, Thomas. 1987b. "Our Revolution Needs a People who are Convinced, Not Conquered." On fourth anniversary of revolution, August 4.

Sankara, Thomas. 1987c. "The Revolution Cannot Triumph without the Emancipation of Women." On International Women's Day, March 8.

Sankara, Thomas. 1987d. "A United Front against the Debt." At Organization of African Unity conference, Addis Ababa, July 29.

Sankara, Thomas. 1987e. "We Can Count on Cuba." Interview with Radio Havana, August.

Sankara, Thomas. 1987f. "You Cannot Kill Ideas." A Tribute to Che Guevara, October 8.

Sautet, F. 2013. "Local and Systemic Entrepreneurship: Solving the Puzzle of Entrepreneurship and Economic Development." *Entrepreneurship Theory and Practice* 37, no. 2: 387–402.

Seely, B. 2005. "Patterns in the History of Engineering Education Reform: A Brief Essay." In *Educating the Engineer of 2020: Adapting Engineering Education to the New Century*, 114–130. Washington, DC: National Academies Press.

Senghor, Léopold Sédar. 1966. "Prayer to Masks." In *Modern Poetry from Africa*, ed. Gerald Moore & Ulli Beier, 49–50. Harmondsworth: Penguin

Senzanje, A., N. Moyo, and I. Samakande. 2006. "Relevance of Agricultural Engineering Education against the Background of a Changing World: The Case of Southern Africa." *International Journal of Engineering Education* 22, no. 1: 71–78.

Shambare, R. 2013. "Barriers to Student Entrepreneurship in South Africa." *Journal of Economics and Behavioral Studies* 5, no. 7: 449–459.

Sheppard, S. D., K. Macatangay, A. Colby, and W. M. Sullivan. 2008. *Educating Engineers: Designing for the Future of the Field*. New York: Springer.

Shepperson, George. 1960. "Notes on Negro American Influences on the Emergence of African Nationalism." *Journal of African History* 1, no. 2: 299–312.

Siamonga, Elliot. 2014. "Rhodesian Landmines Stalk Ba Tonga." *The Patriot*.

Sibanda, Mkhululi. 2014. "War Vet Sibasa Speaks on War." *Sunday News*.

Simbi, D. J., and S. Chinyamakobvu. 1997. "Quality Assurance and Relevance of Engineering Education in Zimbabwe." In *Quality Assurance and Relevance of Engineering Education in Africa*, edited by J. G. M. Massaquoi and F.M. Luti, 44–49. Paris: ANSTI and UNESCO.

Smith, Ian. 1997. *The Great Betrayal: The Memoirs of Ian Douglas Smith*. London: Blake.

Smith, Ian. 2008. *Bitter Harvest: Zimbabwe and the Aftermath of Its Independence*. London: John Blake.

Soudiena, C., and D. Grippera. 2016. "The Shifting Sands of Academic Output: University of Cape Town Research Output in Education and Social Anthropology, 1993–2013." *Higher Education Policy* 29: 495–510.

Sterne, Max. 1967. "Distribution and Economic Importance of Anthrax." *Federation Proceedings* 26: 1493–1495.

Sunday News Reporter. 2014. "Dube's Military Exploits." *Sunday News*.

"Sydney Sekeramayi, ZANU—Student in Sweden—Deputy Secretary for Health, Minister of State for National Security," Interview, Harare, July 27, 1995. Liberation in Southern Africa—Regional and Swedish Voices, Nordic Africa Institute, Aluka.

Tagore, Rabindranath. 1902. "Bharatbarsher itihas." *Bhadra* 1309.

Tebeje, A. 2014. *Brain Drain and Capacity Building in Africa*. International Development Research Center. Available at https://www.idrc.ca/en/article/brain-drain-and-capacity-building-africa.

Tekere, Edgar. *A Lifetime of Struggle*. Harare: SAPES, 2007.

Thebe, T. A., and T. Koza. 2012. "Agricultural Mechanization Interventions to Increase the Productivity of Smallholder Irrigation Schemes in Zimbabwe." Paper presented at the conference Creating a Competitive Edge through Agricultural Mechanization and Post-Harvest Technology in Developing Countries, Valencia, Spain, July 8–12.

"The Grim Reality of War." 1978. *Africa News* (US) (June 26).

Tilley, Helen. 2011. *Africa as a Living Laboratory*. Chicago: University of Chicago Press.

Thomson, Ron. 2001. *The Adventures of Shadrek: Southern Africa's Most Infamous Elephant Poacher*. Long Beach, CA: Safari Press.

Trayler-Smith, A. 2014. "Making Higher Education Work for Africa." *SciDev Net*. Available at http://www.scidev.net/global/education/feature/higher-education-africa-facts-figures.html.

UCT (University of Cape Town). 2010. *Annual Report*. Available at https://www.uct.ac.za/usr/finance/afs/afs2011.pdf.

UNCTAD (UN Conference on Trade and Development). 2007. *The Least Developed Countries Report 2007: Knowledge, Technological Learning and Innovation for Development*. Geneva: UNCTAD.

UNECA (UN Economic Commission for Africa). 2013. *African Science, Technology and Innovation Review 2013*. Addis Ababa: UNECA.

UNESCO (UN Educational, Scientific and Cultural Organisation). 2015. *Towards 2030: UNESCO Science Report*. Paris: UNESCO Publishing.

UNISA (University of South Africa). 2011. *Consolidated Financial Statements*. Available at http://www.unisa.ac.za/happening/docs/AnnualReport_2011.pdf.

UP (University of Pretoria). 2010. *Annual Review*. Available at http://web.up.ac.za/sitefiles/file/publications/2011/UP.

van Onselen, Charles. *Chibaro: African Mine Labour in Southern Rhodesia*. London: Pluto Press. 1976.

Walsh, K. L. 1999. "Curricular Changes in the Geology Department, University of Zimbabwe: Towards a More Vocational Degree." *Journal of African Earth Sciences* 28, no. 4: 879–884.

Walshe, A. P. 1970. "Black American Thought and African Political Attitudes in South Africa." *The Review of Politics* 32, no. 1: 51–77.

Walt, V. R., and V. S. J. Walt. 2008. "Entrepreneurial Training for Human Resources Practitioners and Potential Services Rendered to Small Enterprises." *Southern African Journal of Entrepreneurship and Small Business Management* 1: 21–34.

Washington, Booker T. 1895/2005. "Atlanta Exposition Address, 1895." *Black History Bulletin* 68, no. 1: 18–20.

Washington, Booker T. 1899/1972. "Early Life and Struggle for an Education." In *Booker T. Washington Papers Volume 1: The Autobiographical Writings*, edited by Louis R. Harlan and John W. Blassingame. Chicago: University of Illinois Press.

Washington, Booker T. 1900. "Education Will Solve the Race Problem. A Reply." *The North American Review* 171, no. 525: 221–232.

Washington, Booker T. 1901a. *An Autobiography: The Story of My Life and Work*. Toronto and Atlanta: J. L. Nichols & Co.

Washington, Booker T. 1901b. *Up from Slavery*. New York: Doubleday & Co.

Washington, Booker T. 1902a. "Tuskegee Institute—Annual Report of Principal." *The Journal of Education* 56, no. 17: 292.

Washington, Booker T. 1902b. "The Race Problem." *The Journal of Education* 56, no. 4: 72–73.

Washington, Booker T. 1902c. *Character Building: Being Addresses Delivered on Sunday Evenings to the Students of Tuskegee Institute*. New York: Doubleday, Page & Company.

Washington, Booker T. 1904/1915. *Working with the Hands*. Garden City, NY: Doubleday.

Washington, Booker T. 1904a. "The Education of the Southern Negro." *The Journal of Education* 60, no. 4: 72.

Washington, Booker T. 1904b. *Industrial Training for the Negro*. Tuskegee, AL: Tuskegee Institute Steam Print.

Washington, Booker T. 1905a. "What Education Does For The Negro." *The Journal of Education* 61, no. 15: 397–398.

Washington, Booker T. 1905b. "The Religious Life of the Negro." *The North American Review* 181, no. 584: 20–23.

Washington, Booker T., ed. 1905c. *Tuskegee and Its People: Their Ideals and Achievements*. New York: D. Appleton and Company.

Washington, Booker T. 1906. "Tuskegee: A Retrospect and Prospect." *The North American Review* 182, no. 593: 513–523.

Washington, Booker T. 1907. "Make Me An American Negro." *The Journal of Education* 66, no. 15: 400–401.

Washington, Booker T. 1908. "Negro Education and the Nation." *The Journal of Education* 68, no. 4: 111–112.

Washington, Booker T. 1909. "Relation of Industrial Education to National Progress." *The Annals of the American Academy of Political and Social Science* 33, no. 1: 1–12.

Washington, Booker T. 1910a. "Educational Engineers." *The Journal of Education* 72, no. 6: 149–150.

Washington, Booker T. 1910b. "Training Colored Nurses at Tuskegee." *The American Journal* 11, no. 3: 167–171.

Washington, Booker T. 1910c. "The Negro's Part in Southern Development." *The Annals of the American Academy of Political and Social Science* 35, no. 1: 124–133.

Washington, Booker T. 1910d. "Booker T. Washington on the Country School." *The Journal of Education* 72, no. 9: 243.

Washington, Booker T. 1913. "Industrial Education and the Public Schools." *The Annals of the American Academy of Political and Social Science* 49: 219–232.

Washington, Booker T. 1917. "The Exceptional Man." *The Journal of Education* 85, no. 18: 484.

wa Thiong'o, Ngugi. 1985. "The Commitment of the Intellectual." *Review of African Political Economy* 32: 19–24.

wa Thiong'o, Ngugi. 1986. *Decolonizing the Mind: the Politics of Language in African Literature*. Nairobi: Heinemann.

Wilcox, William C. 1909. "John L. Dube, The Booker Washington of the Zulus." *Missionary Review of the World*, 32 (December, 1909): 915–916.

Wits (University of the Witwatersrand). 2011. *Consolidated Financial Statements*. Available at http://www.wits.ac.za/files/11deq_490532001344327610.pdf.

Wood, J. R. T. 2005. "The Pookie: A History of the World's First Successful Landmine Detector Carrier." Available at www.jrtwood.com/article_pookie.asp.

Woollard, D., M. Zhang, and O. Jones. 2007. "Creating Entrepreneurial Universities: Insights from a New University Business School." Paper presented at 30th Institute for Small Business and Entrepreneurship Conference. Glasgow, Scotland, November 7–9.

Wuta, M., G. Nyamadzawo, J. Mlambo, and P. Nyamugafata. 2016. "Ground and Surface Water Quality along a Dambo Transect in Chihota Smallholder Farming Area, Marondera District, Zimbabwe." In *15th WaterNet/WARFSA/GWP-SA Symposium: IWRM for Harnessing Socio-Economic Development in Eastern and Southern Africa, Physics and Chemistry of the Earth*, 112–118. Amsterdam: Elsevier.

Wynter, Sylvia. 2003. "Unsettling the Coloniality of Being/Power/Truth/Freedom: Towards the Human, after Man, Its Overrepresentation-An Argument." *The New Centennial Review* 3, no. 3: 257–337.

ZSC (ZANU Support Committee). 1976. *Medical Aid to Zimbabwe*. New York: Zimbabwe Support Committee.

ZSC (ZANU Solidarity Committee). 1979. *Zimbabwe Not Rhodesia: The African People's Struggle for Independence and Liberation*. New York: Zimbabwe Solidarity Committee.

Zavale, N. C., and E. Macamo. 2016. "How and What Knowledge Do Universities and Academics Transfer to Industry in African Low-Income Countries? Evidence from the Stage of University–Industry Linkages in Mozambique." *International Journal of Educational Development* 49: 247–261.

Zeelen, J. 2012. "Universities in Africa: Working on Excellence for Whom? Reflections on Teaching, Research, and Outreach Activities at African Universities." *International Journal of Higher Education* 1, no. 2: 157–165.

ZEPARU (Zimbabwe Economic Policy Analysis and Research Unit) and BAZ (Bankers Association of Zimbabwe). 2014. *Harnessing Resources from the Informal Sector for Economic Development*. Available at http://baz.org.zw/resources/harnessing-resources-informal-sector-economic-development.

ZIMA (Zimbabwe Medical Association). 1979. "Report by Publicity Secretary to ZIMA Committee on the Weekend Conference." Leiden, Holland, Saturday/Sunday, June 23–24.

Zimmermann, Andrew. 2001. *Anthropology and Antihumanism in Imperial Germany*. Chicago: University of Chicago Press.

ZimStat. 2012. *Zimbabwe Population Census*. Harare: Population Census Office.

Zippia. 2023a. "Professor Demographics and Statistics in the United States." Zippia. https://www.zippia.com/college-professor-jobs/demographics/

Zippia. 2023b. "African Studies Professor Demographics and Statistics in the US." Zippia. https://www.zippia.com/african-studies-professor-jobs/demographics/

ZMD (Zimbabwe Medical Drive). 1978. "Dumi and the Maraire Marimba Ensemble Benefit Dance." African Activist Archive, Michigan State University.

ZMD (Zimbabwe Medical Drive). 1978–1979. *Zimbabwe Medical Drive* (brochure). Seattle: ZMD.

ZMD (Zimbabwe Medical Drive). 1979. "Benefit Rummage Sale." African Activist Archive, Michigan State University.

INDEX